D1104180

# Shen Gua's Empiricism

HARVARD-YENCHING INSTITUTE MONOGRAPH SERIES 113

# Shen Gua's Empiricism

## Ya Zuo

Published by the Harvard University Asia Center
Distributed by Harvard University Press
Cambridge (Massachusetts) and London 2018

The Harvard-Yenching Institute, founded in 1928, is an independent foundation dedicated to the advancement of higher education in the humanities and social sciences in Asia. Headquartered on the campus of Harvard University, the Institute provides fellowships for advanced research, training, and graduate studies at Harvard by competitively selected faculty and graduate students from Asia. The Institute also supports a range of academic activities at its fifty partner universities and research institutes across Asia. At Harvard, the Institute promotes East Asian studies through annual contributions to the Harvard-Yenching Library and publication of the *Harvard Journal of Asiatic Studies* and the Harvard-Yenching Institute Monograph Series.

Library of Congress Cataloging-in-Publication Data

Names: Zuo, Ya, 1982- author.
Title: Shen Gua's empiricism / Ya Zuo.
Other titles: Harvard-Yenching Institute monograph series ; 113.
Description: Cambridge, Massachusetts : Published by the Harvard University Asia Center, 2018. | Series: Harvard-Yenching Institute monograph series ; 113 | Includes bibliographical references and index.
Identifiers: LCCN 2017060582 | ISBN 9780674987111 (hardcover : alk. paper)
Subjects: LCSH: Shen, Kuo, 1031-1095. | Scientists—China—Biography. | Statesmen—China—Biography. | Science—Political aspects—China. | Empiricism. | China—Intellectual life—960-1644. | Neo-Confucianism—China—History.
Classification: LCC PL2687.S5 Z96 2018 | DDC 181/.11—dc23
LC record available at https://lccn.loc.gov/2017060582

Index by the author

∞ Printed on acid-free paper

Last figure below indicates year of this printing
27  26  25  24  23  22  21  20  19  18

For my parents,
Dong Songjiu and Zuo Yisheng

# Contents

Appendix 3 is available online at https://www.zuoya.org/book or https://bit.ly/2KhɪnHɪ

# Tables and Figures

## Tables

## Figures

Tables 5 and 6 are available online at https://www.zuoya.org/book or https://bit.ly/2KhɪnHɪ

# Acknowledgments

The long journey of writing a book has taught me to value kindness, generosity, and friendship more than ever. When I began this project in graduate school, I was surrounded by inspirational teachers. My advisor, Willard Peterson, has a way to turn the hustle and bustle of a young mind into a love for rigor. He set a lifelong example of clear thinking and intellectual open-mindedness, and his faith in me sustained me through thick and thin. The uplifting words of Ben Elman, Yang Lu, Sue Naquin, Jerome Silbergeld, and Buzzy Teiser also remain with me.

I benefited greatly from the warmth and openness of the scholarly communities I inhabit. Don Wyatt patiently guided me through different stages of my career. Peter Bol and Ron Egan engaged my work with characteristic acumen and generous support. Nathan Sivin shared his publication and valuable insights. Chang Woei Ong generously extended thoughtful counsel and encouragement. Ari Levine and Xiao-bin Ji provided invaluable mentorship in my scholarly work and development as a teacher in the early years of my career. I also benefited from conversations with numerous scholars over the years: John Chaffee, Hilde De Weerdt, Pat Ebrey, Charles Hartman, T. J. Hinrichs, Jeehee Hong, Shigehisa Kuriyama, Eugenia Lean, Peter Lorge, Freda Murck, Gil Raz, Adam Sabra, Pierce Salguero, Anna Shields, Linda Walton, Ping Yao, and Cong Zhang, among many others. I owe special thanks to Steve Angle and Justin Tiwald for sharing their writings with me and engaging my ideas with the probing clarity characteristic of philosophers. The brilliant minds of all these fellow scholars motivated me to expand my intellectual horizons.

Friends and colleagues in the Sinophone world offered unflagging support. My college teachers Deng Xiaonan and Rong Xinjiang introduced me to the field of middle-period Chinese history, and their enduring support has been crucial for my professional growth. Since graduate school, Ge Zhaoguang and Dai Yan have been generous mentors, bestowing fascinating conversations and books, and hosting me at Fudan University during my sabbatical. Chu Ping-I extended a valuable opportunity to a young graduate student to publish her work, and Chu Longfei, Guan Zengjian, Guan Zhenyu, Lei Hsiang-lin, Shi Rui, Shi Yunli, and Wei Bing shared their invaluable insights on sources and possible new inquiries.

My thanks also go to scholars in Europe, where a large volume of fine scholarship on Shen Gua has been produced. I am indebted to Karine Chemla for pointing me to French resources, to Pierre-Étienne Will and Christian Lamouroux for generously sharing the unpublished translations of *Brush Talks* from the workshop conducted at the Centre National de la Recherche Scientifique, and to Alain Arrault for thoughtfully keeping me updated on the seminar series on *Brush Talks* he organizes at the École française d'Extrême-Orient. I thank John Moffett for his warm support and G. E. R. Lloyd for our stimulating conversations at the Needham Institute. Years ago, to my happy surprise, I received newly published materials on the subject from Lincoln Tsui and Dagmar Schäfer, who kindly kept this project in mind.

My peers are nonpareil. Graduate school friends Ian Chong, Jinsong Guo, Brigid Vance, and Shellen Wu repeatedly read parts of the manuscript with unwavering patience. Among the many women friends from whom I drew intellectual inspiration and emotional support, I owe thanks to Sare Aricanli, He Bian, Chunmei Du, Feiran Du, April Hughes, Yajun Mo, Margaret Ng, Megan Steffen, Zhiyi Yang, Ling Zhang, and Xiaowei Zheng. Over the course of writing this book, I also benefited from discussions with Yunju Chen, Fang Xiaoyi, Loretta Kim, Natalie Köhle, Pei-ying Lin, Jiyan Qiao, Fabien Simonis, Lena Springer, and Lawrence Zhang. Old friends from college, Cao Jin, Deng Fei, Gu Liwei, and Zhu Yi, among others, kindly took care of me during my sabbatical in Shanghai and Beijing. I especially thank Chen Hao, for his decade-long intellectual companionship and all the conversations that kept me on my toes.

Bowdoin College has given me a warm institutional home. My colleagues keep my best interests at heart, and my students make teaching enjoyable with their energy and excitement. Dallas Denery, Belinda Kong, Jen Scanlon, Jayanthi Selinger, and Shu-chin Tsui read parts of this book and provided comments that led to significant changes. I also benefited from the advice and support of Connie Chiang, Sakura Christmas, David Hecht, Page Herrlinger, Chris Heurlin, John Holt, Padma Holt, Matt Klingle, Henry Laurence, Meghan Roberts, Arielle Saiber, Matthew Stuart, Rachel Sturman, Susan Tanabaum, Allen Wells, and Peggy Wang, as well as former colleagues Tom Conlan and Femi Vaughan. My heartfelt thanks to them all.

Generous research support from the American Council of Learned Societies, the Luce Foundation, the Andrew W. Mellon Foundation, and Bowdoin College made a paid leave possible and greatly facilitated the completion of the book. I thank Duke University Press for permitting me to use my previous publication from *East Asian Science, Technology, and Society*. Bob Graham, the best of editors, shepherded the publication of this project with impeccable editorial skills and encouragement.

I have been blessed with love and support from a close-knit family. My grandparents, aunts, uncles, cousins, and Manzi and Yezi kept me grounded with their enduring affection despite the geographical distance between us. I thank my parents for their unconditional love. They are my rock, my role models, and the staunchest supporters of my single-minded devotion to academia. I dedicate this book to them with abiding love.

Z. Y.

# *Abbreviations*

The following abbreviations are used in the notes and bibliography.

BBT     Shen Gua, *Bu bitan* 補筆談 (Shanghai: Shanghai renmin chubanshe, 2011)

CXJ     Shen Gua, *Changxing ji* 長興集 (Hangzhou: Zhejiang shuju, 1896)

MXBT     Shen Gua, *Mengxi bitan* 夢溪筆談 (Shanghai: Shanghai renmin chubanshe, 2011)

SHY     Xu Song, *Song huiyao jigao* 宋會要輯稿 (Taipei: Xinwenfeng chuban gongsi, 1976)

SBCK     *Sibu congkan* 四部叢刊 (Shanghai: Shangwu yinshu guan, 1919–1936)

SKQS     *Wenyuange Siku quanshu* 文淵閣四庫全書 (Taipei: Shangwu yinshu guan, 1983–86)

SS     Tuotuo, *Song shi* 宋史 (Beijing: Zhonghua shuju, 1977)

XBT     *Xu bitan* 續筆談 (Shanghai: Shanghai renmin chubanshe, 2011)

XCB     Li Tao, *Xu zizhi tongjian changbian* 續資治通鑒長編 (Beijing: Zhonghua shuju, 2004)

# INTRODUCTION

For anyone with an interest in learning about science and technology in traditional China, Shen Gua 沈括 (1031–1095) is a name that merits notice. In fact, Shen's presence in the scholarship on premodern China at large is inescapable. His treatise *Brush Talks from Dream Brook* (*Mengxi bitan* 夢溪筆談) has become canonical for students of Chinese science and an indispensable reference for a well-rounded view of traditional Chinese culture. In this book, I present Shen as a significant figure in intellectual history and make the case for him as an empiricist. Shen's empirical stance led to the emergence of a new horizon of epistemic praxis and consequently a range of concrete discoveries widely acclaimed for their "scientific" value.

The most convenient place to first encounter Shen and get a quick glimpse of his many doings, interestingly, is the twentieth century, and to that I now proceed.

## Shen Gua in the Twentieth Century

Who was Shen Gua? The mass media in contemporary China has had a lot to say about this household name, calling him "a science giant," a "superstar in science," or more specifically, the "ancient scientist who discovered and named petroleum."[1] A children's picture book published in 1977 applauded Shen as the author of a wondrous book titled *Brush Talks from Dream Brook* (hereafter *Brush Talks*) and recorded a wide range of Shen's amazing "scientific" discoveries. For example, he witnessed a meteor falling to Earth and perceived similarities between the meteorite and iron, an insight achieved "more than 600 years earlier than foreigners" (figure 1).[2] He was the first person in the world to discover astronomical refraction, a concept "not known to Western countries until 500 years later."[3] "More than 600 years ahead of European geologists," Shen became aware of the mechanism of erosion in mountain formation.[4] He made a three-dimensional topographic map, whereas the European counterpart, the Switzerland Topographic Map, did not emerge until the eighteenth century.[5] While traveling in western China, Shen was intrigued by a type of black liquid local people extracted from the ground. When burned, the liquid emitted strong smoke and left a residue on residential tents in the area. Shen named it the "oil of rocks" (*shiyou* 石油) and contemplated its utility, a discovery that later won him the title "father of petroleum" in China.[6] He wrote about the magnetic compass, small-aperture optics, and resonance in physics, among other advanced topics in mathematics, chemistry, biology, and medicine.[7] The picture book concluded Shen's story with a claim by "a foreign scientist who specializes in studying the development of Chinese science" that Shen was a "milestone in the history of science in China."[8]

Though presumably addressed to a juvenile audience, this picture book laid out a reasonable factual ground for a first encounter with Shen. He lived in the eleventh century, a middle-period man who deeply impressed the modern age with his various talents. He traveled extensively and keenly observed the natural world, where he made discoveries bearing striking similarities to modern scientific conclusions. The *Brush Talks*, where Shen gathered his various technical insights, stood out as an unusual old text featuring exceptional technical sophistication and curiously modern overtones.

The media's enthusiasm about Shen was the spillover of a lasting scholarly trend. He has emerged as a subject of fascination in modern China, first to professional historians and then to a mass audience. Starting in the early twentieth century and through the 1970s, generations of Chinese scholars devoted diligent research to uncovering Shen's "scientific" contributions; some had a clear intent to scour the past for precedents of modern science.[9] The "foreign scientist" cited in the closing of

FIGURE 1. Chen Guangyi, *Shen Gua Examines a Meteorite.* From Chen Guangyi et al., *Shen Gua,* 4. Courtesy of the Shanghai Renmin Meishu Publishing House.

the children's book, Joseph Needham (1900–1995), was the founding father of the history of science in China as a disciplinary field in English-speaking academia. In the mid-twentieth century, he chose to highlight Shen as a nodal point in the master narrative of science in China and thus threw in an external catalyst for the interest in Shen among Chinese scholars. Needham's choice became so well received that Shen found his way into the aforementioned children's literature. Over time, scholarly and popular interests in Shen have become commingled.

The modern narratives depicting Shen as an extraordinary medieval man practicing modern Western science reveal more about the present than the past. The intensive interest in Shen emerged as part of two important century-long transformations. First, since the turn of the century, modern Western science came to prevail over traditional learning and served as the measure of China's modernity.[10] At the same time, another movement saw the first generation of indigenous Chinese scientists and newly emerged historians of science make a joint effort to compose a history of science that looked to connect modern learning with China's glorious past.[11] Historical examples like Shen Gua were advanced as proof of China's scientific accomplishments in the remote past, a warrant for the prospect of the nation modernizing yet remaining "Chinese." These developments prevailed over the course of the twentieth century and persisted through the most tumultuous political changes, including the transition from the first republic (ROC, 1912–49) to the second (PRC, 1949–present). The discourse on science in China's past has provided fuel for patriotic sentiments and an important item on the state's propaganda agenda.[12] The interest in Shen, for instance, became heightened during the Cultural Revolution (1960s–70s), when Mao's China, enveloped in an unprecedented surge of ideological zeal, strove to assert its capability to industrialize and modernize amid Cold War hostilities and alienation. Under the government's sponsorship, volumes of new research on Shen's scientific discoveries were produced by natural scientists in collaboration with industrial engineers, constituting a most unconventional type of scholarship.[13] The children's book mentioned earlier, also a product of the 1970s, was a repercussion of this movement. The constant emphasis on Shen's ability to make key discoveries centuries earlier than others in the West bespoke an unmistakable nationalist passion, a resolution to seek modernity on China's own terms, and a message that Chinese adults eagerly wanted to relay to the next generation.

## *Shen Gua in History and in an Intellectual Biography*

The twentieth century's imagination of Shen Gua was more self-reflexive than historical, and Shen was certainly no wondrous prophet who advanced modern science in eleventh-century China. Despite the anachronistic distortion it induced, the modern narrative still rightly highlighted the singularity of Shen and pointed to a significant historical phenomenon that cannot easily be overlooked. After all, people in twentieth-century China chose Shen among a small group of premodern figures to bear the flag of modern science, and they could reasonably ground this choice in Shen's noticeable difference from the majority of the Chinese cultural elite in the premodern era. Modern or not, Shen's accomplishments are nonetheless impressive: among many fields of knowledge, he studied astronomy, optics, chemistry, mathematics, and medicine; he perpetuated his name in history by leaving valuable (if not always the earliest) records of astronomical refraction, geological erosion, petroleum, and movable-type printing. Since any one of these discoveries would warrant historical interest, to find them all within the intellectual scope of one person makes Shen a compelling candidate for a genuine Renaissance man.

Thus Shen Gua merits a judicious historical study. Rather than marveling at even the most peculiar phenomena, a historian poses the questions of how and why. This intellectual historian directs her how-and-why inquiries specifically to the thinking behind Shen's concrete achievements. If the fruits of one's thinking turn out as distinctive as Shen's, we can reasonably assume the thinking itself equally singular. To phrase it more specifically as a problem of knowledge, inquiries that yield knowledge of an idiosyncratic type probably indicate the development of a different way of knowing.

This book, therefore, is a study of Shen's way of knowing, an intellectual biography expositing his epistemological stance. Throughout, I examine Shen's beliefs and assumptions about the conditions of knowing, his decisions on what to know and actions on how to know, and how, on such a basis, Shen formulated an empirical stance.

This is not a book about a genius, however. Although it is a biography of an individual and his ideas, the central question that drives this study is how such a line of thinking came into being under broad historical and conceptual circumstances. As singular as Shen's epistemological stance might have been, what is truly remarkable is what made his originality possible, a question that enlists answers profoundly more complex than Shen's personal genius. Thus, instead of picturing him as an intellectual hero, this book sees the individual Shen as a searchlight which, scanning the vast dim infinity of historical multiplicity, makes visible the conditions, historical and conceptual, broad and intricate, that shaped a significant way of seeking knowledge.

## Shen Gua in the Song Dynasty (960–1279)

This book examines how Shen Gua's thinking intersects, challenges, and enriches two greater historical narratives. The first concerns Song China at large. The Song Dynasty, conventionally divided into the Northern Song (960–1127) and the Southern Song (1127–1279), is a critical age in middle-period China of unprecedented cultural efflorescence and profound social transformations. In textbook narratives, the dynasty is justifiably accredited as the "most advanced society in the world in its day," sometimes analogized as the "Chinese renaissance."[14] This book demonstrates that Shen emerged from this context not as a reclusive hermit, nor was his "science" a stroke of magic. Instead, all of his insights grew organically from his experience as a man of his times.

The first thing to know about Shen in the Song context is that he was a literatus, a member of the newly arisen social elite who ushered in sweeping changes across the empire.[15] As the medieval aristocracy dissolved and their collaborative leadership with the imperial clan diminished, the Song emperors started to recruit people of common origins as officials to staff the new establishment. The recruitment initiative developed into an elaborate system of civil service examinations and an educational curriculum that gradually shifted focus from literary composition to the Confucian Classics.[16] A person who passed the examinations would become an incumbent of the officialdom; if he possessed the examination education and an official appointment, he became a member of the literati. This new ruling class was unprecedented in terms of background diversity and intellectual ambition. Its members created a collective political culture that dominated eleventh-century society and prefigured the so-called neo-Confucianism, the ideology that ruled China for hundreds of years after.

Shen's life did not deviate in a major way from the classic profile of a Song literatus. A writer of elegant prose and a diligent student of old texts, the young Shen passed the civil service examinations with flying colors and earned a position in the imperial government. In the next three decades, he primarily performed administrative duties and formulated policies. His thinking developed through his participation in the literati culture, a flower burgeoning along with many other blooms.

Shen's life as a literatus and Song history illuminate each other in three specific ways. The key political event that shaped his career also happened to be the movement that redefined the Song state and society in every critical way. The so-called New Policies, the most far-reaching reform in the middle period, was an ambitious attempt by literati to reorganize the state and systematically redefine its ideological, economic, and military functions. A crucial participant in this movement, Shen spent the majority of his active years involved in many aspects of the reform. This biography demonstrates that he formulated his intellectual stance, gradually and reflexively, over the long course of serving the reform initiative. The New Policies constituted the most immediate political context of his story.

Shen Gua's life also belonged to a historical moment when the Song intellectual world was at its most dynamic. A critical transitional period, the eleventh century witnessed the rapid decay of the old culture and the surge of the new. Having abandoned the fixation on belles lettres associated with the bygone aristocratic culture, eleventh-century literati quickly established their own goals and intellectual style. As a scholar and a statesman, a Song literatus aimed to envision and implement order in the world, namely, to translate his cultural virtuosity into policy making. Classical literacy and philosophical articulateness, as well as competence in statecraft, constituted the complex enterprise literati venerated as "learning" (*xue* 學).[17]

Due to the central status accorded the Classics, learning in the Song had an antiquarian outlook. During the latter half of the eleventh century, learning meant to envision order through testimonies articulated by ancient sages. The ultimate order was known as the *dao* (conventionally translated as the "Way"); to access the sages' rendering of the *dao* one turned to reading old texts, especially the Confucian Classics.[18] Candidates at the civil service examinations, especially the dominant kind known as "advanced scholars" (*jinshi* 進士), had to demonstrate their ability to articulate core values in antiquity, write about contemporary issues concerning government, and propose solutions, all based on a sound understanding of the Classics.[19]

Somewhat counterintuitively, the antiquarian setting gave rise to an unprecedented explosion of cultural creativity. The new model of learning encouraged every ambitious literatus to contemplate the great issues of the day under the rubric of pursuing the *dao*, develop a total view of unitary order on his own, and put it into praxis if opportunity allowed. Such a total view I call a system.[20] The New Policies was a system most aggressively in action, and the eleventh century witnessed the mushrooming of many other systems and spirited competitions between them. This era was an exemplary "age of systems."

Although systems could vary greatly in terms of detail, the "spirit of system" flaunted by eleventh-century literati gave impetus to the rise of a new mainstream. The commonalities among system builders were already salient during Shen's time: they all aspired to reach a total-view unity (the *dao*), all relied on old texts, and all endorsed the same epistemological stance and specific understanding of knowledge. The further evolvement of system building, along with an intensifying homogeneity, eventually consummated in Zhu Xi's 朱熹 (1130–1200) synthesis of neo-Confucianism in the twelfth century, a legacy that dominated the intellectual scene of China for centuries. In this book, the system-building mainstream provides the most immediate intellectual context for understanding Shen's thinking.

Shen's story adds a new chapter to Song history, specifically in the sense that he was not a system builder. He pronounced a new stance that stood out among "total views" and diverged in profound ways from the system-building mainstream and its long-term evolution leading to Zhu Xi's synthesis. As this book demonstrates, Shen's empiricism was a nonsystem, and it sheds light on the deep and truly complex dynamics in the Song world as well as on the rich possibilities they afforded. In this sense, Shen lends a truer testimonial to diversity in the Song world and necessitates a major revision of the master narrative of Song intellectual history, which features predominantly, if not exclusively, system builders.

Yet the significance of Shen in the Song world goes beyond his life as a gloss on the master narrative. In fact, he ushers in a completely new perspective, one that centers on the problems of knowledge and from which a new analytical narrative of the Song intellectual scene emerges. Shen posed challenges to the concrete decisions of system builders and the epistemological assumptions they complacently endorsed. For instance, how did a literatus know he was on the right track to approach the *dao*, and how did he prove the validity of the system he applied to this purpose? More specifically, when contemplating ideas, how did he advance from an old idea to a new one; in statecraft, how did he advance from existing information to new decisions? In other words, where did he find his epistemic guide? In addressing these questions, this book intends to avoid a simplified answer such as "reliance on classical authority." The flaw of this argument is obvious: if literati were simply dogmatic antiquarians who passively recycled old knowledge as new, there would have been no creativity, much less a renaissance, in the Song. Underlying the system-building mainstream were significant epistemological assumptions that need to be analyzed, and a study of Shen Gua, a voice who turned against these assumptions, provides an excellent opportunity to do so. This epistemological line of inquiry and the new conceptual horizon it creates, I argue, not only can account for "nonmainstream possibilities" such as Shen, but can provide a more thorough explication of the grounds of competition between systems and make better historical sense of the evolution of the mainstream.

In addition to coinciding with the New Policies and the heyday of system building, Shen's life provides a third angle able to illuminate the landscape of Song history: the development of knowledge of the material world and of using the material world to meet practical human needs (in modern parlance, science and technology). Indeed, for decades the Song has remained a fruitful subject for historians of science and technology. In Nathan Sivin's words, the Song was "an important period in the history of every branch of science and technology."[21] The tenth to thirteenth centuries witnessed considerable developments in mathematics, astronomy, and medicine, among other subjects in natural studies.[22] The era was also associated with a range of technological innovations, most famously, the invention of movable-type printing (recorded by Shen), the use of magnetic needles in navigation, and the first established formula for gunpowder.[23] These achievements did not converge into a coherent intellectual movement, as the practitioners emerged from different social sectors and often had no contact, much less intellectual communication, with one another.[24] Even so, the sheer accumulation of concrete accomplishments was impressive.

Besides piecemeal discoveries, other changes in Song society attested to a growing enthusiasm for exploring the material world. State sponsorship, for instance, played a salient role in the production of technical knowledge.[25] The Song government invested significant resources in maintaining institutionalized support for astronomical observation and calendar making. The state machinery actively promoted the publication of old mathematical texts, classical medical works, contemporary pharmacological literature, and agricultural treatises.[26] Conspiring hand in hand with these publication projects was the cultural elite's increasing interest in technical knowledge: as the primary readers, authors, and editors of these publications, literati valued them as cultural resources auxiliary to the Classics and made an effort to engage the technical content in depth. They often sought to distinguish themselves from nonelite practitioners through coopting technical knowledge into existing cultural frameworks, for instance, the pursuit of the *dao*.[27]

The flourishing of science and technology in the Song undoubtedly serves as an important background for the study of Shen. Indeed, Shen's multifarious interests became possible only in the context of broader developments. His cultural literacy enabled him to effectively use the aforementioned textual resources. At some point, his official rank afforded him access to the most comprehensive collection of books at the imperial court. His appointment to the state-run astronomical bureau provided the opportunity to operate and design observational instruments. The growing peer interest in exploring the material world brought informants and interlocutors on technical topics into his social circle. Since the 1970s, historians of science have meticulously considered these connections in their efforts to study Shen against the background of Song science.[28] Indeed, in historiography, Shen's technical achievements and the prosperity of Song science are two narratives egging each other on: the former is deeply enmeshed in the backdrop of the latter, and the latter consistently features the former as a central testimony.

Any new study of Shen, including this one, is deeply indebted to the rich accumulation of scholarship in Song science. The current book nevertheless aims to present Shen's interest in technical knowledge in a different analytical narrative. Instead of first highlighting his technical achievements (as a scientist) and then incorporating him into a broader discussion of literati culture (e.g., Confucianism), I place his thinking in the context of "learning" right away. After all, learning was the overarching framework every Song scholar relied on to navigate his journey of acquiring knowledge and organize his encounters with various subjects, including the Classics, astronomical ephemerides, and medical treatises. Participation in learning afforded Song literati a shared intellectual vocabulary, a robust sense of community, and a most valued common identity. In the world of learning I present here, Shen shared his interests and goals with statesman Wang Anshi 王安石 (1021–1086), literary master Su Shi 蘇軾 (1037–1101), numerologist Shao Yong 邵雍 (1011–1077),

philosophers Cheng Hao 程顥 (1032–1085) and Cheng Yi 程頤 (1033–1107), and scientist Su Song 蘇頌 (1020–1101).[29] Not only does the learning-based analytical framework bring out the relevance of Shen in intellectual history, a meaningful task in itself, it also introduces an inclusive perspective that avoids partitioning the Song cultural sphere into ahistorical subdivisions.

## Shen Gua in the History of Knowledge

To enable an in-depth discussion of the epistemological divergence between Shen and the mainstream, this book annexes another greater narrative concerning the history of knowledge in China. A responsible investigation of epistemology requires attention to a longer historical period, in this case, from antiquity through the Song, as basic epistemological assumptions are habits of mind which take long periods of time to formulate and to dissolve. Multiple features of Song epistemic praxis had strong connections with ancient thinking. As important contexts, a number of continuities in the history of knowledge will be demonstrated in detail in this book.

The "history of knowledge" I mobilize in this study is a complex narrative synthesizing a number of different strands of scholarship. "Knowledge" was an extraordinarily broad category, certainly not limited to scientific knowledge. My discussion of the history of knowledge focuses on ways of knowing while striving to stay true to the empirical inclusiveness of "knowledge." The central strategy is to use philosophical inquiry into the meaning of "to know" as the linchpin to connect a range of empirical discussions concerning government, morality, science, and cosmology, among other pertinent topics.

Against the broad background of the history of knowledge since antiquity, Shen's story poses two questions that have yet to be fruitfully answered. First, is there a homogeneous "Chinese way of knowing"? In attempting to distinguish Chinese epistemic praxis from the logocentric Cartesian tradition, scholars from different fields have come to define a distinctive Chinese way of knowing with a few interlocked cardinal features. More accurately, such a "way" is a "framework of taken-for-granted," a horizon that directs concrete choices to common commitments.[30] The framework, which may or may not be consciously acknowledged by the knower, provides tacit directionality rather than articulated guidance. The designation of a framework is consistent with the first cardinal feature scholars have rightfully identified for the Chinese way of knowing: virtually no formal theory of knowledge, in a systematic, modern (Cartesian) fashion existed in premodern China.[31]

The second feature concerns orientation. As some philosophers argue, in seeking knowledge the Chinese are often not interested in pursuing substance, essence, or an attribute; rather, they are primarily concerned with relations and processes.[32] Knowledge takes the form of "how to be adept in relationships, and how, in optimizing the possibilities that these relations provide, to develop trust in their viability."[33]

The third feature defines the way of knowing. Scholars realize that in many contexts "to know" is not a simple act of cognition, as modern readers might assume. Instead, it meant to participate and realize, and it clearly involved a bodily aspect.[34]

The fourth feature of the Chinese way of knowing most starkly sets it apart from the Cartesian model: the Chinese knower admits no division between subject and object, nor does she assume a distinction between an object and an idea. Her way of knowing thus involves "presentation" instead of "representation."[35]

These basic features and their combinations aptly account for a variety of more concrete Chinese characteristics of epistemic praxis across different empirical contexts. For example, features two and three, when applied to a social context, characterize the central task of "to know" as to appreciate all interhuman relationships one needs to know and become able to handle all interactions harmoniously. To acquire knowledge is thus a "moral, prudential skill" in action.[36]

Features two, three, and four in combination explain the working principles of the so-called correlative thinking with which the Chinese organized the universe. The system classifies all entities, material things, living creatures (including humans), and phenomena in cultural and natural realms into "categories" (*lei* 類), stipulating that one entity would resonate with another when both belong to the same category. Categories do not stand alone; they annex one another and form relations (feature two). The most exemplary relations include those between yin and yang and the Five Processes, among other numerical sequences. Things that belong to different categories thus interact according to these relations. Take yin and yang, for example. In certain contexts, both a man and the sun belong to the yang category, while a woman and the moon belong to the yin. A man may resonate with the sun because both are yang, and a man may have a contrasting yet complementary relationship with a woman, precisely as yang interacts with yin. Such category-based resonance and interaction constitute a correlation.[37] A human agent participates in networks of correlation as part of them rather than observing them as objective phenomena (features three and four). Modern scholarship often compares a correlation to a cause with an emphasis on polar distinctions.[38] The difference is commonly taken as what distinguishes Chinese thinking (sometimes a more inclusive "proto-scientific" thinking rather than exclusively Chinese) from modern scientific thinking.[39] Some scholars call a correlation "mysterious" with a strong implication of irrationality.[40]

In the context of science, features three and four logically entail the argument that the Chinese knower is indifferent to truth and thus to the classical questions raised by Western epistemologies. To know is not an act of cognition directed at a reality. The knower is thus not worried about whether her idea is a truthful representation of a putative reality; as a result, she is not concerned with certainty or justification, among many other key issues that define the Western epistemological tradition.[41]

Though a number of studies cited here focus on the classical period, these fundamental features, as I discuss in the book, accurately characterize the mainstream way of knowing in the Song. For instance, acquiring knowledge often involved a performative component beyond cognitive activities. System building, with its eventual consummation in Zhu Xi's neo-Confucianism, was essentially a "relation-oriented" line of thinking. While pursuing the *dao*, the Song knower did not see himself as a subject or the *dao* as an object.

Does the versatility of this framework mean that there was practically only one way of knowing in premodern China? To address this question, Shen Gua posits extremely interesting challenges. As will be laid out step by step in this book, his epistemological stance concerned the essence and substance of things no less than their relations. The "knowing" activity in his practice and description was often a clear instance of

cognition. He started to contemplate the idea of representation and im-
plicate a division between reality and idea. Though he justifiably made
no mention of absolute or total truth, Shen was methodically concerned
with the reliability of knowledge, a concept in the same family with
truth and certainty. Last but not least, in his way of learning, a subject-
object division started to emerge. Shen provides a firmly negative answer
to the aforementioned question: there is certainly not just one way of
Chinese knowing, and a paradigmatic distinction between a monolithic
"Chinese" epistemology and a Cartesian tradition is barely viable.

In addition, Shen's deviance from the basic parameters of the mainstream model was not an idiosyncratic monologue. Various hands-on
practitioners such as artisans, technicians, and physicians were attentive
to the essence and substance of objects, from which they sought guidance to navigate their productive activities. They also developed justificatory procedures and assessment criteria to monitor the quality of their
work, thus maintaining an awareness of epistemological reliability. The
concern with reliability applied to some literati, too, when they sought
to ensure the accuracy of the textual knowledge they produced in reading and writing. Shen Gua provides a good individual case and effective
textual sources (*Brush Talks*) for us to peruse these epistemological sub-
streams, which productively operated in diverse contexts across Song
society.

## The Issue of Empiricism

Another key question Shen poses in the history of knowledge is: what
does "empiricism" look like in the Chinese context?[42] This inquiry is central to the thesis of this book and deserves a separate section of discussion.
Let's look first at how this concept has been construed in philosophy. In
the European tradition, empiricism has conspired with rationalism, and
the two are defined in relation to each other. Some philosophers state
reservations regarding the use of empiricism in the Chinese context,
because "China lacked this rationalism to react to."[43] A knower in the
Chinese tradition has no grounds (or philosophical interest) for claiming
"innate knowledge" completely independent from experience.[44] To know

is "irreducibly experiential."[45] Without a rival, empiricism seems unable to claim the significance it enjoys in the West.

Another reason that potentially deters scholars from using empiricism in the Chinese context regards perception. Despite the long contestation among Western philosophers to define empiricism, the source of knowledge has remained a persistent concern, and reliance on sensory perception as the source of knowledge constitutes a basic parameter of many different versions of empiricism. Historical examples are numerous. In antiquity, for instance, Epicurus (341–270 BCE) regarded "incorrigible" acts of perception as the foundation of knowledge.[46] In the sixteenth century, the forefather of the so-called British empiricists, Francis Bacon (1561–1626), claimed that "we have served men better than those who have dealt with natural history in the past, for we use the evidence of our own eyes, or at least of our own perception, in everything, and apply the strictest criteria in accepting things."[47] Following his lead, John Locke (1632–1704) claimed "in that (experience), all our Knowledge is founded; and from that it ultimately derives itself."[48]

Sensory perception in the Chinese tradition, however, does not enjoy a role as central as it does in Western epistemologies, nor does it entail the same degree of complexity. Most distinctively, the so-called problem of perception barely bothered any premodern Chinese thinkers. In Western discourse, the sensory modalities are not always to be trusted because of their capacity for illusion or hallucination. As a result, the discussion of perception always involves a relationship between the senses and a mind-independent, public object, and this relationship is neither open nor transparent.[49] In contrast, the Chinese knower handles the issue of perception with what her Western counterparts might regard as naïveté. For her, sensory knowledge is a direct and unobstructed awareness. Given her tendency not to isolate herself (including her senses) from the surrounding circumstances, she does not see her act of perception as subjective; moreover, she hardly doubts the capacity of this act to enable an undistorted access to the entity under gaze.

Sensory knowing did attract philosophical interest, however, and remained a theme in intellectual discourse from antiquity through the Song period.[50] "Knowing from hearing and seeing" (*wenjian zhizhi* 聞見之知), or "hearing and seeing" (*wenjian* 聞見) for short, was a popular term in the Song times, and the core meaning of the concept resided in sensory knowing. Moreover, the theoretical awareness of the senses led to a discrimination of different types of knowledge. Another Song term that often appeared in a dichotomous relationship with "knowing from hearing and seeing" was "knowing from virtuous nature" (*dexing zhizhi* 德性之知).[51] Although the invocation of "virtue" might invite readers to assume that this type of knowing primarily concerned morality, such was not the case. Instead, this type of knowing, which I render more generically as "modeling" to avoid potential misunderstanding, addressed deep orders beyond the sensory facade of the phenomenal world. The dichotomy gave rise to a two-tier epistemology frequently discussed by Song literati: the majority believed that sheer reliance on "hearing and seeing" constituted one type of knowing, albeit one often construed as trivial and inferior; to reach the *dao* (the most profound order of reality), one needed to rely on his heart-mind (*xin* 心) and engage "modeling," a superior type of knowing.[52] While modeling, the knower needed to seek epistemic guidance via the connection between his heart-mind and the deep orders, a practice saliently distinguishable from sensory knowing. This means that the Song knower did have his own share of distrust of the senses, not for fear of illusion but out of an aspiration to achieve intellectual depth.

Here emerges a conceptual structure that renders possible a kind of empiricism comparable to its European counterpart: at least two models of knowing existed, the distinction between them resided in the sources of knowledge, and among the two possible sources, one was sensory knowing. The demarcation was analogous to that between empiricism and rationalism in the European tradition, except that the source of the other type of knowing was not innate knowledge precisely, nor was it a mode of absolute transcendence. Indeed, rationalism did not exist in premodern China. The Chinese experienced similar feelings of uncertainty while wading through the myriad particulars in the universe and wanted to find ways to contain them. This impulse drove them to produce rich systematic knowledge without having to depend on a god or other forms of transcendental idealism.

The foregoing comparison should certainly be subject to more refined analyses in diverse historical contexts. The relationship of Shen's empiricism with the two-tier epistemology, for instance, needs a more nuanced elaboration. Shen's empirical stance was not a simple endorsement of "knowing from hearing and seeing" backed by a renunciation of modeling. In fact, most Song scholars, including Shen, did not translate the conceptual distinction between the two types of knowing into a categorical demarcation in praxis. For some, the two models often worked in sequence, with "hearing and seeing" preceding and preparing for modeling. The distinction between an empiricist and a devotee of modeling (also a system builder in this context) resided in the assignment of priorities. In Shen's praxis, he often prioritized the epistemic guiding power of "hearing and seeing," a tendency he demonstrated in contexts sometimes involving deep orders and sometimes not. His empiricism was the result of subtle negotiations over epistemological commitments, a phenomenon I engage in detail over the course of the book.

In addition to a general consideration of philosophical compatibility, the current project also highlights some characteristics of Shen's empiricism from the vantage point of a historical study. First, in Shen's case, the movement of epistemological commitment toward "hearing and seeing" not only generated a reliance on sensory knowing, it also put emphasis on the epistemic autonomy of an individual knowledge seeker who actively exercised critical skills. These skills, aiming at epistemological reliability, were not limited to the sensory aspects of the world in a rigorous philosophical definition.

The second point concerns the methodology of studying empiricism. Unlike the European philosophers I cited already, Shen did not define his empiricism with normative articulation. He did not, for instance, coin a concept that can be rendered into empiricism in one way or another, nor did he muse elaborately on what sensory perception was and how it worked. His reluctance to propagate general epistemological norms was salient when examined in his own times as well. Unlike his system-building peers, Shen's philosophical musings were disproportionately laconic, a sharp contrast to the volume of his other writings, especially his descriptive accounts of concrete activities.

The lack of normative articulation nevertheless does not invalidate the historical study of empiricism. From a historian's point of view, Shen's empiricism should be understood as a conceptual entity distinctive from a philosophical position defined through normative articulation. It was, instead, an empirical stance, which Shen demonstrated in the form of attitudes, commitments, and approaches in concrete inquiries.[53] The rich description of his epistemic praxis in texts such as *Brush Talks* was the main form he used to textualize this stance. As a matter of fact, the emphasis on praxis—instead of solely on propositional articulation—is not foreign to historical studies of empiricism worldwide, primarily due to the fact that empiricism was (and still is) difficult to define a priori. Scholars who study practitioners of *historia*—the version of empiricism in early modern Europe—also notice that the "epistemology of the empirical" could sometimes only be implicit in praxis and did not always merit much normative verbalization.[54]

Let me further accentuate the methodological message by carrying it beyond the specific case of empiricism. The study of epistemology has remained a broad, inclusive horizon on which various approaches differ from and complement one another. While analytic philosophers focus on theories of knowledge presented in formal propositions, historians venture into concrete processes of knowledge production embedded in cultural and historical contexts. This methodological tradition, known as "historical epistemology," is an important source of inspiration for the current book.[55]

As a result, I do not intend to limit the source of my discussion of Shen's empiricism to formal propositions, which were lacking in historical sources to start with. I will use existing Song concepts, such as learning, pursuit of the *dao*, and the two-tier epistemology, as important organizational components of my analytical narrative. In addition, this book explores Shen's epistemic praxis in depth. I pay attention to what Shen did as well as the assumptions, implications, and tacit commitments involved in his deeds. For example, chapter 4 invokes individuation to reveal the ontological implication of Shen's use of sensory knowing, and chapter 10 identifies the subject-object division as the underlying assumption behind his critical pursuit of epistemological reliability. A grasp of these conceptual entities is essential for understanding Shen's empiricism, as they delineated the directionality of his thinking and motivated his concrete decisions in praxis. By illuminating his "framework of taken-for-granted" as part of my account of his empiricism, I intend to offer an in-depth comparison between Shen and the mainstream Chinese way of knowing.

## *A Note to Readers*

This book is a biography of a person and an idea, and thus a complex account containing two intertwined narratives. The first narrative, encompassing chapters 1, 3, 5, 6, and 8, delineates Shen's life experience as a scholar-official in the context of the Song world, and the second, comprising chapters 2, 4, 7, 9, and 10, analyzes his thinking. Each chapter in the first narrative finds a correspondent section in the second, forming a "life and thought" pair.[56] A commentary on its matching "life" installment, each "thought" chapter highlights the key conceptual issue(s) arising from the particular set of experiences and contextualizes them within the broad themes of Song intellectual history. Such a format serves two purposes key to the thesis of this book. First, the double-narrative structure juxtaposes a detailed account of Shen's life with a thorough analysis of his thinking, fulfilling the mission of an intellectual biography. Second, the independent thought narrative affords ample space for contextualization, allowing me to build the overarching conceptual framework (system versus empiricism) by making extensive connections with key issues in the broad cultural sphere of premodern China. This way, readers will see not only Shen's thinking—the searchlight—but also the teeming landscape of Chinese thought it scans.

Now begins the tale of Shen Gua.

# CHAPTER I

## Peripateting the World (1051–1063)

In the last month of 1076, Shen Gua was appointed as an imperial adviser—a Hanlin academician—and thus joined the top echelon of the Song state. The forty-six-year-old Shen responded to the imperial grace with a confession of his humble past:

> Having risen from extreme obscurity, I have barely led a moral life. Without a fine gift for learning, I have gone low and become increasingly defiled. Having studied with teachers for twelve years, I have failed to make any achievement. Having served in the bureaucracy for ten years, I have come close to being demoted thrice but fortuitously survived. [I] have not yet terminated this ignominious life and returned to the tomb. [I] am ready to be diligent and self-motivated, rising to observe the prosperity of ritual and music.
>
> 伏念臣起身至微,涉德未幾,無良質以受學,從下習以日汗。一紀從師,訖無一業之僅就;十年試吏,鄰於三黜而偶全。未能捽茹苟生,歸老壙埌之下。尚將壹浴自勵,起觀禮樂之興。[1]

Formulaic as this statement might sound, Shen's conflicted feeling on the contrast between his past and present was genuine. At this moment, he had reached the peak of his career. But for a prolonged period in his early life, he struggled and suffered. Shen had been a sensitive, intelligent young boy who lived a relatively privileged life in a literati household until 1051, when his father died. Afterward, he endured career stagnation and financial hardship for more than a decade. During the time of struggle, Shen remained perched between two worlds—one the dignified life of a literatus, hazy in his memory, and the other his own experience of the harsh reality of Song officialdom.

This chapter starts with an account of Shen's family and early life, which planted in his mind an attachment to the literatus ideal, and follows with an analysis of the first decade of his professional life, which witnessed his constant effort to sustain and redefine this ideal while struggling to survive in the immense bureaucracy.

## Son of Literati

Shen often recalled his early life with his family as one of privilege. The most prominent aspect of this privilege, as he repeatedly recollected, was the membership in the officialdom his forebears had earned and the ideal of a cultured life they passed on to him. Shen was born into a household of bureaucrats and degree holders, where he learned at a young age the significance of an official title and cultural prestige. Though most of his relatives served in the middle reaches of the bureaucratic hierarchy and bore merely dim reputations, they sowed in his mind a solid identification with the literati class and an ambition to realize a life germane to this status.

Along with many other literati families at this time, the Shen family did not have a prestigious pedigree, a fact that did not seem to bother them much. Their long history bore only sporadic illustrious moments. The earliest forebears came from Qiantang, an affluent southern town in present-day Zhejiang Province. They constituted a branch of the Shen clan in Wukang (present-day Deqing in Zhejiang Province), which rose to power in the turmoil of warfare toward the end of the Han Dynasty (206 BCE–220 CE).[2] The Shen clan was not one of the aristocratic families that dominated the political scene of early medieval China, but it did produce a series of capable military generals with access to the center of power.[3] In 1051, at the death of their father, Shen Zhou 沈周 (978–1051), Shen Gua and his elder brother, Shen Pi 沈披 (fl. ca. 1050s–1070s), visited Wang Anshi and asked him for an epitaph.[4] Wang did a dutiful job in assembling all aforementioned details into a lofty eulogy, boldly praising that "the [Shen] clan of Wukang has been uniquely prominent under Heaven."[5]

Shen Gua took the matter of pedigree lightly and demonstrated little effort to gild his family history. In the tomb inscription he composed for a relative, Shen Qi 沈起 (fl. ca. 1050s–1070s), he presented the history of the Shen clan in an objective tone: in early times people with the same surname but lower origins migrated from other localities to join the clan in Wukang, so that "the Shen [people] in Wukang do not firmly know the exact origin of themselves."[6]

Shen's disinterest in remote forebears reflected the general identity politics of literati in his times. After the end of the ninth century, the attachment to pedigrees fast dissipated along with the decay of the old aristocracy.[7] To the newly emerged literati, the promise of professional success steadily resided in the meritocratic civil service examinations and an officialdom that awarded examination winners with rank and title. To pursue that prospect, family connections and personal merits were much more relevant than claims of remote, aristocratic blood.

For Shen, the more recent past of his family set the goal for his life and career: to become an official. He was born into generations of government service, beginning with his great grandfather, Shen Chengqing 沈承慶 (fl. ca. 960s–980s), a low-ranking bureaucrat in the Wuyue Kingdom (907–978) prior to the Song unification.[8] Following Chengqing's example, most of his male descendants—including Shen Gua's grandfather, father, uncle, and paternal nephew—held official titles of one kind or another.[9]

Among the male relatives of Shen Gua, Shen Gou 沈遘 (1025–1067), the nephew, was the most accomplished as an official. Though one generation junior to Gua, Gou was six years older and made a name as a renowned official a decade sooner. When compilers of later histories grouped Shen Gou, Shen Gua, and Gou's younger brother, Shen Liao 沈遼 (1032–1085), together under the name "Three Masters of the Shen Clan" (Shen shi san xiansheng 沈氏三先生) in this descending sequence, they not only mistook Gua for the cousin of the other two but also erroneously ranked him below Gou.[10]

Shen Gou claimed an important presence in the Shen family and was the first in generations to access high-ranking political power. Had he not perished prematurely in his forties, he might have become a political luminary. At the age of twenty, he attained the leading place in the *jinshi* examinations, the highest honor a man of his age could aspire to. Soon after, Gou made his name known to Emperor Renzong (r. 1023–1063) by presenting a set of insightful essays on government.[11] This was an early demonstration of an ability to blend "cultural virtuosity" (*wenxue* 文學) with "competence in statecraft" (*zhicai* 治才) his contemporaries credited him with.[12] The emperor promoted him into the Imperial Libraries and then appointed him to draft edicts in the Hanlin Academy. By securing a position in the Imperial Libraries, Shen Gou entered the de facto candidate pool of policy makers; at this moment, he was just a step away from the summit. This precocious man represented the most privileged type of "fast starters" in the Northern Song bureaucracy.[13] By comparison, his uncle Shen Gua did not manage to reach this height until his forties. After eight years serving as an editor and drafter, in 1061 Shen Gou was demoted to Yuezhou (close to present-day Shaoxing) and then to Hangzhou as prefect because of an undisclosed mistake his father had committed. But as soon as Renzong's successor, Yingzong (r. 1063–1067), came to the throne in 1064, he called Shen Gou back to the capital and restored him as an edict editor and later prefect of Kaifeng. Just as Gou was about to ascend to a new height in his career, he died in 1067 at age forty-three while mourning his mother's death.[14]

Shen Gou's influence on Shen Gua was intricate. Their careers overlapped, and each had its ups and downs. But one's rise always seemed to coincide with the other's fall. During the eight years (1054–1061) Gou served in the prestigious post in the Imperial Libraries, Gua struggled alone in local bureaucracy. Later, as Gou temporarily fell out of favor with the emperor and stayed away from the capital until 1064, Gua made a substantial step upward in 1063 with his newly attained *jinshi* degree. Gou's early achievement surely brightened Gua's path, but his influence was indirect. In a world whose pace gradually came to be measured by the progression of civil examinations, an academic degree often appeared to outweigh a personal connection. For Shen Gua, a relative in high power raised a light of hope and remained a source of inspiration, but he still had to work hard for the *jinshi* certificate, an institutional promise of a career breakthrough.

The person who cast the greatest influence on Shen Gua and directly introduced him to an official's life was his father, Shen Zhou. Zhou had followed the career trajectory of a typical midranking bureaucrat in the early Northern Song. His own father died when he was still a boy, but Zhou managed to complete his education and secure the *jinshi* degree at the age of thirty-seven (1015).[15] He then embarked on an itinerant journey in local service across the empire.

During his thirty-year career, Shen Zhou traveled in all three major geographical macro-regions of Song China: he started in the southeast, sojourned in the southwest, and eventually made his way to the north.[16] First a supervisor of the wine trade in Suzhou in the lower Yangzi delta, he then moved westward into the Sichuan basin and became magistrate of Pingquan County. Shen Zhou was apparently well received in Pingquan, and after his departure the locals erected a stone stele to honor his work. He swung back to Suzhou, and in 1039 traveled to Runzhou (present-day Zhenjiang in Jiangsu Province) and became prefect (*cishi* 刺史) there. In 1040 he was appointed to Quanzhou, a town on the southwestern coast.[17] Except for the first post, which engaged specific fiscal responsibilities, Shen Zhou spent almost two decades serving as a local administrator, comprehensively tending to issues of all kinds.

In 1043 Shen Zhou received a critical promotion: his growing reputation as a capable administrator brought him to the capital, Kaifeng. Most Northern Song officials longed to come to the capital city, the nexus of political networks, cultural resources, and magnificent urban scenes.[18] Shen Zhou spent five years as assistant prefect of Kaifeng (*Kaifeng fu pan-guan* 開封府判官). In 1048, at age seventy and perhaps no longer in good health, Zhou returned to the south for his final official appointment, surveillance commissioner in Jiangnan East Circuit (*Jiangnan dong lu an-cha shi* 江南東路按察使). He retired after three years in this post and died in his hometown of Qiantang in 1051.[19]

Shen Zhou devoted his entire official career to concrete statecraft, including fiscal and general administration, and he earned a reputation for competence.[20] In the capacity of an administrator, he developed his own working style: a soft approach blending moral persuasion with legal punishment. While he was prefect of Quanzhou, a legal case brought two wrangling locals to his court. Shen Zhou simply sent them home and ordered them to perform moral reflections on their own. The case thus simmered down without a need to assert formal judgment or enact legal punishment.[21] When the authorities were considering whether to assign him to the position of prefect of Kaifeng, this case specifically caught their attention and eventually convinced them in favor of the promotion.[22] Although never known for brilliance of any kind, Zhou remained competent and reliable at work. The adult Shen Gua took great pride in his father and constantly referred to his achievements as a standard against which to measure his own.

In addition to a deeply experiential understanding of the officialdom, Shen Gua acquired a solid elementary education from his family. The first person who introduced him to the fascinating world of learning was his mother, Née Xu 許 (986–1068).[23] She was a daughter of an established official lineage in Suzhou, an affluent town in the lower Yangzi delta.[24] As a girl she demonstrated a notable affinity for books and an impressive ability for memorizing texts. Like many educated mothers of this time, Née Xu shouldered the responsibility of imparting an elementary classical education to her son and supervised his reading at home.[25] Coming from a highly cultured family, she also introduced Shen Gua to the heritage of his maternal relatives. Née Xu grew up closely with a brother, Xu Dong 許洞 (ca. 976–1015), who had a sound classical education and a strong interest in military affairs.[26] A typical *jinshi* degree holder, Xu Dong served as an official and was known for his expertise on the classic *Zuo's Commentary on the Spring and Autumn Annals* (*Chunqiu Zuo shi zhuan* 春秋左氏傳). A nontypical civilian man, he was interested in martial arts and military skills. In 1005, he completed a book titled *Treatise of the Tiger Seal* (*Huqian jing* 虎鈐經) on the subject of military tactics, which was a significant item in the corpus of Song military writings.[27] Although Gua was born after his uncle's death, he inherited his interest in the classic *Spring and Autumn*, on which he later composed extensive commentaries, and he became fascinated with the military, a proclivity he likely developed from flipping through the *Tiger Seal* on his own.[28] The influence from the maternal side added to that imparted by the degree holders in the paternal clan, and the young Shen Gua experienced a rich and stimulating environment for reading and learning.

Growing up in such a family, Shen Gua became acquainted with the basic meanings of being a literatus in his times: he remained sheltered under the wings of a bureaucrat father, hearing stories about his great-grandfather, grandfather, and uncles all serving in the grand state machinery; he read aloud old texts under his mother's supervision and witnessed the moments of jubilation when his older nephew acquired his *jinshi* degree. In the small world he inhabited, official title and cultural prestige claimed evident significance, and talk of them infiltrated his everyday experience. This milieu allowed him to intuitively grasp the two basic parameters of a literatus's life even before he acquired the ability to articulate them.

## Son of the South

In addition to the experience that applied to literati households across the Song empire, the Shen family had distinctively southern qualities. Situated in the southeast, the Shens struck roots in the heartland of the lower Yangzi delta. Such an origin permanently marked Shen Gua as a southerner. In addition, Shen Zhou's iterant career brought his young son to southern localities far beyond their home place.

The south was bustling, and Shen Gua's eyes were bound to open early. First he saw Zhejiang and Jiangsu, where he was born and his father served the longest. Local kilns made the finest green-glazed ceramics, and weavers produced quality silk commissioned by the state.[29] Shen may have retained some memory of the central Sichuan basin in the southwest, where numerous waterways nurtured the land, grains thrived, and tea grew in abundance in foggy valleys.[30] In late adolescence, Shen arrived on the southeastern coast. Urban Quanzhou, where legal and illegal maritime trade flourished and thousands of immigrants huddled, undoubtedly offered a world of fascination for a boy to admire.[31]

Being a southerner also had consequential implications in the political world. During this period, educated men from the south were still slightly underrepresented in the examination compound and officialdom, and the reason was obvious: the Song erected its capital in the north, the heartland of Chinese civilization since antiquity, and the emperors—northerners themselves—vested more interest in men from their home region.[32]

By the 1040s, when Shen Zhou came to the north and served as the vice head of the capital, the new generation of southerners—himself included—were growing quickly and striving to tip the balance of political power. A steady stream of fellow southerners came to serve in the capital, and some of them had penetrated the top echelon.[33] The capital city was alive with the winds of change due to their arrival. In 1043, when the twelve-year-old Shen Gua stood in front of the city gate of the magnificent Kaifeng and admired the most vibrant city in the world, the setting was right for him to vividly imagine following his father's path and building his own career in the heart of the empire, as an adventurous southerner.

## A Displaced "Clerk"

The glowing dreams of the young Shen Gua, however, met harsh reality in 1051. In the following years, he identified himself as a "clerk" (*li* 吏) and constantly lamented his inability to keep up the literatus lifestyle he once enjoyed.[34] Although he did find some silver linings, such as taking on new skills on bureaucratic posts and extending his path into the vicinity of the capital, this decade primarily witnessed his struggles with sluggish promotion, financial hardship, and a profound sense of displacement.

The year 1051 was most certainly a hard one for the Shen family. Shen Zhou, the sole provider, died near his hometown, leaving his wife and two sons adrift. No remarkable signs indicated that the Shen family enjoyed an opulent life while Zhou was alive, and he certainly left behind little personal wealth. The loss of his father and a stable source of income became Shen Gua's first traumatic memory. It was imperative that the brothers, now in their early twenties, leave the house to seek a new life.

Owing to the privileges their late father had acquired, they soon found their way. Both entered the officialdom through "protection" (*yin* 蔭) without first having to earn an academic degree. The Song state conferred the privilege of protection on its mid- and high-ranking officials, in which case their relatives (mostly direct descendants) could take offices without having to attend civil examinations.[35] As such, Shen Gua quickly detached from his life as a schoolboy and became registrar of Shuyang (*Shuyang zhubu* 沭陽主簿), a county in the northern part of modern Jiangsu Province.[36] Shen Pi also obtained a post. His earliest appointment that we know of was magistrate of Ningguo (*Ningguo xianling* 寧國縣令, in present-day Anhui Province).[37]

This was when and how Shen Gua's decade as a lowly "clerk" started. From 1054 through 1062, he held a number of local posts without receiving any significant promotion. After finishing his tenure as registrar of Shuyang, in 1055 he was appointed surrogate magistrate of Donghai (*She Donghai xianling* 攝東海縣令), in a neighboring county.[38] In the following years, repetitive assignments recurred monotonously.

When Shen Gua recollected this peripatetic career trajectory and called himself a "clerk," he was clearly bitter. The basic meaning of the original term for "clerk," *li*, designated an assistant who served below the normal officialdom. A clerk was a stranger to the cultured world: he received no formal education and worked through drudgeries with no prospect of joining the elite ranks.[39] With neither educational credentials nor an office, a clerk was not considered a literatus. In truth, Shen Gua was never a clerk, and given his family background, he never had been. Institutionally speaking, he was a member of the formal officialdom from his very first appointment.

Besides elements of whining and self-pity, Shen's frustrating self-address exposed the hard reality that confronted all low-ranking Song officials: gaining the membership in the officialdom did not necessarily guarantee status or wealth. In this decade, everything Shen associated with a literatus's life—decent rank, cultural prestige, and financial security—eluded his reach. First, he faced an extremely dim prospect of gaining promotions and rising above the bottom stratum of the officialdom. This problem was especially acute for those who entered the officialdom through "protection." The promotion rate for the low and middle reaches of the Song bureaucracy was already slow in general, and the situation was only worse for those without a proper degree. A seemingly generous privilege, "protection" also imposed a glass ceiling. Without having to go through the examinations, protection-based bureaucrats enjoyed a relatively easy start, yet had to endure longer rotations and more restrictions in future promotions.[40] If a protection recipient followed the regular pace and did not gain access to the trajectory of degree holders, he could expect to spend his entire life in the base stratum. That was precisely why Shen Gua found himself mired in repetitive low-ranking appointments for a decade; he had to take the examinations before he could lift himself above the morass.

With slow promotion came financial hardship, a real issue that troubled all low-ranking bureaucrats in his times. The early Song state granted their officials high social prestige but low financial compensation. Even after a few systematic raises in the 1010s and 1050s, the salaries bureaucrats received were far from impressive, especially for those in the bottom strata.[41] The monthly payment Shen Gua received in this decade fell in the range of 6,000 to 8,000 *wen*.[42] Around this time, the average price for a unit of rice (approximately twelve pounds) in his region fluctuated between 70 and 100, and a pound of pork would cost at least 30.[43] A quality brush was worth 100, and an inkstone of supreme quality could easily exceed 10,000.[44] At this time, Shen Gua was married to his first wife, who had given birth to their first son.[45] In his repeated complaints, Shen gave the impression that he had great difficulty supporting a family of three plus his aging mother with such a limited income. As he once explained to a colleague why he stuck with his poorly paid magistrate job: "Clothing, rice, and salt, none of them can be missed even for a single day so as not to let [me and my family] starve. How can I act differently from others?"[46]

In addition to being impecunious, Shen Gua barely found anything associated with the cultured status of a literatus in these jobs; the daily routine was tedious and lacked dignity. In a long, grievance-ridden letter to a friend named Cui Zhao 崔肇 (ca. eleventh century), he grumbled that his job was "most inferior and grueling," featuring a schedule overloaded with endless responsibilities, including "socializing interactions and condolence visits, seasonal festivals and rituals, public duties and private errands."[47] The swirl of chores drove him around in an incessant flow of motion, as he recalled: "one moment I am in the upper area and the next in the lower; one minute I am in the south and the next in the north."[48] Eventually, he no longer felt his mind, as it was "dull and numb."[49] He lost contact with his sensations, so that "Heaven and Earth no longer appear to me as Heaven and Earth, nor do snow and wind seem dark or chilly."[50]

The current situation contrasted sharply with the life Shen used to live with his father, or at least with the version in his own recollection. In deep frustration, he reminisced about the past in glowing terms. The bona fide literatus's lifestyle he previously experienced was dignified. But now he could no longer "make [his] usual broad strides with [his] head held high."[51] Also, a true literatus lived immersed in culture. When his father was alive, Shen enjoyed the company of "literati from all under Heaven" (*tianxia shidafu* 天下士大夫).[52] Now, in the tiny and cramped world he inhabited, he only sought to "complete one bureaucratic appointment" (*wan yu yi guan* 完於一官) and "exhaust a lone self" (*jin yu yi shen* 盡於一身).[53]

Despite all this distress, Shen Gua nevertheless exhorted himself to remain diligent at work and focus on learning new skills. "Since I *am* now serving as a clerk," he persuaded himself, "I should compare myself to other local clerks (*xiangren zhi wei li zhe* 鄉人之為吏者) in terms of competence (*neng* 能) and skills (*yi* 藝), and I am not entitled to any difference."[54] "Competence and skills" referred to all kinds of practical techniques a local official employed to perform his many duties. They ranged from highly technical expertise in hydraulics to general administrative aptitude. Shen proved himself to be a fast learner in this area, and, although he acquired these skills in a regrettable situation, they remained a consistent aspect of his professional persona and later took on important intellectual meanings.

A southerner who grew up among rivers and rice paddies, Shen revealed his talents first as a capable hydraulic engineer. In his observation, the Jiangnan area (lower Yangzi delta) was prominently mountainous, so arable lands were all low and wet.[55] In this natural environment, effective supervision of agricultural activities required an ability to handle matters concerning water. Shen, along with his brother, Shen Pi, seemed to have cultivated some expertise in hydraulics at a young age, and they made use of this skill in their earliest appointments.

Shen first applied his acumen in hydraulic engineering in the north. Except for his short visit in Ningguo (a county in the Eastern Circuit of Jiangnan and south of the Huai River), all known jobs he received were located in the coastal region below the Shandong peninsula, an area that belonged to the macro-region Qi-Lu 齊魯. It was one of the few neighboring areas that supplied the capital, Kaifeng; thus its agricultural productivity was crucial at the local and national levels.[56] Since the inception of the dynasty, the court kept investing extravagant funds in the area's hydraulic infrastructure.[57] While serving in Shuyang, a county in the Qi-Lu region, Shen built a flood control system with a series of dams and levees over the Shu River. The system enabled farmers to reclaim a large area of arable land, a boon to the local economy.[58]

Later in the early 1060s, Shen joined his brother in conducting hydraulic projects in his home region. While Shen Pi served as magistrate of Ningguo, he made a joint effort with two other officials, Zhang Yong 張顒 (1008–1086) and Xie Jingwen 謝景溫 (1021–1098), to revamp an embankment system known as the Wanchun Dike (*Wanchun wei* 萬春圩). The old dike was destroyed in a flood in the late tenth century and had been abandoned for eight decades. Shen Pi and his colleagues carried out a detailed survey of the geographical conditions and eventually persuaded the state to authorize the project by refuting five opposing arguments. The new dike proved to be well designed, and it protected large tracts of land from extraordinary floods.[59] During the construction, Shen Gua was visiting his brother and taking a sojourn in Ningguo. Although scholars debate whether Gua participated in the construction, he did witness the event and documented it in great detail.[60] Given the depth of engagement he demonstrated in producing the record, it is likely that he played at least a consulting role in the project.

In addition to hydraulic expertise, Shen honed his leadership skills while tending to the small rural populations. He strove to be a benevolent official receptive to the concerns of local people. In his opinion, the close relationship he cultivated with his subjects was precisely the reason he received his first promotion, one that advanced him from a local registrar to a magistrate. In a letter to an acquaintance, possibly another local official, Shen spoke of the promotion with a mixture of pride and modesty: he denied that it had much to do with any outstanding talent he possessed, so to speak, but that it certainly testified to his bond with the local population, his deep understanding of their needs, and his flexible spirit to accommodate their requirements in policy making.[61]

For Shen, the true challenge in developing leadership abilities on these jobs came with the cultural shock he personally experienced. The culture and geography of the Qi-Lu stood in sharp contrast to the southern regions he knew. When Shen first landed here, he found many local customs foreign and difficult to manage.

To meet the challenge, Shen cultivated a habit of observing. He maintained a keenness for particular details in the places he served and recorded them in writing. He noted that residents in Qi-Lu tended to be "headstrong and recalcitrant."[62] In his (not so respectful) description, the locals "frequently slaughtered cows and skinned pigs at home," and when they conducted daily affairs, they were "strong-hoofed" and "stretched their bows to the full."[63] To rule a population like this, he would be "unable to rule effectively without some techniques (*shu* 術)."[64] One central technique was to assert the authority structure and streamline the organizational form. For instance, after Shen arrived in Shuyang to supervise the hydraulic project, the local laborers launched at least two revolts within a short time. Shen sensed pent-up frustration and identified the cause of unrest as ineffective communication. Before his arrival the previous authority had issued a total of twenty-one orders since the start of the project, sometimes multiple orders in a single day. "The orders were not assertive, and the boundary [between fields] not established, hence the riots."[65] Shen's initial curiosity in analyzing the Qi-Lu customs certainly owed to his motivation to perform the job well (especially as a southerner). As his peripatetic life across different localities extended to a decade, curiosity developed into a habit of observation, and this trait assisted him in adapting quickly to new environments.

## *A Reluctant and Eager Official*

Years into working as a "clerk," Shen started to grasp the true meanings of being a literatus in his situation, and on that basis he developed new insights regarding this cultural identity. Among several thoughts, one seemed to occupy him most: how to coordinate a literatus's professional ambition with his moral status. Many people in his times believed that the chase of rank and title was a sign of moral weakness, because it revealed a greedy attachment to "profit" (*li* 利), whereas detachment from career advancement was a virtue, because it signaled invulnerability in guarding moral righteousness (*yi* 義). The righteousness-profit dichotomy was the source of long-running and intensive debates in Song times, and morality was an indispensable criterion when evaluating the intellectual qualifications of a literatus.[66]

Such a formula posed a dilemma for a bottom-level official like Shen. As a literatus of his age, he had no other means of financial support, so taking office and seeking promotions constituted the only way to maintain status and dignity. But excessive eagerness in pursuing career advancement would invite the aforementioned moral suspicion and tarnish his honor. Accepting his situation of having to serve, Shen, like any peer, was a "reluctant" official. Craving dignity (in the form of rank and title), he nevertheless also acted as an "eager" striver who thus constantly felt the need for moral self-defense.

To Shen, "reluctance" accurately characterized the situation of all literati in his times. He saw clearly that he and his peers depended on the state for their livelihood. As he exclaimed to an unnamed colleague: "How can it be my own desire to stay [on this job] for four years! Being as impoverished as I am, I have no choice but to settle for it."[67] He explicitly expressed that he settled for the current low placement because he had no other choice (financial means). Scholar-officials were professional bureaucrats whose livelihood primarily (if not always completely) depended on their office holdings. For most of them, the earnings they received as government officials constituted the sole source of income, which made such employment mandatory for self-sufficiency. In a letter addressed to a colleague named Xu 徐, Shen commented on this situation with palpable despair:

> Literati nowadays do not own farmlands to grow food, nor mulberry and flax to make garments, but they cannot thus give up on providing for their flesh and blood and die.
>
> 今之為士，無田畝以爲之食，無桑枲以爲之衣，不可遂棄骨肉之養而死亡之是蹈。[68]

Therefore,

> If one does not take office, are there other more deserving options [at all]?
>
> 不出於仕祿，則將何適而不枉?[69]

Such reluctance was a relatively recent phenomenon, closely associated with the new social characteristics of the literati class, especially their nonaristocratic origins. Many members of the bygone aristocracy were independently wealthy and lived off inherited land assets, which contrasted with the living condition of Song literati as dependents of the state.[70] The difference was a key point in Shen's imagination of the past;

as he once emphatically pointed out, "scholar-officials (*shidafu* 士大夫) from antiquity possessed land and received salaries."[71] Such double financial security was a great privilege absent from his own experience. In theory, the moral problem was a less critical issue for aristocrats, because they often enjoyed financial independence from employment and thus could afford to detach themselves from career advancement.

Regarding the "eager" part of his professional persona, Shen was full of self-defense. In the first place, he indeed seemed to be competitive. Judging by the defensive tone of a number of letters between him and colleagues, we can safely assume that criticisms for greed and moral deficiency had already been directed at him.[72]

To put a positive light on his actions and avoid appearing to be a striver, Shen suggested a dissociation of morality and career building, and he raised this point repeatedly in discussions with peers. In effect, he wanted office holding to be seen as a purely professional matter with little moral burden. On the one hand, he exhorted his peers to accept realistic expectations and "not to whine about being treated by superiors in inappropriate ways," because "literati have no choice (*bu de yi* 不得已) and have to take office."[73] This was to exempt bureaucratic authorities from their "moral" obligations.

On the other hand, Shen strongly opposed imputing moral meanings to one's career plans, whether to actively seek advancement or to lie low in the spirit of detachment. In his own words, "a literatus should not view 'retreat' (*tui* 退) as being morally worthy (*xian* 賢); those who consider 'retreat' as being worthy in fact fixate on profit."[74] Here he was speaking with mild cynicism, implying that an apparent surrender of ambitions (retreat) could be a circuitous strategy for seeking advancement, as the moral high ground accorded to self-effacement might open doors to further opportunities. He then asserted in the name of the "superior man" (*junzi* 君子), the role model for all literati, that the superior man did not make decisions to "advance" or "retreat" based on moralistic labels people might unjustifiably attach to him.[75]

In place of moral dogmatism, Shen suggested the "heart-mind" as a more reasonable standard in judging the validity of an action: "do something when your heart-mind is at peace (*an* 安) with it, and relinquish it when your heart-mind is not."[76] He specifically emphasized that the natural impulses of one's heart-mind provided more accurate guidance for one's behavior: "if one does something which incurs distress in his heart-mind, even though this undertaking may be righteous, he is still harming himself (*zizei* 自賊)."[77]

This was not, however, a modern call for "following your heart." What enabled the heart-mind to become a gauge of morality was its connection with the cosmic force, Heaven (*tian* 天). As the ancient thinker Mengzi (372–289 BCE) famously claimed,

> To exhaust one's heart-mind so that one knows his nature. To know one's nature, one knows Heaven.
>
> 盡其心者，知其性也。知其性，則知天矣。[78]

Shen endorsed Mengzi's stipulation of the heart-mind, as he repeated verbatim this point in his own commentary of *Mengzi*, a text on which he spent considerable time at this stage of his life.[79] Following this line of argument, the impulses one sensed in his heart-mind were surely more illuminating than existing moral precepts because they were live manifestations of the perfect cosmic order.[80] By invoking the fundamental mechanism that governed the generation of right beliefs, Shen gently and firmly refuted current moral labels and thus defended his aggressive self-promotion with an intellectual counteraction.

Though his line of argument was clever and reasonably persuasive, Shen did not succeed in clearing his name. In fact, long after he climbed out of the quagmire of serving as a "clerk" and secured a foothold in the top echelon of the state, peers still questioned his moral stature from time to time. The anxiety and self-defense at this early moment foreshadowed his future reputation as someone who combined exceptional competence and moral ambivalence.

Let's return to the speech Shen made at the opening of this chapter, where at the most glorious moment of his career he could not help but recollect the struggles of the early years. The time he spent in "extreme obscurity" remained a powerful memory in his heart for complex reasons. It was the first time he left the haven of the family to plunge into the real world on his own, only to find that the literatus lifestyle he once enjoyed was not guaranteed. It was also the first time that Shen, with his southern disposition and sensibilities, started to explore the northern part of the empire. During the decade he endured poverty and felt twinges of self-doubt, but he acquired new practical skills and developed new insights into the pursuit of an official career. One component critical to the life design of a literatus nevertheless demands further discussion: what about learning? That is the central topic I explore in the following chapter.

# CHAPTER 2

# *Envisioning Learning*

Whenever "Clerk" Shen Gua had a moment of solitude away from daily drudgery, he turned to thinking about "learning," a therapeutic word that was soothing to his mind. For any Song literatus, "learning" defined what they did and who they were. In learning, a literatus sought cultural credentials and career success. Driven by a clear understanding of its significance, Shen Gua demonstrated a salient devotion to learning, and more important, he set out to actively discover his own way of engaging it.

This chapter starts with a discussion of the general features of learning in the Northern Song and then examines an important concept, "things" (*wu* 物).[1] I discuss the role of "things" in the definition of learning, with a focus on a consequent metaphysical issue. In the last two sections, I discuss how Shen developed a robust interest in "things" and how he sought to incorporate it into his vision of learning.

## *Learning and Its Two Features*

Shen embarked on his journey of learning during an exciting time. The culmination of two long-term changes characterized mainstream learning in the mid-eleventh century. The first regarded the pursuit of the *dao* as the goal of learning.[2] Starting in the ninth century, a group of revolutionary-minded scholars renounced the obsession with prose and enlisted literary composition to serve larger purposes, the central one being to understand the *dao*. This movement was known as the Ancient-Style Prose Movement (*guwen yundong* 古文運動).[3] In the following centuries, the *dao* became the central subject that occupied almost every Song thinker as well as the conceptual foothold from which all varieties of "learning" derived.

Generally speaking, the *dao* was an ideal order that governed all entities (including humans) and processes in the world. As A. C. Graham aptly puts it, the *dao* laid "down the lines along which everything moved."[4] By definition the *dao* bore unique ontological features. For one thing, it did not exist in a transcendental realm and thus could not be spoken of independently from experience. Song thinker Cheng Hao characterized the *dao* as inextricably woven into the world: "There is not a thing external to the *dao*, and there is no *dao* external to things."[5]

For another thing, the *dao* was neither an entity nor an object. Graham's metaphor correctly prompts readers to conceptualize it as a modality or process. A number of eleventh-century thinkers described the *dao* as a process with no static reality. For example, according to Zhang Lei 張耒 (1054–1114), "the movement [of the *dao*] has no trace; it encompasses the utilities of the ten thousand things albeit bearing no fixed name; it tracks the transformations of the ten thousand things albeit having no fixed form."[6]

A consensus arose among eleventh-century literati that the *dao* was not meant to be subject to verbal articulation or confined to any normative definition. For example, Wang Anshi asserted that "conceptions cannot reach all places the *dao* exists, and words cannot exhaust all places where conceptions reach," and, on a related note, that "the *dao* resides in the void, the desolate, and the invisible."[7] Many Song thinkers characterized the inexpressibility of the *dao* as having no "name" (*ming* 名). Shen Gua stated that the word *dao* itself was an imposition, that "the *dao* is invisible; the ancients imposed a name ("dao") on it."[8] Su Shi pointed out that to invoke names (verbalization) was an erroneous means to approach the *dao*; in his judgment, "among those who speak of the *dao* in the world, some name what they see, and some speculate about what they cannot see; both pursue the *dao* in wrong ways."[9] Cheng Yi, another renowned peer, deemed the *dao* "too large to be named."[10]

The *dao* nevertheless remained an ultimately attractive aspiration for the literati. Many Song thinkers believed that anyone fully drawn to the *dao* qualified to be a sage.[11] Without having to resort to self-control or willpower, a sage would respond appropriately to any situation and achieve a satisfactory outcome.[12] Thus, despite its challenging elusiveness, the *dao* remained at the center of scholarly discourse and the learning agenda of the literati.

To render the ineffable *dao* as an accessible subject in discursive learning, Song literati envisioned various orders at different levels in the cosmos and often invoked the ancient sages as guiding authorities in exploring these orders.[13] The sages were taken as precedent observers of the *dao* in high antiquity, and their versions of this cosmic order, the so-called *dao* of sages (*shengren zhi dao* 聖人之道), became the most prominent motif in Song intellectual discourse. For the purpose of understanding social, political, and cultural orders in antiquity, Song literati turned to old texts, especially the Classics, for testimonies of sages' words and deeds. Classical exegesis became the dominant praxis of learning. In this sequence of reasoning, textual learning was just a means, and pursuit of the *dao* remained the ultimate end.

In addition to the prominence of the *dao*, the second feature of learning culminating into great salience at this time was its active disposition. Literati practiced philosophical contemplation and essay composition as the basic procedures of learning, yet they produced discursive knowledge for an active aspiration: to create order in the world they inhabited. More specifically, they aimed to build a better government in the capacity of scholar-officials.[14]

The active temperament of learning led to several substantive consequences. For instance, it was both cause and effect of important reforms in civil examinations in the mid-eleventh century. In the 1040s and 1050s, the leading minister Ouyang Xiu 歐陽修 (1007–1072) campaigned to highlight "policy questions" (*cewen* 策問) in the *jinshi* curriculum and gave the new subject priority over old ones, such as poetry.[15] To appropriately respond to a policy question, an examinee was supposed to deliberate contemporary statecraft issues in light of classical references.[16] The young Shen Gua was right in the middle of this development. The moment he swore his devotion to learning, he joined a Classics-based, contemporary ideological enterprise that actively advanced discourse into political solutions.

## *The Dual Nature of "Things"*

Both characteristics of learning were closely related to "things," a concept that enjoyed unprecedented prominence in the Song. In the Song discourse, *wu* referred both to what we call "objects" and "affairs" in combination. This concept designated all entities (sometimes including humans) and processes in the phenomenal world.[17] "Ten thousand things" (*wanwu* 萬物) was often synecdochic with the experiential world in its entirety.[18]

The philosophical significance of the ten thousand things escalated particularly in the Song period. The advocates of the Ancient-Style Prose Movement waged a war against Buddhism, which viewed the phenomenal world as impermanent and transient, thus deluding people from grasping the Buddhist truth. In an effort to revive Confucianism—an enterprise grounded in the phenomenal world—Song literati actively worked to dismantle the Buddhist denial by imputing new meanings to the ten thousand things.[19]

Two peculiarities of "things" are worth notice. First, beyond the primary designation of "objects," "things" also connoted affairs.[20] Another term, *shi* 事, which more exclusively referred to affairs and undertakings, was often used by Song literati interchangeably with *wu*. As Cheng Yi pointed out, "speaking of *wu*, any *shi* one encounters is *wu*."[21]

The second characteristic of *wu* was its dual nature. A "thing" was distinctive; being a *wu* was to be an object/affair among other objects/affairs.[22] A "thing" had some durable reality and maintained a limited, individuated identity. Generally speaking, an individuated object can possibly claim possession of an independent existence, a substance, and definable properties such as shape and color. The Song thinkers often relied on sensory qualities to define this level of reality of a "thing."

A "thing," however, was not limited to a sensory appearance. More importantly, it was the reification of a place in a larger order. To define a "thing" in terms of its "placement" enabled Song thinkers to connect it to the *dao*, the ultimate order of reality. By positing a philosophical connection between every single "thing" and a unitary order, literati built a meaningful and intelligible cosmos out of the particulars.

Song literati expressed the dual nature of "things" in terms of a binary between "knowing from hearing and seeing" and "knowing from virtuous nature" (what I render as "modeling"). The most famous scholars who spoke of this dichotomy were Zhang Zai and Cheng Yi. To know from hearing and seeing in a rigorously philosophical sense (as Zhang and Cheng defined it) was to derive knowledge regarding the sensory qualities of a "thing" via one's visual and auditory modalities, thus corresponding to the first aspect of a "thing." To know from "virtuous nature," or to "model," was an inquiry into the placement of a "thing" in terms of relationality, which addressed the second and deeper aspect of a "thing."[23]

Song thinkers imputed much intellectual significance to the placement of a "thing," and they concocted an elaborate philosophical scheme to frame the issue. First, they posited a generative thesis to relate a "thing" to the *dao*. The thesis featured a famous dyad, "origin" (*ben* 本, "fundamental" when used as an adjective) and "branch" (*mo* 末, "trivial" as an adjective). The *dao* was the origin, and the ten thousand things were the branches. As Wang Anshi asserted: "In the *dao* there is an origin and there are branches. The origin is by which the ten thousand things are born, and the branches are the ten thousand things in their completion."[24] Wang ascribed an "origin" (the *dao*) which gave birth to the "branches" (the ten thousand things).[25] The origin, to which the branches owed their existence, presumably emerged first and claimed the ultimate "fundamentality" (*ben*-ness). The branches appeared later and thus stood for triviality (*mo*-ness).

In the stipulation of some other Song thinkers, this generative thesis acquired a more elaborate look by incorporating more orders—what I call intermediate stages—between the *dao* and "things." A most famous formula came from Shao Yong, who presented it in different versions. The first version read: "The Great Ultimate (*taiji* 太極) is one. Unmoving, it gives birth to two (duality). Duality is the numinous. The numinous gives birth to number. Number gives birth to figure. Figure gives birth to objects (*qi* 器)."[26] An alternative version followed: "*Yin* and *yang* separate and generate the two modes. The two modes interact and generate the four figures. The four figures interact and generate the eight trigrams. The eight trigrams interact and generate the ten thousand things."[27]

Both versions came from Shao Yong's numerological work based on the *Classic of Change* (*Yijing* 易經).[28] Together they stipulated a generative scheme more complicated and yet perfectly consistent with Wang Anshi's simpler thesis. First, just like Wang's stipulation, this scheme began with the *dao*, with which the Great Ultimate in the first version was synonymous (in Shao's diction).[29] Second, Shao's formula ended with the "ten thousand things" in the second version, with *qi* (objects) as its alternative phrasing in the first statement. The two versions thus can be combined as follows:

*dao* → the numinous (*shen* 神) → number (*shu* 數, e.g., yinyang) → figure (*xiang* 象, e.g., hexagrams) → ten thousand things.[30]

In between the *dao* and "things," Shao stipulated multiple developmental stages—the numinous, number, and figure—each standing for a different order of reality. In coordination with the origin-branch ascription, these stages sequentially decreased in fundamentality and increased in accessibility and concreteness. The numinous was, according to multiple Song scholars' descriptions, an enigmatic state that was "without a definite form" (*wufang* 無方), "not verbalizable" (*buke yan* 不可言) and "could not be accessed through reflection" (*buke zhi si* 不可致思).[31] The next stage, number, was more concrete: it encompassed a rich body of numerical relations, such as the most famous yinyang dyad and the Five Processes. "Figure," which followed, included basic images, such as trigrams and hexagrams in the system of *Change*.

While this formula might look like part of Shao's esoteric numerological invention, it was in fact a popular theme in Song thinking, though not always in the complete and systematic aspect Shao posited. Given its provenance in the *Classic of Change*, the generative thesis permeated Song writings of many kinds. Let's return to Wang Anshi, for instance.[32] Although Wang spoke only of the *dao* and "things," he was well aware of the intermediate orders and would recount the connection between number and "things" in fastidious detail:

As I see it, when the Five Processes become things: (they generate the aspects of things such as) the time, the place, the material, the *qi*, the nature, the shape, the engagement, the condition, the color, the sound, the smell, the flavor. Each has its complementary half, and so do [things] extend and disperse.[33]

The Five Processes constituted a prominent example of number. Therefore, number gave rise to many an object with a life span, a physical location, a certain shape, color, sound, and smell—in a nutshell, "things" with sensory qualities.

Of special note was the nature of this generative process: it was a relation-oriented scheme accounting for the divisibility of the world and interactions within it. The formula was not, I emphatically point out, a process based on the transformations of substance, such as how atoms form molecules and then generate matter. The numinous, number, and figure—like the *dao*—were modalities instead of stuff. They structured reality instead of illuminating its substance. Number was supposed to divide the cosmos according to numerical ratios, and figure ordered the world through illuminating connections among particular cases via a repertoire of basic images.[34]

What the generative process generated in regard to "things" were their places in larger orders. These orders might consist of numerical relations or imagery correlations, among other possibilities. A single "thing" had its place among these relations, and a correct understanding of the "thing" meant to seek this place. When all "things" find their due slots and move on the right tracks, it is called the *dao*. Thus to deal with a "thing" in accordance with the *dao* was to ensure its "well-placed-ness."[35]

One should note, however, that such well-placed-ness did not concern only cosmological orders such as number and figure, which many eleventh-century literati considered to be prior to the birth of the ten thousand things.[36] Particular "things" were also supposed to find their proper places in orders that presumably originated in the phenomenal world and claimed weaker ontological priority to the ten thousand things. Orders of morality and governance were the most prominent examples of this kind. In chapter 7 I revisit this issue with a richer repertoire of intermediate orders extracted from Song writings.

In sum, "things" were dual in nature. A particular "thing" always had a distinctive appearance accessible to human perception; more importantly, it was supposed to have a place in larger orders of the world. For an aspirant literatus, the latter attribute bore much more significance, because it assisted him in dealing with the particular "thing" in accordance with the *dao*. A *dao* seeker deemed the sensory properties of a "thing" trivial, because they did not reflect the place of the "thing" amid the fundamental structures of the world. Unsurprisingly, the seeker was less keen on knowing from hearing and seeing, a practice that would potentially hamper his understanding of deep relationality.

## Learning vis-à-vis "Things"

The dual nature of "things" extended beyond philosophical stipulation. When "things" joined learning, far-reaching consequences shaped the practice of the latter. In the eleventh century, one could not speak of learning without paying attention to "things."[37] The previously mentioned characteristics of learning—a focus on the *dao* and an active disposition— should be understood in a world of "things." In the first place, "things" constituted the inevitable environment of learning because they stood for the phenomenal world, which Song literati actively defended in the face of Buddhist rejection of it. Instead of seeking the *dao* in other possible worlds, Song scholars pursued it among real entities and processes.

Second, the active agenda of learning exhorted a literatus to engage objects and affairs, often (though not exclusively) in the context of statecraft. "Things" thus often appeared in the discourse regarding government. To cite one famous example, Wang Anshi applauded the fact that Emperor Shenzong (r. 1067–1085) had "intentions of treating the population with humaneness and things with care" (*renmin aiwu zhi yi* 仁民 愛物之意).[38] The invocation of "things" in this statement was formulaic rather than singular.[39] At the time Wang was writing, "things" as the context of ordering the world (statecraft) had already been established as a discursive convention.

While "things" constituted an indispensable context of learning and influenced its content, every *dao*-seeking literatus understood that while they dealt with "things," they should constantly endeavor to stay in accord with the *dao*. Therefore, they had a distinctive preference when facing the dual nature of "things": the place of a "thing" in larger orders was much more important than its properties as an individuated object. Knowing from hearing and seeing, which focused on the sensory properties of "things," was thus regarded as an inferior type of knowing in the scheme of learning.

This was precisely why in some contexts the intellectually important "things" became intellectually distracting—an apparent oxymoron caused by their dual nature. "Things" in their individuated existences, especially when identified with their sensory qualities, presented distractions to the goal of seeking the *dao*. "Things" in their obstructive capacity were often referred to as "external things" (*waiwu* 外物) and as such were considered external to the *dao*-seeking mission.

The reservation regarding "external things" arose within the Confucian tradition since antiquity. Originated in a famous passage of *Mengzi*, this line of argument admonished readers that one should engage "things" with the heart-mind instead of the senses. The sensory organs were "things," too. A superficial thing–thing interaction would only lead one astray. Mengzi advised as follows:

> It is not the function of the ear or the eye to reflect, and so they can be deluded by things. Being unreflective, when they come in contact with other things, they are led astray. The function of the heart-mind is to reflect. When it reflects, it gets things right; if it does not reflect, it cannot get things right.[40]

"External things" frequented the writings of Northern Song literati with the same dubious reputation. For example, Wang Anshi resonated with Mengzi by pointing out that the misconceptions of "external things" induced by appealing to sensory perception posed a serious challenge. To solve the problem, Wang suggested that one should hold his heavenly endowed nature (*xing* 性) steadfast.[41] When one remained in a lucid and stable connection with his nature, "things under Heaven will not be able to rattle my hearing" and "things under Heaven will not be able to disrupt my sight."[42] Su Shi made a similar comment that a superior man could "lodge" (*yu* 寓), or put up temporarily in "things," but should never "abide" (*liu* 留) in them, because "the five colors make one's eyes blind; the five notes make his ears deaf; the five tastes injure his palate."[43]

Wang's and Su's discussions provided concrete examples of how "external things" became a problem and provoked ongoing philosophical contemplation in the eleventh century. Scholars focused attention on illuminating fundamental orders in which "things" were properly placed and on such basis formulated several famous philosophical propositions, such as the Cheng brothers' *gewu* 格物 (investigating things) and Shao Yong's *guanwu* 觀物 (observing things). Whether to investigate or to observe, vis-à-vis "things," these thinkers were motivated by a concern to reveal the larger orders in which "things" were deployed in accord with the *dao*. The Cheng brothers (and later Zhu Xi) named their focal order *li* (Coherence), and Shao Yong named his number.[44] The concern with the placement of "things" served as the implicit background of learning in general.

Together, the Song literati directed their philosophical gaze to orders that contained a "thing" rather than setting the foothold in its perceptible properties. In praxis, this philosophy corroborated the mainstream methodology of learning which relied primarily on textual studies: literati contemplated and discoursed about fundamental orders, structures, and patterns, which then provided orientation in dealing with particular "things."

The formula of learning should be understood with a caveat, however: the problem of "things" did not entail a total renunciation of a sensory understanding of them. The dual facets of a "thing" were not segregated or mutually exclusive, nor were the two correspondent ways of knowing. In learning, the two sides (in both dichotomies) were often connected; the difference lay in the attribution of priorities. A devoted learner should surely focus on the placement of "things" by not indulging too much in the sensory facade of the world, but he should not completely withhold epistemological commitments to sensory knowledge. Conventionally speaking, when one contemplated on the placement of a "thing," he often could not help possessing some sensory knowledge of the "thing." In a way, knowledge from hearing and seeing provided a preparatory springboard for the higher form of knowing.[45] "External things" would only become distractions if one lingered too long in the preparatory stage. Effective navigation of learning required a proper deployment of priorities between the two epistemic practices, not the choice of one to the exclusion of the other.

## Shen Gua's Interest in "Things"

When the young Shen Gua eagerly envisioned his participation in learning, the whole scheme of a dual approach to learning vis-à-vis "things" was available as the most important background. Although not prominently vocal in the metaphysical discussions, he developed a strong interest in "things," a robust proclivity to appreciate the intricacies of distinctive objects and processes. Shen's devotion seemed to imply a new possibility of learning, one that presumed desirable intellectual outcomes from engaging "things" in a hands-on way, a hypothesis he had yet to articulate as a concrete plan at this early stage.

Judging from the results, Shen managed to excel at learning. He successfully passed the examinations and received his *jinshi* degree in 1063. He studied the Classics extensively despite a busy schedule at work, a diligence attested to by the number of classical commentaries listed in his oeuvre. Shen authored two chapters of commentaries on the *Classic of Change*, a few chapters of work on the *Spring and Autumn*, a systematic reorganization of the *Zuo's Commentary* in fifty chapters, and a chapter's worth of commentary on *Mengzi*.[46]

Although most of these writings were lost and left no direct clues for precise dating, circumstantial evidence strongly implies that Shen produced much of the work early in his career. For example, his interest in the *Spring and Autumn* probably blossomed early in his life, before he started working under Wang Anshi in the 1060s. Wang was known for his dislike of this particular classic; he expunged it from the examination curriculum he designed in the 1070s.[47] A follower in Wang's cadre, Shen would not have risked offending the strong-minded authority with openly defiant choices. Shen's commentaries on *Mengzi*, too, reflected thoughts and concerns closely related to his experience serving in low-ranking administration (see chapter 1).

The grim decade of wandering in the local bureaucracy placed Shen in circumstances that were awkward for pursuing textual learning. A practical issue he repeatedly complained about was his inability to concentrate amid the swirl of incessant daily drudgeries. In a letter addressed to Cui Zhao, Shen articulated the predicament: "It is not possible for people to pursue learning with no concentration."[48] "My misfortune," he lamented, "is that I have been excessively multitask-ing."[49] Too much distraction was detrimental to work of any kind. Even a bearer of an "utterly trivial" task needed a clear focus to succeed, "not to mention the *dao* of the superior man!"[50]

Interestingly and unsurprisingly, Shen conceptualized such distrac-tions as coming from "things." In his own words, "while I was young, even though I had a firm devotion to learning, I was still unable not to let external things approach."[51] In the rest of the letter Shen described the distractions of "external things" in specific detail: how he had to run around all day and perform myriad trivial tasks with a numb heart, how he became so jaded that nothing in the world could provoke any response in him, a situation I briefly discussed in chapter 1.

In some ways this statement resonated with the problem of "external things," yet did so with subtle differences. In this very specific context, Shen viewed the trivial tasks he performed in the bureaucratic routine as "external things." These "things" included both activities and objects: fix-ing hydraulic infrastructure, managing the local population, and attend-ing rituals and funerals, as well as the dams, hydraulic tools, and ritual paraphernalia involved in these activities. These "things" were distracting in the sense that hands-on involvement with them kept Shen away from quiet immersion in textual studies and disrupted the bliss of reading and writing. The issue was primarily practical—Shen was overburdened by drudgery and thus spent less time with books—and secondarily intellec-tual, that in these desultory and clamorous activities he found no access to the *dao*.

Without going down the metaphysical road of repeating the philo-sophical rejection of "external things" in terms of deep orders, Shen kept the discussion at a practical level. In response to the situation, he suggested exploring alternatives: since the presence of "external things" was inevi-table and time for quiet study scarce, why not embrace them and see what follows? He invoked Kongzi (551–479 BCE) to justify the consideration of other possibilities: "The way in which Kongzi seeks it, isn't it different from how ordinary people seek it?"[52] Without specifying the subject of Kongzi's pursuit, Shen strongly implied a reference to learning (and the *dao*), as he further stated, "although it is truly beyond my capability, I am eager to learn."[53] He exhorted himself: "As remiss as I am, I cannot ex-pect to emulate that (Kongzi's spirit); yet, within the limits of [my] talents, I dare not resist my master's (Kongzi) advice."[54] Finally, he stated his resolution: on the journey of learning he resolved to "examine it with heed, ponder on it with caution, and practice it with dedication."[55] After all the diligent work, if the *dao* still did not illuminate his horizon, he should accept it as his own "destiny" (*ming* 命).[56]

At this early stage of his thinking, Shen perhaps did not have a clear conception of his alternative way of learning. Instead, he fixated on one simple point: that "things" were important and he wished to learn from extensively engaging "things" in concrete praxis. He used a simple strategy to make this point: he drew widely on classical references to reinforce the positive presence of "things" in the world and in one's intellectual pursuits. He called for expanded attention to "things" in an upbeat and robust tone, which laid the foundation for his future thinking.

The statement of resolution was an example of such a strategy. Despite its appearance as a terse, somewhat ambiguous claim, it was in fact carefully packaged in the classical language and drew on famous claims in old texts to link "things" with intellectual goals. The term *ming*, which Shen used to frame his lifelong journey of learning, was famously associated with Mengzi's philosophy. In the Mengzian discourse, this term denoted two intertwined meanings, "destiny," the course of one's life charted by an external agency, and "mandate," a decree one received from the same agency—Heaven. Mengzi designated the process of fulfilling one's destiny as follows: one came to understand his nature by fully developing the heart-mind, and by understanding his nature he understood Heaven (in the form of feeling unobstructed impulses in his heart-mind). To diligently cultivate one's self toward this goal was to "establish his destiny (and simultaneously, fulfill the heavenly-decreed mandate)" (*liming* 立命).[57] The three-step program was the conceptual framework in which Shen projected his vision for the destiny of learning.

Shen also managed to integrate Mengzi's discussion of the intellectual significance of "things" into the mandate/destiny scheme. He accomplished this goal in his commentary on a passage of *Mengzi*, where the master claimed that "the ten thousand things are available in me" (*wanwu jie bei yu wo yi* 萬物皆備於我矣).[58] Shen commented on this line with an emphasis on the significance of the ten thousand things: "Without thoroughly fathoming the patterns of the ten thousand things, one is unable to discern values under Heaven."[59] The concept "pattern" (理) came from "Explaining the Trigrams" (*Shuo gua* 說卦), an early commentary attached to the *Classic of Change*. The original line encouraged one "to exhaust the patterns and to complete one's nature, and thereby to fulfill one's destiny" (*qiong li jin xing yi zhi yu ming* 窮理盡性以至於命), which Shen cited in an abbreviated form ("to exhaust the patterns and to complete one's nature").[60] This formula bore a very similar structure to the three-step program proposed by Mengzi, involving one's nature as the middle step and ending with destiny. Shen blended the first step in the second formula, "to exhaust the patterns," into the first (Mengzian) scheme; then, by ascribing patterns to "things," he successfully incorporated the discussion of "things" into the goal of fulfilling one's destiny. In this line of argument, Shen assigned "things" indispensable significance in the whole intellectual agenda.

Shen made similar efforts when he captured other references to "things" in old texts. For instance, again in commenting on *Mengzi*, he linked the *dao* of the "great man" (*daren* 大人) to "things." In his own words, "to [act] in accordance with the destiny and with righteousness, and to rectify oneself so that things becoming rectified—such is the *dao* of the great man."[61] The great man designated an exemplary human being, a role model considered even higher than the "superior man."[62] To "rectify oneself" meant to cultivate one's self, and to "rectify" things (handling things with propriety) constituted the major actions that culminated in the *dao* actualized by a great man.

This statement was meant as a creative paraphrase of a Mengzian quotation, that "there is the great man, who rectifies himself so that things become rectified."[63] In the original context, Mengzi identified four types of exemplary humans. The first type was known for his devotion to the ruler, and the second for his loyalty to the state. The third type, "Heaven's subject" (*tianmin* 天民), had a good understanding of the *dao* and showed readiness to put it into practice. The fourth kind was the great man at issue. Mengzi pronounced no preference for any of the four, although in tone he implied some favored attention to the latter two.

In his commentary Shen emphatically applauded the great man, and according to how he set up his own interpretation, such a preference was due to the intellectual significance he saw in rectifying things. Shen placed the great man in comparison with Mozi 墨子 (ca. 470–391 BCE) and Laozi 老子 (ca. fifth century BCE), two renowned ancient thinkers. According to Shen, Mozi "constantly worried that all under Heaven (the world) cannot be ordered," and Laozi "stood aloof and invested no attention to the ten thousand things."[64] In contrast, the great man traced out his *dao* through rectifying his self and the "things" he encountered. "If one can live up to the great man in action," Shen exclaimed, "that would be perfection."[65] Once again, "things" boasted a positive presence and, in this formula, a direct link to the *dao*.

Beyond the classical realm, Shen made frequent references to "things" in a variety of positive ways. Although it was indeed a common term in his time, the frequency with which Shen used the word was distinctive, and his almost habitual invocation of "things" revealed a deep philosophical attachment to it. The following examples do not all come from his early writings, yet together they demonstrate a lifetime consistency that is hard to ignore. For example, when applauding the intellectual competence of fellow bureaucrats, Shen described one as being able to "fathom the subtleties of the patterns of things" and the other as "having no ambiguous understanding of things."[66] When exalting the achievements of Ouyang Xiu, the top political luminary in his times, Shen praised him as "not missing one single thing—small or big—when it comes to statecraft."[67] In a celebratory essay he wrote for a newly built city gate in Chizhou (in present-day Anhui Province), Shen envisioned that the "administration [of the locality] has its schedule, and things have their order; here (under the gate) [people gather] to hear [orders and policies]," implying that "things," such as the new gate, asserted a significant role in articulating political power.[68] When singing a eulogy to the New Policies reform, Shen said to the emperor that "your majesty has issued orders to renew the ten thousand things; [the orders are] to be passed on [to later generations] infinitely."[69] All in all, in Shen's conceptualization, the appropriate handling of "things" constituted a primary achievement for individual officials and the state machinery.

Shen's interest in "things" was accompanied by a certain degree of philosophical naïveté. None of his arguments displayed an originality to speak forcefully to his own experience. If we isolate the two examples from Shen's oeuvre and compare them with his contemporaries' writings, we see that Shen shared similarities with the work of many other thinkers. For instance, to link the *li* of "things" to the *dao* was a claim much more famously associated with the Cheng brothers (see chapter 10).

In the second example, any literatus who advocated for active governance vis-à-vis Buddhist and Daoist retreatment would not have disagreed with Shen.

Shen did touch on the dual nature of "things," though with a tinge of vagueness. In his commentary on *Mengzi*, he made sporadic references to the metaphysical issue in relevant language. He invoked the term *li*, yet spoke no more whether he was headed in the same direction with the Chengs, who employed the concept to address the place of a "thing" in a larger order. Later in *Brush Talks*, Shen used *li* in a manner visibly distinctive from the Cheng brothers', a difference that took time to formulate and reveal (see chapter 10).

Shen echoed Mengzi in praising the heart-mind over the ears and eyes when commenting on *Mengzi* 6A:15. In his opinion, the heart-mind could "select" (*ze* 擇) between the "right" (*shi* 是) and "wrong" (*fei* 非) and was thus "greater" (*da* 大) than the ears and eyes.[70] By choosing vocabulary like right and wrong, Shen highlighted the judicious capacity of the heart-mind in moral contexts. When discussing 2A:2, Shen repeated the identical language regarding the choice between right and wrong and associated it with a heart-mind filled with "flood-like *qi*" (*haoran zhi qi* 浩然之氣).[71] In *Mengzi*, the flood-like *qi* was the "energy of moral courage," which enabled humans to perform challenging moral tasks.[72] To follow Mengzi on this point, Shen envisioned a fully cultivated superior man who "gathered righteousness in his self and thus had *qi* filling up his heart-mind" and "fully extended his determination and held no grudge between Heaven and Earth."[73] The thesis was particularly useful for Shen to resolve moral controversies in his personal experience; for instance, he invoked the heart-mind's superior guiding power in moral choices in an effort to justify his competitiveness (see chapter 1). In contexts other than the moral, Shen demonstrated an intensive interest in using the ears and eyes, for he deemed perception a key epistemic guide, in sensory inquiries or search for deep orders (see chapters 4 and 9).

The lack of resolute clarity in these metaphysical arguments surely had to do with the fact that Shen was still in the early stages of formulating his thought. His mellow presentation of interest in "things" in classical exegesis was also part of a larger consistency in his general discursive persona. It is important to notice that Shen rarely stirred up polemics when it came to discussing the Classics. Throughout his lifetime, he was not known for pronouncing noteworthy ideas in conventional forms, such as in commentaries or political essays.[74] His words in these genres often bordered on clichés, yet his ideas in less politically sensitive writings (including *Brush Talks*) and, more importantly, his actions, frequently betrayed the differences concealed under ideological compliance. In the rest of the book, I rely primarily on Shen's nonpolitical writings and his praxis to trace the development of his thinking.

## Making Music as a "Thing"

I conclude this chapter with a concrete example of how Shen applied his new vision of learning in his bureaucratic career. Unlike his nephew Shen Gou (and many other peers), who submitted essays on statecraft to attract the imperial court's attention, Shen Gua decided to make a distinctive "thing" for the same purpose: music.

The genre of music Shen worked on was known as the "ceremonial music" (*yayue* 雅樂).[75] Closely associated with state rituals, ceremonial music bore an ideological significance far beyond the sonic paraphernalia of imperial politics. Good ceremonial music was supposed to be the instantiation of the perfect order a monarch had achieved in ruling the empire, and thus was a vital testimonial to his legitimacy.[76]

Due to its political importance, music promised unconventional career opportunities for ambitious literati. The two hundred years of the Northern Song witnessed six music reforms, many supervised by literati who had special talents in musicology but came from lower ranks. A music reform was almost always the imperial monarch's initiative. In all Northern Song cases, the emperors personally received the reformers, regardless of their origins.[77] Participating in such activities thus offered the literati a favorable channel for cutting through the bureaucracy and making a name in the inner circle of the imperial court.

Although looking on from the margins of the empire, Shen was well aware of the potential opportunity music promised. In the 1050s, Ruan Yi 阮逸 (1023–1053) and Hu Yuan 胡瑗 (993–1059) launched the hitherto most elaborate music reform in the Song.[78] This was precisely when Shen started writing on music in the countryside. As soon as he completed a treatise titled *On Music* (*Yue lun* 樂論), Shen presented it to a number of high officials, including Ouyang Xiu, Sun Shi 孫奭 (962–1033), and Cai Xiang 蔡襄 (1012–1067), expressing unconcealed eagerness for self-promotion.[79]

*On Music* is no longer extant, so we rely only on bibliographical records to deduce its content. According to the *History of the Song*, it had a one-chapter main body, to which a chapter of illustrations of musical instruments, a chapter of three scores, and a chapter of harmonic theories were attached.[80]

Viewing this scarce information against the background of Song music suggests that Shen intended his work to be immediately useful in a potential music reform. All six music reforms in the Song involved the three major tasks: to recalculate the harmonics (primarily the pitch standards), to build new sets of instruments (in accordance with the new harmonics), and to compose new melodies (or revive old ones).[81]

The pursuit of utility emerged simultaneously with an intensive attention to "things" (in music) as distinctive objects/processes. Although most Song literati were able to and felt obligated to discourse on the ideological significance of music in statecraft essays, Shen was clearly more drawn to the "thing" aspect of the subject: the dimensions of instruments, the melodic progressions, and high and low pitches. What he wished to present to the authorities was his ability to manage all relevant objects and processes involved in music making and bring a useful product to life.

Shen revealed his interest in distinctive "things" when crafting self-promotion letters to authorities. In the letter to Cai Xiang, he reviewed the history of music as a process fraught with and driven by "things" in fluctuation. When Shen described the golden times of ceremonial music, the Western Zhou Dynasty (1045–771 BCE), he did so by naming the melodies performed at the time: the "Lofty Shrine" ("Qingmiao" 清廟) and "Great Brightness" ("Daming" 大明) proffered to deities; "Military" ("Wu" 武), "Figure" ("Xiang" 象), "South" ("Nan" 南), and "Pole" ("Xiao" 簫) performed at the imperial court; and "Fish Trapped" ("Yu li" 魚麗), "Deer Belling" ("Lu ming" 鹿鳴), "Osprey Chirping" ("Guan jiu" 關雎), and "Raccoon's Head" ("Li shou" 貍首) circulating in the world.[82] When Shen discussed the ending of the golden age, he narrated the scene by describing the flight of musicians, who scattered in all directions with "instruments held in their arms."[83] In the letter to Ouyang Xiu, Shen argued that not only was the general model of music making established by the sages significant, the material forms made by "various craftsmen, bureaucrats, and people from markets and fields," such as "technics and implements, sizes and measurements, and [colors such as] black, yellow, blue, and red," were also indispensable.[84] In Shen's conception, melodies, instruments, and the handlers of all these "things," constituted what was vital to music, and this is where he wished to make his own contributions.

Although Shen Gua's investment in music did not return the kind of political payoff he expected, a few years later he successfully garnered substantial promotions as a degree holder. The hard work he devoted to textual learning in the first decade of his career brought him to the precipice of a new phase. As he began to enjoy happier days, Shen did not forget the lessons of the difficult times. He continued to learn through "things" and turned his fascination with distinctive objects and processes into career successes. We now proceed to that period.

# CHAPTER 3

## *Measuring the World (1063–1075)*

The arrival of 1063 brought auspicious news. After placing in Suzhou as a prefectural-level candidate, Shen Gua passed the metropolitan examination and got his *jinshi* degree.[1] A couple of years later, in his early thirties, Shen stood before the magnificent city gate of Kaifeng again; this time he made it on his own. During the years from 1063 to 1075, Shen worked in a series of positions in Kaifeng, gradually moving from an Imperial Libraries associate into the policy-making nucleus of the empire. Shen returned to the more prestigious world he had been longing for, and more importantly, he managed to use skills and orientations he had cultivated in the past to open fresh avenues for advancement. His adeptness with "things," for example, remained his most stellar quality. During these ten or so years, Shen conducted a series of quantitative projects, ranging from his old trade in hydraulics to a new engagement in astronomy. In conducting these activities, he had crafted a stentorian statement of his interest in distinctive objects and processes, most of which he explored with hearing and seeing.

This chapter begins with Shen's appointment to the Imperial Libraries, where he expanded his social and intellectual milieu and gained access to a rich repertoire of new knowledge. Then I introduce his projects in hydraulics, astronomy, and the Song empire's border survey. This chapter focuses on Shen's independent quantitative work during this time. Some of his other important activities—all closely supervised by reform leader Wang Anshi—appear in chapter 5, where I offer a focused narrative of the New Policies reform.

## *The Imperial Libraries*

While the promising new time in Shen's career began with attaining his degree in 1063, it reached its first high point three years later, when he was selected to join the Imperial Libraries. The opportunity came by virtue of Shen's hard-earned degree, new connections, and a bit of luck. As a new degree holder, he received an appointment as administrator of order in Yangzhou (*Yangzhou sili canjun* 揚州司理參軍). The prefecture of Yangzhou provided an attractive geographical foothold for an ambitious young man. It claimed a central status in the Circuit of Huainan, the key locality joining the north to the south, and enjoyed national recognition for its critical influence over the transportation of supplies to Kaifeng.[2] Shen delightfully observed that numerous boats and carriages filled with goods glided through Yangzhou on their ways in and out of the capital city.[3]

In Yangzhou, Shen met someone who changed the course of his life. A senior official called Zhang Chu 張蒭 (1015–1080) had just finished mourning his deceased father and came to serve as financial commissioner in Huainan (*Huainan zhuanyun shi* 淮南轉運使).[4] On Shen's first visit to Zhang, the two found mutual admiration and had day-long conversations.[5] They had much in common—financial hardship in childhood, a deep investment in education and learning, and an aptitude for practical affairs concerning agriculture and finance, on foundation of which they quickly built a close mentor–disciple partnership.[6] In his capacity as a circuit-level official, Zhang had the prerogative to recommend a junior person to the capital, and he generously extended the opportunity to Shen. Zhang held a decade-long tenure at the Imperial Libraries as editor at the Jixian Library (*Jixian xiuzhuan* 集賢修撰), and on his recommendation, Shen managed to become an associate of the Zhaowen Library.[7] Four years later, Zhang gave his daughter to Shen in marriage.[8] Shen's first wife had died in the early 1060s and he joyfully remarried.

The Zhang family was an official lineage featuring generations of civil servants. Zhang Chu's grandfather, according to the family's recollection, was a wartime hero who risked his life to communicate messages between the Song monarch and neighboring Liao (907–1125) while the two states were at war in the early eleventh century.[9] By all accounts, Zhang Chu was a successful official. At the time he met Shen, Zhang was in charge of transporting rice from the Huainan region to the capital city, a business key to the operation of the Song state. In the twelve years after Shen joined his family, Zhang was promoted to vice commissioner for the Salt and Iron Monopoly (*yantie fushi* 鹽鐵副使) and then prefect of a number of localities close to Kaifeng, continuing to exercise his competence in the fiscal sector at the state level.[10] From the mid-1060s through the mid-1070s, the critical period of Shen's career, his father-in-law assumed considerable influence in fiscal affairs in the pan-Kaifeng region, thus lending robust support to Shen. Zhang had four other sons-in-law who served in the officialdom; among them, Zhang Yuanfang 章元方 (fl. eleventh century) became erudite of the National Academy (*guozi boshi* 國子博士) and Li Ji 李稷 (fl. eleventh c.) served as fiscal commissioner in Shaanxi (*Shaanxi zhuanyun shi* 陝西轉運使).[11] Support from his in-laws instilled in Shen a renewed confidence in pursuing his professional ambitions.

In 1066, Shen set foot on the grounds of the Imperial Libraries, where he found the change of scene dramatic. The library complex sat in the heart of the capital, protected by the high walls of the imperial palace. Emperor Taizong (r. 976–997) relocated it close to the Gate of Grand Celebration (*Daqing men* 大慶門), the main entrance that overlooked every visitor who entered the imperial city.[12] Artisans presented the library buildings in lavish architectural motifs: vermilion-lacquered doors with golden studs, layered brackets, and murals of dragons and phoenix floating in clouds.[13] The collection of books had flourished, with the number of *juan* (scrolls or chapters) exceeding 36,000 by the time Shen arrived.[14]

Shen began as an intern and moved up to official editor. For institutional reasons, working conditions were flexible and protocols adaptable at the Imperial Libraries. Except for top-tier executive posts, editorial positions were not regular appointments, because higher authorities created them ad hoc whenever they spotted desirable candidates. As a result, editors enjoyed a considerable degree of liberty in determining work schedules and choosing texts to work on.[15]

The flexibility suited Shen well. By exploring the most extensive collection of books in the empire, he imbibed knowledge of diverse kinds, some of which he put to immediate use. From 1071 through 1072, he supervised a reform of the Southern Suburb Ritual (*Nanjiao li* 南郊禮)

and composed a treatise on it.[16] Efforts to reconstruct state rituals had always relied on historical precedents, for which the imperial collection provided the optimal sources. In 1072, Shen launched a six-year project to overhaul the old calendar and the imperial observatory. He could undertake the task because he had taught himself astronomy while editing calendrical texts.[17] The production of astronomical knowledge had long been a state monopoly because of the paramount significance of Heaven in symbolizing imperial legitimacy. The Song government kept most texts concerning astronomical observation and calendar making out of circulation and put them under rigorous control at the Imperial Libraries.[18] Access to this exclusive collection of texts was undoubtedly Shen's best opportunity to learn the subject, and he capitalized on it by reading ferociously.

The institutional flexibility Shen enjoyed was largely the result of the dual functions the Imperial Libraries served: to "gather books and cultivate the worthies" (*jushu yangxian* 聚書養賢).[19] The "worthies" referred to policy makers, and the primary means of cultivation was to let them browse freely among the finest repertoire of books. Later in their lives, many library incumbents applied their enriched minds in policy-making capacities, using the editorial jobs as a springboard leading to the top echelon of the Song officialdom. Some of them had attracted attention from the emperor and taken up other responsibilities while officially still in the waiting pool. For these people, the librarian titles had become half-time obligations and sometimes even titular honors.[20] Such was the case with Shen. After spending a few years reading quietly between the shelves, he ventured into various projects that took him outside the imperial city; all this time he held an official title as an editor, an emblem of cultural virtuosity and a promise of a bright career.

## *Measuring the Bian*

The first few tasks Shen performed as an Imperial Libraries associate concerned the quantitative management of "things." In 1073 he came to work on the Bian Canal, a transportation artery of vital significance for the Song regime. The canal connected the capital to two major river systems, the Huai in the south and the Yellow in the north, supplying grain and coal to thousands of people residing in the greater Kaifeng area.[21] During Shen's time, the canal was also linked to almost all major waterways within Kaifeng and received a large amount of water from them.[22] A persistent issue impaired the utility of the Bian, however. Despite the early Song monarchs' efforts to set up cleanup routines, the waterway remained excessively silted and incurred massive flooding. By the beginning of Shenzong's reign, the river had not been properly dredged for two decades and the riverbed stood meters higher than the ground outside the dikes.[23]

In the 1070s, Wang Anshi rose into power and resolved to tighten the management of the canal system. He became interested in the so-called method of silt fertilization (*yutian fa* 淤田法) and intended to apply it to the Bian.[24] The method aimed to channel the silt-laden, nutrient-rich waters of the canal to irrigate farmlands in suburban Kaifeng by installing sluice gates. If successful, it could be a one-stone-two-birds solution that tackled silt dredging and irrigation at the same time. Wang recruited Shen to turn the idea into practice.[25]

Although extant sources afford meager details as to how Shen conducted the work, circumstantial evidence shows that he carried out careful surveys of the surrounding environment, which earned him further opportunities to supervise hydraulics. In these activities, Shen's acumen in quantitative work shone. For example, the imperial court assigned him to study the possibility of channeling the Luo River, a waterway lying west of the capital, into the Bian Canal. For this purpose, Shen had to inspect the change in altitude of a certain section of the canal, starting from Shangshanmen, a place close to the capital, to Sizhou, where the canal joined the Huai River. The two spots were a considerable distance apart, as great as 840 *li* and 130 *bu* (399.37 kilometers/248.11 miles), and the terrain in between was not flat, which rendered the measurement task particularly thorny.[26]

To accomplish the goal, Shen had the following instruments at his disposal: a water-level (*shuiping* 水平), a sighting board (*wangchi* 望尺), and a graduated pole (*ganchi* 幹尺).[27] A water-level, as shown in figure 2, consisted of a trough and three floats. Each float carried a measure, the top of which was supposed to come into a straight line with the other two when the water-filled trough was placed on an even surface. A sight-

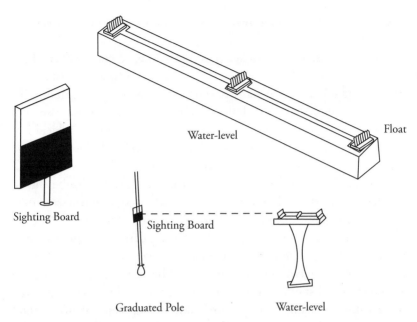

FIGURE 2. Water-Level, Sighting Board, and Graduated Pole. Adapted from Wuhan shuili dianli xueyuan and Shuili shuidian kexue yanjiuyuan, *Zhongguo shuili shigao*, vol. 2, 54, with permission of Zhongguo Shuili Shuidian Publishing House. Drawing by Kelly Maccioli.

ing board was a rectangular board with a handle. The board (excluding the handle) was 4 *chi* long (1.27 meters/4.16 feet) and evenly divided into two vertical halves. The upper half was painted white, and the lower was black. A graduated pole, 2 *zhang* 丈 in height (6.34 meters/20.78 feet), had a scale divided into *fen* 分 (1/1,000 of a *zhang*).

A team of two surveyors could use the three instruments in combination to implement a standard procedure for measuring altitude differences. Surveyor A set up the graduated pole, holding the sighting board; surveyor B placed the water-level at the first location to be measured, making sure the three floating measures aligned in his sight. Surveyor A then adjusted the position of the sighting board in relation to the pole until the black-white division aligned with the water-level line (formed by the three floats) in B's sight and read the height of the division on the graduated pole. Afterward, B moved the water-level to a different location and the team repeated the same procedures. A comparison of the readings revealed the altitude difference between the places.[28]

Though reasonably convenient, the procedure did not suit Shen's Bian project. The method relied on the sight of a surveyor, and a distance as great as 248 miles was obviously beyond any human visual capability. To address this problem, Shen designed a new method that overcame the constraints of existing tools by wedding them to the topographical features of the land. From surveys of local conditions, he noticed some ditches from which laborers used to take earth for the construction of canal dikes. These ditches existed extensively along the embankments of the Bian and intersected the canal. Shen let water into the ditches and then built dikes within, turning them into a succession of reservoirs. Not too distant from one another, these reservoirs rendered the 248-mile long slope into manageable sections. The next step was to determine the altitude difference between each pair of neighboring reservoirs, using the aforementioned standard procedure. The sum of all results yielded the overall altitude change of the entire 248-mile-long slope. Shen's method produced a result more accurate than any previous efforts, which provided a solid foundation for the canal–river connection project.

## Measuring the Firmament

In addition to water and dams, Shen also dealt with "things" in the sky with a quantitative approach. In 1072, he launched a calendrical reform, making good use of the astronomical knowledge he had acquired in the Imperial Libraries. The calendrical reform lasted six years, during which Shen more or less achieved his two goals: to overhaul the imperial observatory and to make a new calendar. As with ceremonial music, calendar was a technical field subjected to frequent reform initiatives in the Song. Shen's new calendar—the so-called Oblatory Epoch System (*Fengyuan li* 奉元曆)—was the sixth among nine calendrical systems made during the Song.

Before discussing the details of Shen's work, I first clarify the particular meanings of a "calendrical reform" in the Chinese context. First, a Chinese *li* 曆, which we insufficiently translate as "calendar," was an astronomical system featuring predictions of eclipses and planetary phenomena.[29] Sivin suggests "astronomical reform" as a more accurate designation when we discuss any reform initiative directed at a Chinese *li*.[30] Second, the computational system of a Chinese calendar consisted of elaborate arithmetical rules, in contrast to the distinctive presence of kinematic geometry in the Ptolemaic tradition. Third, a calendrical system was evaluated based on its accuracy in predicting celestial events, especially eclipses. Over time, a computational system could gradually lose its accuracy as sources of errors accumulated—hence the necessity of a reform.[31] To build a new astronomical system, a calendrical expert would typically collect observational data and revise existing arithmetical rules toward finding a better match between data and calculations.

When Shen was appointed to lead the Bureau of Astronomy in 1072, he was on a mission to conduct a reform in the face of turbulent political and technical circumstances. On the political front, his patron, Wang Anshi, was contesting a foe, Sima Guang 司馬光 (1019–1086), for control over imperial astronomical work. To Wang, Shen's appointment was a victory he earned from factional fights. He surely expected to seize partisan control of the Bureau of Astronomy via Shen, and the most efficient way of doing it was to start a reform.[32]

At the same time, technical challenges abounded. Controversies regarding the accuracy of the calendar currently in use had been going on for a while. In 1068, the Resplendent Heaven (*Mingtian* 明天) system demonstrated major discrepancies in eclipse prediction; for the lack of a better solution, officials replaced it with an old system, the Reverence for Heaven (*Chongtian* 崇天) calendar, which had its own inaccuracies.[33] Some calendrical conventions were alarmingly out of date. The lunation factor in use was the legacy of Yixing 一行 (683–727), a calendrical expert centuries ago, and it had accumulated considerable error.[34] What upset Shen most was shoddy practice—particularly blatant negligence of astronomical observation—in relevant institutions. At the time, two offices were responsible for making routine observations: the Bureau of Astronomy and its auxiliary counterpart in the Hanlin Academy. The Song state specifically created the second office to double-check observational and calculation work done by the main bureau.[35] Shen was dismayed to see that many incumbents in both bureaus failed to fulfill their basic responsibilities and simply copied observational reports from one another.[36]

It was clear that a comprehensive reform had to be the first and foremost item on his agenda.

To implement a reform quickly and effectively, Shen made a controversial decision: he recruited Wei Pu 衛樸 (fl. 1070s) to lead the initiative. Wei was a curious choice in many ways: he was an outsider to the civil service, a commoner with an obscure background, and visually challenged (likely blind). Nevertheless, the new computational system he devised deeply impressed Shen. According to Shen, with the new system Wei was able to correctly calculate thirty-five out of thirty-six eclipses recorded in the *Classic of Spring and Autumn*, an achievement surpassing even that of Yixing—the legendary Tang astronomer—who managed only twenty-nine.[37] Unable to rely on vision, Wei had developed an uncanny talent in mental calculation. He could immediately recite critical data from a computational procedure after hearing an assistant read it aloud once, and he could detect mistakes in calculations when someone recited the computational treatise to him. Wei's fingers flew across the counting rods when he calculated, and "others could barely catch up with their eyes."[38] Wei considered the new Oblatory Epoch System as merely a partial manifestation of his capability, yet Shen proudly judged it more accurate than any other existing system.[39]

In addition to recruiting an outsider as the new leader, Shen waged a war against the negligence of observation. During his tenure, he repeatedly and adamantly emphasized the decisive significance of observation in calendar making. He had a specific plan in mind. Unsatisfied with existing data on planetary movement, he decided to launch a five-year observational program that would generate more detailed records of the apparent movements of the planets.[40]

To prepare for the new initiative, Shen commissioned a set of new instruments, including an armillary sphere (*hunyi* 渾儀), a gnomon (*guibiao* 圭表), and a clepsydra (*fulou* 浮漏), and for each he composed an elaborate introduction. An armillary sphere was a model of the structure of the celestial sphere, with multiple "armillaries" standing for major astronomical circles such as the equator and ecliptic. The instrument had a long history in China. After carefully reviewing historical precedents, Shen pronounced that he endorsed the Tang model designed by Li Chunfeng 李淳風 (602–670) and ameliorated by Yixing as the foundation of his new design.[41]

Like the Tang model, Shen's instrument featured three layers consisting of multiple rings. The outer layer, which Shen named "framework" (*ti* 體), comprised four concentric rings: the prime meridian double ring (*jing* 經), which passed through the zenith, nadir, and two celestial poles; the celestial equator ring (*wei* 緯), which presented the equatorial plane; and the horizon ring (*hong* 紘), which marked the cardinal points. All four rings were fixed to one another and rested together on a four-leg pedestal (*fu* 趺).[42]

Shen named the intermediate layer "figure" (*xiang* 象), which referred to varied phenomena associated with celestial bodies. All four rings in this layer were movable. The "Red Road" (*chidao* 赤道), the equator ring, and the "Yellow Road" (*huangdao* 黃道), the ecliptic ring, were pegged onto a polar-mounted declination double ring (*ji* 璣) in such a way that they could rotate about their centers. The double ring could also revolve around the polar axis. The equator and ecliptic bore calibrations and had small holes evenly distributed, one in each "degree" (*du* 度).[43] The two rings thus could be connected to each other at different points (with a copper nail through the holes) to accommodate changes entailed by the precession of the equinoxes along the ecliptic.[44]

The inner layer included a sighting tube (*heng* 衡) mounted on another polar-mounted declination ring (also named *ji*).[45] The ring could rotate on the meridian plane, and the sighting tube could revolve about the center of the ring. When pointing the sighting tube to a celestial body, an observer was able to read important data on correspondent rings, for example, the object's distance from the north pole and its position vis-à-vis the equator and ecliptic.

Comparing to the Tang precedent, Shen's new instrument featured a number of improvements. The most important was the removal of the lunar ring, which Li Chunfeng included in the intermediate layer. The lunar ring could not be fixed on the ecliptic because of periodic retrogradation of the lunar nodes. Li had to drill as many as 249 holes on the ecliptic and move the lunar ring every month between these holes.[46] Even a design as complicated as that, however, failed to capture the moon's movements accurately. Shen decided to completely remove the lunar ring, and astronomers in later ages endorsed this change by never restoring it.[47]

Because the instrument needed to point to true north to yield accurate results, Shen invested serious attention to locating the so-called celestial pole (*tianshu* 天樞), the center around which the stars revolved. Observers conventionally chose a visible-to-the-naked-eye star near the pole and designated it the pole star. The distance between the celestial pole and the pole star was minimum and yet perceptible; more importantly, it changed over time due to the precession of the Earth's rotational axis. Shen noticed this phenomenon while observing that the pole star sometimes drifted outside the scope of the sighting tube on an old sphere. He then increased the diameter of the tube to the extent that the pole star remained well within the view for three consecutive months and proposed that the pole star was over three "degrees" away from the celestial pole.[48]

Shen also improved the design of the sighting tube to help the observer's eye focus better. The old tube was an open cylinder with two identical ends, and Shen turned it into a flat cone with a smaller eye end. With some trigonometry, he reduced the radius to 0.16 inches/0.40 centimeters, which approximated the size of the pupil and thus prevented any influence the observer's movements might incur.[49] All in all, Shen's armillary emerged from his careful study of precedents and became known as the one of the four major armillary spheres of the Northern Song.[50]

As the redesigned armillary sphere turned out to be a success, Shen moved on to another instrument: the gnomon. Since antiquity, the Chinese used the gnomon to determine the seasons of a year either by measuring the changes of the sun's shadow by day or by tracking the movements of stars by night. The instrument normally consisted of two major parts: a vertical pole and a calibrated template (see figure 3). In daytime, the shadow of the former would project onto the latter. Conventionally speaking, the shortest shadow would appear on the summer solstice and the longest on the winter solstice. The instrument was simple and convenient, but computations based on it involved a number of uncertainties—the date, the place, the height of the gnomon, and the weather.[51]

To maximize the gnomon's efficacy, Shen developed innovative ways to use it. First he moved it from open space into a closed chamber, where the sun's rays entered through a small slit and thus left a more distinctive shadow. He designed a double-pole method, in which he added an auxiliary pole to the main gnomon for the purpose of reducing

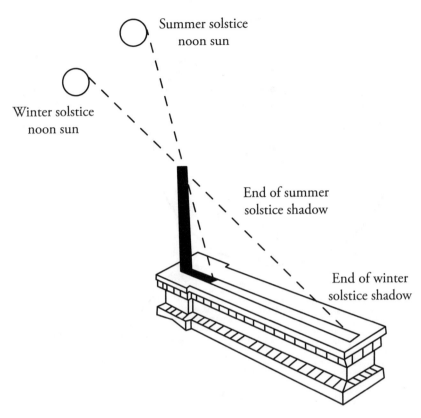

Summer solstice
noon sun

Winter solstice
noon sun

End of summer
solstice shadow

End of winter
solstice shadow

FIGURE 3. Gnomon. Drawing by Kelly Maccioli.

the shadow's penumbra. The main gnomon was 8 *chi* (2.53 meters/8.31 feet) tall; at this height, a penumbral zone of considerable size appeared at the end of the pole's shadow, which made it difficult to locate the exact tip of the shadow. The addition of a second, shorter gnomon, only one twentieth the height of the main pole, provided a solution. The second pole was supposed to be placed in a location where its head was just in contact with the sunbeam and thus helped render the shadow of the main pole clearer.[52]

Shen was also concerned with the method of setting up the gnomon, particularly the challenge of determining the cardinal directions. To more accurately locate the north and south, Shen proposed a triple-gnomon system (see figure 4). First, line up three poles, G1, G2, and G3 in a row,

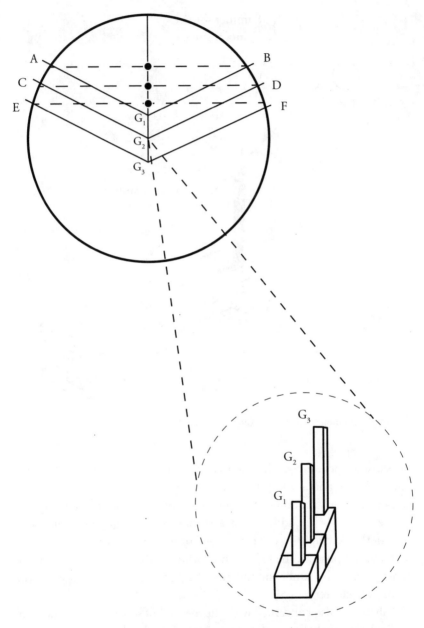

FIGURE 4. Triple-Gnomon System. Adapted from Chen Meidong, *Zhongguo kexue jishu shi*, 474, with permission of Yu Lishang and Kexue Publishing House. Drawing by Kelly Maccioli.

each 2 *chi* (0.63 meters/2.08 feet) apart from another and following an approximate north–south direction. Second, draw a circle from where G2 stands as the central point. Mark the intersections of the shadows and the circle at sunrise and sunset, generating six points in total; join the ends of each pair and make three lines (AB, CD, and EF). To follow, adjust the direction of line G1-G2-G3 to the extent that AG1 and BG1, CG2 and DG2, EG3 and FG3, as well as AC and BD, CE and DF all become symmetrically equivalent pairs. The line that connects G2 and the midpoints of AB, CD, and EF should be the north–south axis.[53] This system generated more accurate directions than a single gnomon did, and did not require the assistance of other equipment (for instance, a clepsydra).[54]

The last item on Shen's list was a clepsydra. Also known as a water clock, a clepsydra kept time by a regulated discharge of water. Shen's clepsydra was what Needham categorizes as the overflow type.[55] The instrument comprised four major vessels (see figure 5). Water flowed from the reservoir on the top into the overflow tank, from which it dripped into the inflow receiver with a calibrated ruler on a float. The ruler bore one hundred "quarters" (*ke* 刻), which encompassed all twelve "double-hours" (*shi* 時) during a day.[56] Any excessive water from the overflow tank would exit into the extra liquid receiver, the tank on the bottom.

Shen designated each component with an elegant name. The reservoir, which upheld the source of water in the entire system, was the "tank of pursuit" (*qiuhu* 求壺), and the overflow tank was the "tank of duality" (*fuhu* 複壺). The receiver with the ruler was called "tank of establishment" (*jianhu* 建壺) for instituting numerical readings, and the bottom vessel was the "tank of otiosity" (*feihu* 廢壺), for accommodating superfluous liquid.[57]

The key to a properly functioning clepsydra was the steady drip of water, an issue Shen attended to in several creative ways. The overflow tank featured two horizontally juxtaposed compartments, a structure that stabilized the motion of water from the reservoir. Liquid entered the left chamber, which Shen named "origin" (*yuan* 元), and then flowed into the right chamber, "medium" (*jie* 介), through an aperture in the division wall. The aperture was called "arrival" (*da* 達), and it reduced the flow from the reservoir to half; the other half left the overflow tank through the two "branch canals" (*zhiqu* 枝渠) into the extra liquid receiver.[58] This was a design Shen adopted from Yan Su 燕肅 (961–1040), who made the first overflow type of water clock in Chinese history, the Lotus Clepsydra (*Lianhua lou* 蓮花漏).[59]

Shen managed to further control the water discharge from the overflow tank into the receiver. He placed the exit pipe at a relatively high position on the wall of the overflow tank—about an inch beneath the opening.[60] This arrangement had a number of benefits. It forced water up so that only half of the liquid in the right chamber would be released into the inflow receiver. The higher the outlet was placed, the less it would be affected by sediment deposits.[61] At the orifice of the pipe Shen placed a "jade regulator" (*yuquan* 玉權), a plug with a hole, which made flow regulation as easy as adjusting the position of the small opening.[62] A number of other attempts Shen made enhanced the accuracy of the apparatus. He suggested verifying readings on the graduated ruler with observations made by the gnomon.[63] He carefully considered building materials in terms of durability, choosing gold for the ruler and jade for the regulator.[64] Although not all of his choices achieved desirable results, Shen showed his devotion to hands-on particulars by tending to as many details as he could.[65]

Despite the elaborate preparations Shen made regarding human and material resources, the Oblatory Epoch System was in use on and off for just three years. In 1076, immediately after its launch, the new calendar failed to predict a lunar eclipse and aroused open attacks on the reform. Because of the extreme scarcity of sources, it is difficult to comprehensively evaluate the technical strength of this system. The only components of the Oblatory Epoch calendar available in textual records are two key constants, on the basis of which scholars have attempted limited reconstructions.[66]

Nevertheless, the causes for the failure were perhaps more political than technical to start with. The computational system indeed harbored innate defects, which Shen was aware of but unable to avoid. Ideally, he would have completed the five-year observational program before incorporating up-to-date data into the calculations. In reality, he received the commission from the emperor in 1072, completed the construction of the instruments in 1074, and launched the new calendar in 1075. With this timeline, Shen barely had a chance to make use of the new observational results in his calculations.[67] Reluctant to relinquish his original goal, he

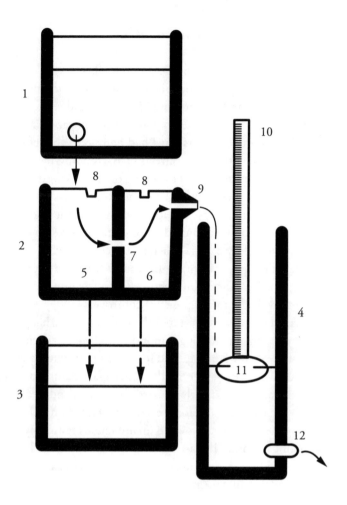

1. Tank of pursuit
2. Tank of duality
3. Tank of otiosity
4. Tank of establishment
5. Origin
6. Medium

7. Arrival
8. Branch canals
9. Exit pipe and jade regulator
10. Ruler
11. Float
12. Water exit

FIGURE 5. Clepsydra. Adapted from Li Zhichao, "Fulou yi kaoshi," 34, with permission of Li Zhichao and University of Science and Technology of China. Drawing by Kelly Maccioli.

strove to continue the observational program while working on the new calculations, maintaining the processes separately until the former concluded in 1077. That year, Shen proposed to correct and revise the calendar, an effort never carried out due to political complications.[68]

Factional infighting over personnel had hampered the reform since its inception. Incumbents at the Bureau of Astronomy naturally held grudges against Wei Pu, an outsider intruding on their turf. After repeated attempts to oust Wei, they managed to have him convicted of malfeasance and sent to prison in 1075.[69] In 1077, when Shen felt ready to revise the system with new observational data, the emperor responded to his earnest entreaty and released Wei.[70] Shen and Wei returned to the project, this time with little support from the Bureau of Astronomy and an even more pressing timeline. They managed to effect a few minor corrections, one of which provoked another political controversy. They proposed to push New Year's Day forward by a month, a decision they had to defend with a public experiment with a gnomon in the face of vehement criticism.[71] Soon after, Shen was impeached for other political reasons and fell out of imperial favor (see chapter 6). The court again banished Wei from the imperial city and permanently aborted the new calendar.[72]

Although the calendrical reform eventually ended on a despondent note, until 1075 the movement provided unambiguous testimony to Shen's rise in the political world. The astronomical instruments and computational system stood out as exemplary products of his quantitative work during this decade of his career.

## Measuring the Border

Shen's talents in the quantitative management of "things" earned him the ultimate reward a probationary policy maker could expect: the emperor's favorable attention. In 1075, he was assigned by Emperor Shenzong to redraw the border between the Song and the Liao and settle a major interstate dispute. Although slightly different in nature and perhaps less technical in practice than his earlier work, this project enabled Shen to exert a more substantial influence on the day-to-day transactions of the empire.

The background of the Song–Liao border controversies traced to the unique geopolitics of middle-period China. As much as the Song emperors aspired to inherit the Tang legacy, they had to acknowledge that the Song was not the only regime that ruled the territory the Tang empire once occupied (the territory we conveniently call "China" in this historical period). To the north of the Song lay the formidable Liao, a state run by a people known as the Khitans. The Khitans had established their empire prior to Song unification of the central plains and the south. Over the entire course of Northern Song history, the two states posed a constant military threat to each other and maintained an uneasy relationship characterized by reluctant diplomatic parity.[73] From the perspective of a Song emperor, dealing with diplomatic tensions vis-à-vis the Liao was an ongoing concern, a task so important that he relegated it only to officials of trustworthiness.

At the moment Shen arrived at the court, a border dispute between the Song and Liao was escalating into a crisis. In 1074–1075, the Liao sent envoys to the Song court twice, complaining that the Song's expansion of fortifications in three border prefectures violated terms stipulated in the last Song–Liao peace treaty (signed in 1042).[74] In the meantime, large numbers of Khitan troops started to gather at the border. Liao's aggressive move alarmed Emperor Shenzong. Intimidated by the idea of an imminent war with the Khitans, he decided to recruit a capable diplomat for negotiations of peace and a new border. After some consideration, he chose Shen Gua.[75]

As the emperor and Shen both understood, the greatest challenge of the issue arose from the Khitans' intentional use of vague language in their request. The leading emissary insisted that the border should be drawn along the "watershed" ( *fenshui ling* 分水嶺), but he refused to specify where it was. A watershed, by definition, is a high ridge dividing two river systems. In theory, almost every mountain can mark a watershed.[76]

Shen decided to take the matter into his own hands. After extensive fieldwork, he located all four watersheds in the region, three of them extremely close to the old borderline; it would not cost the Song state much to cede these small parcels of land to the Liao. The fourth watershed, formed by Mount Huangwei, posed a problem. At this point, the Song emperor was willing to draw the border at the northern foot of Mount

Huangwei. But if the line were drawn where Mount Huangwei parted the river, according to the criterion of "watershed," the new border would move further south by another 30 *li* (14.36 kilometer/8.86 miles). To Emperor Shenzong, that was not an acceptable condition.[77]

Therefore, when Shen set out on his mission, he had a clear, well-defined goal derived from his meticulous research beforehand: to protect the 30-*li* span of land around Mount Huangwei. In the Khitan capital, Shen engaged in a series of contentious debates staged as public events. He finally persuaded the Liao to accept his provision before an audience of over a thousand.[78]

The textual and field research Shen conducted prior to the trip made possible his persuasive performance at the negotiation table. According to Shen's own records, he demonstrated excellent debating skills (the sole extant source of the debates is the report he prepared for the emperor).[79] Despite a self-congratulatory mood that might color the credibility of some details, what Shen chose to include and highlight in this text indicate that the skills of persuasion he took pride in depended on his mastery of textual and experiential evidence. During one short debate, as Shen reconstructed it, the two sides used the word "evidence" seven times, six of which came from Shen's questioning. Following is a section of the conversation between Shen and two Liao officials, Yang Yijie 楊益戒 (ca. eleventh century) and Liang Ying 梁穎 (ca. eleventh century).

[Yang and Liang dutifully insisted that Mount Huangwei was part of the Liao soil from the founding of the dynasty.]

SHEN: What kind of textual evidence does the Northern Dynasty (the Liao) have?

LIANG: What kind of evidence does the Southern Dynasty (the Song) have?

SHEN: The Southern Dynasty has collected plenty of evidence about the Northern Dynasty, which includes evidence from a decade ago, evidence from this year, evidence from prefectures and counties, and evidence in imperial edicts. For starters, in the eleventh year of the Chongxi reign in the Northern Dynasty (1042), the Northern Dynasty sent commander Wang Shouyuan, vice inspector Zhang Yong, and seal holder Cao Wenxiu to a negotiation with Di Dianzhi, commander-in-chief of the Yangwu

Stockade, and Wu Jie, magistrate and commander of the Guo County. They reached the consensus that the borderline lay at the foot of Mount Huangwei. Since then, the official documents issued by the Shunyi Commandery all addressed the foot of Mount Huangwei as the borderline. Isn't it perfectly clear!

臣括答云："不委北朝有何文字照證?" 穎云："南朝有何照證?" 臣括答云："南朝收得北朝照證甚多,亦有十年前照證,亦有今年照證,亦有州縣照證,亦有聖旨照證。且說最先北朝重熙十一年,北朝差教練使王守源、副巡檢張永、句印官曹文秀,南朝差陽武寨都監瞿殿直、崞縣令教練使吳岊,同行定奪,以黃嵬大山腳下為界,自後順義軍累有公牒,皆稱黃嵬大山腳下為界,豈不分白!"[80]

Shen's long response was hard to refute, and such persuasive force owed to the fact that behind every claim of "evidence" he held well-researched information unavailable on the other side of the negotiation table.

The three measurement projects discussed in this chapter were distinct from one another, and they required a broad spectrum of concrete skills. Nevertheless, behind the array of actions stood one man demonstrating an essential preoccupation: to remain steadfast in observing "things" as distinctive objects and processes and to engage them with hearing and seeing. The hydraulic project required meticulous attention to the topographical characteristics along the Bian, and the heart of the calendrical reform lay in systematic observations of the firmament. Success in the border negotiations required the exploration of every watershed in the border region. Shen's sensory engagement with "things" had become even more intensive in practice, and his epistemic decisions revealed a number of interesting proclivities in accordance; one was to individuate "things," a point I discuss in the following chapter.

CHAPTER 4

# Individuating Things

During his decade of intensive work on quantitative projects, Shen Gua was deeply immersed in the sensory world he engaged through hearing and seeing: the banks of a meandering river, the vault of the heavens, and the vast ridges along the borderland. To know from hearing and seeing was an action he productively explored. While measuring the world in quantitative terms, Shen also mused on the orders underlying the sensory properties. To him, the two paths of engagement with the world were certainly not mutually contradictory, but neither were they constantly coherent. Juggling between sensory knowing and the plumbing of certain deep orders, Shen kept a steady emphasis on the utility of the former and embraced it as an epistemic guide. This raises the issue of the individuation of "things," the central topic I explore in this chapter.

In the opening section I define individuation in connection with the major epistemological issues this book has thus far examined. In the following three sections, I discuss the concept "number" and its ontological and epistemological implications. I then explore Shen's act of individuation in separating "things" from number in an astro-calendric system. Finally, I analyze a further ramification of "individuation"—Shen's conceptual grasp of the idea of representation—in his astronomical work.

## *Individuation*

The idea of "individuation" was closely related to the dual nature of "things" discussed in chapter 2.[1] A "thing," on one hand, was a distinctive object/process with its own properties and, on the other, a reification of a place in a larger order. This order could be the *dao*, number, or figure, among other possibilities, and the place of the "thing" was determined by specific relations present in each order. Viewed from the second perspective, a "thing" was nothing but a "knot in a net of relations," a metaphor I develop in this chapter. The subordination of a "thing" to a larger order rendered its own properties—substance and appearance, among others—philosophically insignificant.

Individuating a "thing" meant to accentuate its independent existence and thereby partly release it from the order that determined its place. In this chapter, I demonstrate how Shen engaged in and spoke about this procedure. The specific context of his thinking was "number," and what he did was to individuate "things" from numerical relations. Given the connection between the dual nature of "things" and the two-tier epistemology comprising modeling and knowing from hearing and seeing, individuation was closely associated with the knower's privileged use of hearing and seeing, as the individuated existence of a "thing" was often captured through its sensory qualities.

An important reason I use the concept of individuation is its utility in elucidating the ontological implication of knowing from hearing and seeing and avoiding potential pitfalls caused by modern/Western ontological assumptions. As mentioned in chapter 2, an inquiry into the placement of a "thing" in larger orders did not preclude an understanding of its sensory properties. An important ontological reason underlay this nondemarcation: unlike in a conventional two-tier ontology, the sensory aspect of a "thing" was not segregated from the deep order. For instance, in the current case, number remained in the same realm with concrete perceptible objects/processes. "Individuation" was an act to separate an entity from the relations in which it was placed, during which process one did *not* travel from an ontologically transcendental world down to the experiential realm.

My invocation of individuation will also help answer a question read-ers will likely ask: if Shen were indeed an empiricist, would he not have steered clear of superstition such as yinyang and observed natural phe-nomena in an undistorted manner? Posing the question another way within the field of Chinese history, wouldn't an empiricist in the eleventh century have circumvented the entire yinyang correlative cosmology as an outdated model associated mainly with the Han period? My discus-sion under the rubric of individuation addresses the question in a histori-cal manner.

## *Number*

Before discussing Shen's action of individuating "things" from number, it is important to have an accurate understanding of the nature of the problem from his viewpoint. To begin with, what was number? What was the relationship between number and "things"? To Shen, number was nei-ther superstition nor a disposable theory; in his world it claimed a preva-lent and real presence, which made the task of separating "things" from number more difficult than it sounds.

Number was a construal with an ancient pedigree. I introduce it in three key aspects: function, basic components, and a fundamental opera-tional principle. "Number" was synecdochic with "numerical relations," whose function was to order reality by dividing the cosmos into intelligible patterns. The basic components of number included a set of commonly used numerical relations, among which the yinyang dyad and the Five Processes were the most famous. Yinyang stood for a relationship between an alternating two, and the Five Processes was a relation of a cycling five (normally known as wood, fire, earth, metal, and water). Last, the opera-tional principle of number resided in the so-called correlative thinking, an issue I discuss after presenting concrete examples.

From antiquity through the Song, generations of thinkers developed an extensive body of number-related schemes on the foundation of the three key basics. Contrary to conventional modern understanding, many schemes did not concern a narrowly construed numerology or divination. These systems served a variety of purposes, such as in calendrical and medical practices, as well as in political and moral realms. They also took different forms. Some of them predominantly featured elaborate numerical designs, and others used numerical basics as an auxiliary theme only. The most prominent example of Song number systems was the one designed by Shao Yong. Shao intended to organize the world with an encompassing scheme of numerical relations, an effort reasonably accredited as numerology. Yet this numerological system reached deeply into the development of moral philosophy in his time, a good example of the complexity associated with number.[2]

It is thus important to keep in mind that number was not just in the possession of a few specialists and applied to certain highly technical/ esoteric practices only. Instead, number was prevalent in the Song world: literati widely endorsed its function to organize the world, and they routinely employed yinyang and the Five Processes as basic "conceptual grammars" in intellectual discourse.[3] The users of number included "cosmologists" such as Shao Yong, Zhou Dunyi 周敦頤 (1017–1073), Zhang Zai, and many other practitioners of the *Change*, and people like Wang Anshi, who enjoyed a reputation as being apathetic to cosmology yet comfortably employing numerical basics to delineate the order of the world (see chapter 2). Even more counterintuitively, empiricist Shen found that deep engagement with number provided him an indispensable background for his inquiries into the experiential world.

A key to understanding the significance of number in the Song World was its ontological status. Number was undoubtedly a construal, abstract in nature if subject to modern analysis. Yet it claimed an experiential presence in the Song world; to the predominant majority, basic numerical relations such as yinyang were simple, immediate reality rather than concepts susceptible to skepticism. A thorough understanding of this reality is necessary for grasping Shen's situation. In the next section, I flesh out this point with a central example, yinyang.

## *What Yinyang Was (and What It Was Not)*

What was yinyang? As a concept, it denoted a binary relation between an alternative pair (yin and yang), and such a relation could be contradictory, complementary, or interdependent, among other possibilities. Some scholars suggest that yin and yang can be simply understood as "x and y," or "this and that," and yinyang serves as an "organizing concept."[4] To cite a famous example from the ancient text *Huainan zi* 淮南子:

> A furred mammal or a feathered bird, which runs or flies, is yang.
> An animal with a shell or scales (e.g., a turtle or fish), which lays low
>     and hides, is yin.
> The sun is yang.
> The moon is yin.[5]

The relation between a bird and a turtle obviously contrasted high and low; the relation between sun and moon was contradictory (bright and dim) and complementary (alternating in the sky).

This explanation should be understood with a major caveat regarding the unique ontological traits of yinyang. Put simply, yinyang was supposed to have an experiential existence in the phenomenal world. To explicate this point, I demonstrate what yinyang was not. In a nutshell, it was not akin to any type of "universal" found in the Greco-European tradition.

To serve this study of a medieval Chinese man, I draw a quick comparison between yinyang and the scholastic conceptualization of universals, namely, "universals before the thing" (*universalia ante rem*) and "universals after the thing" (*universalia post rem*).[6] First, yinyang was not a "universal before the thing," of which a typical example is a Platonic form. The Platonic forms were supposed to be universal archetypes of particular things; the phenomenal world, in all its particularity, was the result of modeling after these ideal forms.[7] They were thus construed as "before" things. For example, "brightness" as an ideal form imposed on passive matter resulted in all kinds of bright things, such as the sun. Yinyang, in contrast, did not claim to be perfect, immutable archetypes, or to constitute an immutable realm independent from the experiential world. In the *Huainan zi* example, yang was not a model the sun followed, and it did not exist in a world other than the one the sun inhabited. Although in the generative scheme yinyang (as part of "number") indeed came prior to "things," such temporal priority was more of a metaphor for "fundamentality"—a weak ontological distinction—than a demarcation between transcendence and immanence.

Nor was yinyang a "universal after the thing," a concept derived from particular phenomena. The universal "brightness," for instance, was abstracted from all bright things and "caused" by them. In this scheme, a universal was the product of human thinking. Yang, however, was not a logical concept extrapolated from the commonality between sun and bird. Yang's existence had nothing to do with the active human mind. Yinyang certainly can be used as an "organizing concept" such as "x and y," but in the minds of premodern Chinese, it was not simply a concept created for human convenience.

Another major difference between yinyang and many Western universals is that the former was not a concept of substance or essence. Instead, it denoted relations. In the example of furred/feathered animals versus shelled/scaled creatures, for instance, it is incorrect to assume that a deer was yang because yang was a fixed attribute/essence of this animal. Rather, a deer, when placed in a binary relation with a turtle, *functioned* as yang, while the turtle functioned as yin.

In a different relation, the yinyang assignment might change completely. Another reference to furred mammals from a Song thinker is a telling example. Su Shi once commented on medical recipes employing powdered deer antlers as an ingredient. He distinguished the antlers of a regular deer (*lu* 鹿) from those of a Père David's deer (*mi* 麋) in terms of effect, because "A regular deer is a yang animal, [ . . . ] and a Père David's deer is a yin animal."[8] In this case, although both were furred mammals (members of the same family, in fact), one of them was yang and the other yin. Clearly, yang was not always associated with furred mammals, much less a permanent essence of them.

The focus on relation enabled yinyang to account for changes of all kinds as the result of interdependence. Returning to the *Huainan zi* example, the fact that both the sun and furred animals were yang set a framework for explaining why deer shed in the spring and summer and threw off their antlers on the winter solstice. As the sun rose higher and daylight lengthened in the spring and summer, fur on the deer thickened and began to shed. As the sun waned to its weakest state on the winter solstice, the deer would correspondingly shed their antlers. Actual (unseen) interactions occurred between the sun and deer, because both functioned as yang, as in the same "category." This constituted the famous thesis of "resonance between [things of] the same category" (*tonglei ganying* 同類感應), the central mechanism of the so-called correlative thinking.[9]

For the sake of explicating Shen's case, one more important concept related to yinyang requires an introduction. Since the Han period, in most contexts the characterization of a yinyang relation involved another concept, *qi*. Often left untranslated, *qi* can be understood as both the texture and dynamic of the underlying reality of the world. Regarding its nature, Ted Kaptchuk aptly states that "*qi* is somewhere in between, a kind of matter on the verge of becoming energy, or energy at the point of materializing."[10] The appearance of new concoctions such as yin *qi* and yang *qi* in the Han signaled that *qi* (as a concept) had become a new interpretative framework of yinyang: yin and yang now circulated in the phenomenal world as "dynamic and natural forms of flowing energy and manifestation of the primordial potency of the universe."[11] For example, a Han text, *Luxuriant Gems of the Spring and Autumn* (*Chunqiu fanlu* 春秋繁露), stated: "As the yin *qi* of Heaven and Earth rises, the yin *qi* of humans rises in resonance. When the yin *qi* of humans rises, the yin *qi* of Heaven and Earth also rises in resonance. The *dao* is one."[12] As soon as *qi* joined yinyang, the resonance between two "things" could be understood as an interaction by way of *qi*. Therefore, *qi* provided a unified material base to mutual resonances and rendered them more palpable as an experience.

In sum, yinyang maintained a real, experiential existence in the world, as did the rest of number. "Realness" implied, first, no existence in a realm other than the phenomenal world. Unlike the Platonic forms, yinyang did not necessitate a two-tiered ontology. Second, yinyang was not just an intellectually derived concept whose existence depended on the workings of the human mind. The concept denoted actual, though unperceivable, connections between things. Its later alliance with *qi*—the universal energy-matter—further strengthened its experiential existence. Last, yinyang stood for relation rather than substance. The circulation of yin *qi* and yang *qi* functioned as the dynamic forces that mapped the links between "things" and eventually connected all into an interdependent cosmos.

## *"Things" as Knots in a Net*

Now let me return to the core question of this chapter: the relation between number and "things." Number connected "things," and according to the generative formula discussed in chapter 2, they also gave birth to "things," in the sense of granting them proper places. Thus in this scheme "things" did not stand as distinctive entities. Then what kind of existence did "things" have vis-à-vis number (or for that matter, any other larger orders)? In response to this question Zhang Zai, a Song thinker with a strong interest in number, offered an answer that characterized the nature of "things" in terms of *qi*'s movements. In his characterization, "*qi* has to coalesce into the ten thousand things, and the ten thousand things have to dissipate into the Great Void."[13] As such, Zhang defined "things" as the products of contractions of *qi* and thus "dynamic configurations" of relations.[14] What truly identified a "thing" always resided in relations rather than in any self-contained static reality. Zhang made this point as follows:

> There is no such thing that a thing stands on its own (*guli* 孤立). If a thing does not come into illumination (*faming* 發明) through [processes of] identification and differentiation (*tongyi* 同異), condensing and extending (*qushen* 屈伸), as well as beginning and ending (*zhongshi* 終始), it is not a thing albeit named so.[15]

Without participating in larger scales of relation, a "thing," even with distinct, self-possessed qualities, did not qualify as a "thing." This also explains why the static reality of a "thing" was not important: that reality was merely the manifestation of ever-changing interactions.

The metaphor I proposed at the beginning of this chapter helps us visualize the relationship between "things" and number. The cosmos, as a large whole, was divided according to a body of numerical ratios, which in their entirety constituted an extensive net. The (patterns of) cords were the relations, and the knots, where the cords met and coalesced, were "things." The existence of a knot depended entirely on the interactions of the cords as well as the overall structure of the net. The nature of the knot itself, tight or loose, had little relevance because it asserted no structural significance in the whole picture. What I emphasize with this metaphor is that to a Song person, the cords were as real as the knots, both present in the same net with no ontological segregation. In the mind of the Song, the cords were not only real but also more important.[16]

## Individuating "Things" in the Course-Qi System

When Shen came to grapple with the relation between number and "things" as a problem, he was working on the so-called Five Circulative Courses and Six *Qi* (*wu yun liu qi* 五運六氣, hereafter the Course-*Qi* system), an astro-calendric system primarily used for making weather forecasts.[17] In Song times, the system was also applied extensively in medical practice, and in various ways exerted influence in the philosophical work of Zhou Dunyi and Shao Yong.[18] Shen's engagement with the system started while he served in the Bureau of Astronomy. The basic premise of the Course-*Qi* scheme held that any weather condition resulted from the movements of yinyang *qi*. Predicting the weather involved identifying the paths of these movements through arithmetic calculations and making appropriate readings. The numerical patterns in the scheme, as I show later, were much more elaborate than a simple alternation of yin and yang or a progression of the Five Processes. Nevertheless, the patterns represented variations of the basics and, most importantly, inherited all the ontological features I identify in the discussion of yinyang.

Before plunging into the technical details of the Course-*Qi* system, let me first summarize the problem Shen confronted: when a number system failed to produce an accurate forecast, how should one address the discrepancy between number and actual weather conditions ("things")? My analysis of the ontological status of yinyang lays the ground for a clear understanding of Shen's situation. He faced a net with complicated patterns, each cord as real as each knot. The numerical formulae in the Course-*Qi* system formed the patterns of the cords, and weather conditions at particular temporal points occurred at the knots. Making a forecast required calculating the location of a knot, identifying the surrounding patterns enmeshing it, and transposing the structural reading into weather conditions. Humans, however, did not possess complete knowledge of all the intricate details of the net, and thus their predictions sometimes failed to match the actual meteorological phenomena. Shen's solution was to revise current numerological rules under the guidance of sensory phenomena, by which he accentuated the importance of the individuated existence of "things" and granted them epistemic guiding power.

Now let me proceed to explain the workings of the Course-*Qi* scheme.[19] The system featured two primary sets of variables, the Five Circulative Courses—earth (*tu* 土), metal (*jin* 金), water (*shui* 水), wood (*mu* 木), and fire (*huo* 火)—as well as the Six *Qi*: wind (*feng* 風), cold (*han* 寒), heat (*shu* 暑), dampness (*shi* 濕), dryness (*zao* 燥), and fire (*huo* 火).

Each *qi* was supposed to have a yinyang association, which was marked by a modifier conjugated from yin and yang. The six modifiers included three *yin* modalities—the old yin (*taiyin* 太陰), the young yin (*shaoyin* 少陰), the diminishing yin (*jueyin* 厥陰)—and three yang modalities: the young yang (*shaoyang* 少陽), the mature yang (*yangming* 陽明), and the old yang (*taiyang* 太陽).

Both courses and *qi* operated in cycles: one followed another in a certain sequence, and then the last joined the first to form a loop. The two sets corresponded to more than one cycle, and the cycles differed in duration. For instance, the Five Courses, according to one rule, operated in a five-year cycle, in which case each course covered a whole calendar year. In a different scenario, the Five Courses ran on a reduced scale of one year, so that each course covered merely one fifth of a year. A complicated host of rules guided the coordination between cycles of different scales. The same complexity applied to the Six *Qi*. By following these arithmetic rules, one was able to calculate which course encountered which *qi* (that is, the intersection of a certain cycle of courses and a certain cycle of *qi*) at a certain point in time.

The "encounters" between courses and *qi* provided the basis for making a forecast. To explain it in an extremely simplified fashion: such an encounter was either a conquest or a defeat. A course could either overwhelm a *qi* or surrender to it, and vice versa, depending on the specific associations the two agents at issue had. The associations of *qi* were yin and yang in different phases, whereas the associations of courses included designations such as "dampness" and "dryness." After determining the nature of the encounter, one could translate it into meteorological phenomena.

For instance, a conquest could bring down heavy rain; a defeat could entail protracted gloominess. Generations of practitioners deposited possible readings into diagrams. In an ideal scenario, a user of the system could make a forecast based solely on arithmetic calculation and diagram reading.

Although already rich in complexity, this basic structure was still too finite to cover all possibilities, which resulted in prediction errors. A conventional solution to this difficulty required further complicating the patterns by specifying the behaviors of numerological agents. For instance, the Circulative Course fire is supposed to arrive in 2016 to supervise the year, but the agent may show up late, or it may refuse to exit on time. The erratic behaviors of the course will subsequently affect the operation of the entire cycle and thus alter interactions with *qi*. Some practitioners endorsed certain aberrations so as to offset the rigidity of the original rules.

Shen presented himself as a master of the Course-*Qi* system and claimed that his excellence derived from a sound understanding of aberrations in addition to the standard rules. He took pride in one successful prediction in particular. During the Xining reign, a prolonged drought persisted and caused widespread anxiety in the capital. One day the sky suddenly turned overcast, which people applauded as a positive sign of rain. But to their disappointment, a scorching sun soon returned. The worried emperor consulted Shen about the chances of rain in the near future. Shen responded with confidence that rain would fall the next day. Although no one believed him at the time, heavy rain indeed came down as he predicted.[20]

The key to his success, as Shen explained, was an understanding of so-called overwhelming (*sheng* 勝), a deviance from the standard patterns. During the few days when the weather had turned cloudy, the course at work was earth, which reigned with dampness. In the meantime, the *qi* associated with earth was young *yin*, which means that the power of *yin* was on the rise. The encounter of earth and the rising *yin* in theory should have caused precipitation. However, there was a more complicated cause—a deviant one—for the overcast weather. The young *yin qi* failed to function properly because the young *yin* was "overwhelmed" by the diminishing *yin*, the previous agent in the *qi* cycle. The clash lasted for days and produced gloomy weather. Later, the course shifted from earth to metal, associated with dryness. In the meantime, the *qi* moved from the young *yin* to the old *yin*. The old *yin* marked the peak of *yin* strength and thus easily canceled out the constraining influence of the diminishing *yin*. At this point, the course and the *qi* had resumed the normal paces and returned to their due lots. When dryness met *yin* of maximum strength, abundant rain fell.

Shen went on to articulate his systematic thinking on aberrations. In his view, a proper reading of any single numerological rule should be accompanied with some consideration of deviance. Apparently, he had built a whole repertoire of aberrations, from which he showcased as examples the possibilities in the situation when the diminishing *yin qi* was in charge:

"Following" (*cong* 從): Windy weather and lush vegetation.

"Opposition" (*ni* 逆): Dry, calm, and windless weather.

"Excess" (*yin* 淫): The sky is windy and dark, and yet rivers do not freeze over.

"Overabundance" (*yu* 郁): Gusty winds break trees and heavy clouds make the air murky.

"Overwhelming" (*sheng* 勝): Rivers and mountains dry up and vegetation withers.

"Retaliation" (*fu* 復): Unusual heat and a pest disaster.

"Major excess" (*taiguo* 太過): Natural disasters as severe as mountain sliding and ground cracking, along with an often murky sky.

"Insufficiency" (*buzu* 不足): The weather remains gloomy all the time. Clouds hang heavy in the sky and it is dark even during the day.[21]

The diminishing *yin qi* was normally associated with windy weather. When this *qi* was active, if it was performing normally ("following"), "windy weather and lush vegetation" should occur. But if the *qi* reversed its movement ("opposition"), the weather would turn dry, clear, and windless. All other aberrations applied accordingly. As such, Shen expanded one pattern into a variety of possibilities, which greatly enhanced the flexibility of his readings.

Let's further explicate Shen's story beyond his explanation. First, why was a good understanding of aberrations helpful for better use of this system? The answer resides in Shen's philosophical understanding of the Course-*Qi* scheme. On one hand, the system—the net—was as ontologically real as anything that could be captured in sensory perception. On the other hand, the current basics concerning the regular movements of the numerological agents did not include all details of the net. Shen's tacit argument was that the standard rules reflected the current human understanding of the net, which was by no means perfect.

This implicit claim was consistent with Shen's long-term contemplation of number. In one passage in *Brush Talks*, he rendered the point in more explicit terms: "As I see it, those who speak about number in this world only capture the coarse vestiges (*ji* 跡) of it."[22] Here "coarse vestiges" referred to sporadic traces left by the operation of number, which, according to Shen, were insufficient to account for number's "intricate" (*wei* 微) reality.[23] "Vestiges" might not even provide a proper access to number in its most profound aspect. To illuminate this point, Shen offered some examples. For instance, conventional knowledge held that the scheme of the Five Processes was the deep order underlying the circulation of the seasons: spring corresponded with wood, summer with fire, autumn with metal, and winter with water. The current human understanding of this pattern was presumably the product of "coarse vestiges," and, unsurprisingly, it lacked exactitude, because within a month, even within a day, different phases of time also had correlations with the Five Processes.[24] The scheme presumably provided structure to time by infinite divisions, which constituted the so-called intricacy.

Returning to the Course-*Qi* problem, we see clearly why Shen deemed the standard rules insufficient. Even though humans might never be able to fully fathom the perfect intricacy of number, a responsible practitioner of number systems should exhort himself to approach the goal as closely as possible. To this end, Shen suggested the study of "aberrations," the deviances that revealed subtle details of the cosmos beyond the basics.

Thus, Shen proposed a dichotomy between "constancy" (*chang* 常) and "aberration" (*bian* 變), and attributed to the latter a kind of significance no less than the former's:

Of the patterns of things, some are constant and some aberrant. What the courses and *qi* determine is a constancy, and what differs from what they determine is an aberration.

大凡物理有常、有變。運氣所主者，常也；異夫所主者，皆變也。[25]

In his definition, a constancy was a basic numerological pattern. One example of a constancy he gave was that at a certain temporal point, young *yin* was the reigning *qi*. An "aberration," by definition, was any pattern that contradicted a constancy. In Shen's story, an exemplary aberration was the overcast weather during the young *yin* period, when precipitation was supposed to be abundant. In his opinion, such aberrations occurred often and thus deserved serious attention:

A constancy is just like original *qi*, and yet aberrations reach into everywhere. Each thus merits its own prediction.

常則如本氣，變則無所不至，而各有所占。[26]

In other words, a practitioner of Course-*Qi* had to pay heed to both to predict effectively, hence the long list of aberrations Shen prepared.

The next question concerns how Shen further used these aberrations within the framework of constancies. To apply the net-knot metaphor here, the aberrations were the new knots he identified. In Shen's understanding, the existence of these knots proved that the current understanding of the net was not "intricate" enough. Therefore, he assigned more varied behaviors to numerological agents and created more complicated cord patterns to accommodate these new knots. In the case of a nonraining young *yin* period, he acknowledged the overcast weather as a new knot, and matched it with a new numerological mechanism, "overwhelming." In future readings, an operator of Course-*Qi* should pay heed not only to the existence of young *yin* but also to its possibly deviant interactions with other agents. With an open mind to aberrations, one would be in command of a more comprehensive (understanding of the) net.

In this process, the way Shen defined and used "aberrations" was of specific epistemological interest: he saliently relied on sensory knowing.[27] All aberrations were defined with concrete meteorological phenomena, which further monitored the editing of numerological rules. For instance,

"overwhelming" was defined with drying waterways and withering vegetation. Speaking in practical terms, Shen was indeed revising numerological rules on the basis of sensory phenomena. Such a procedure was
no surprise given the aforementioned philosophical reasoning. In theory
the Course-*Qi* framework commanded the vicissitudes of meteorological phenomena (as the deep order underlying them); in practice, humans
did not have full access to all subtleties of the framework, and sensory
phenomena provided them with key (if not the only) evidence of the insufficiency of their understanding. After revision, the resultant new net
would better map onto concrete sensory experience, because the alterations were made on the foundation of observed phenomena. This was how
Shen successfully managed the Course-*Qi* scheme.

Shen's operation and discussion of the Course-*Qi* scheme provides a
good platform for examining his handling of "things" within the framework of number; specifically, his proclivity to value "things" in individuated existence. In this case, concrete meteorological phenomena—gusty
winds, plagues of insects, withering vegetation—were "things" in their
distinctive existence, perceived by humans through "hearing and seeing."
The Course-*Qi* scheme provided a framework in which these "things"
were supposed to find their places.

Shen demonstrated his propensity to value individuated "things" in
two ways. First, in operating the system, he highlighted the existences of
"things" recognized by their sensory qualities and, in so doing, individuated them from the net that was supposed to determine their placement.
In identifying aberrations, Shen incorporated extensive actual weather
conditions—products of hearing and seeing—and granted them an iden-
tifiable status. These actions pointed to an assumption at work on the
horizon of Shen's thinking: while a "thing" was a transient configuration
of *qi*, its individuated existence, mainly perceived through the senses, was
also important.

Second, Shen accorded "things" epistemic guiding power in revising
numerological rules. This was a substantial challenge to the epistemological hierarchy embedded in the system, where number was considered to
hold normative power. Shen made it official, writing that constancies and
aberrations were nearly equals in terms of guiding one's use of the system. The new numerological rules he devised were based on observations
of concrete meteorological phenomena; therefore, although sensory phe-

nomena eventually returned to the number system, they did so as an epistemic guide through a process of active negotiation with the established rules.

In sum, in Shen's hands, a "thing" was no longer just a knot recognized by its place in a net. It was now known by its "knot qualities," tight or loose, simple or complicated. Even in the world of number, which was heavily regulated by normative rules, a "thing" did not necessarily vanish into relations; instead, it merited observation in its individuated existence.

Let me conclude this section by responding to the empiricist question raised at the beginning of the chapter. Over the course of my analysis, it should have become clear that number was not simply superstition or a world picture susceptible to skepticism. Instead, number was real. I make this argument for all three basic aspects of number: the omnipresence of yinyang and other basic numerical ratios, the belief that they were the real structures of the world, and the central mechanism that enabled all numerical operations: correlative thinking. All three constituted a factual aspect of the lived experience of Song people, including Shen Gua. From antiquity through this time, when people criticized number-related systems, they typically did so by attacking concrete aspects or particular details—understood as the result of erroneous/incomplete human understanding—rather than these basics.[28] The realness of number constituted the key background of Shen's action of "individuation."

While my previous analysis has covered the first two basics of number, the third element—correlative thinking—requires further elaboration. This explanation also serves to answer the second version of the question: why didn't an eleventh-century empiricist simply renounce correlative thinking, given that the Han correlative cosmology was already under attack?[29] In my definition, correlative thinking was not an equivalent of the so-called correlative cosmology in the Han; the former was the basic working principle of the latter. In other words, correlative thinking was the philosophical foundation on which historical thinkers built varied themes, such as the Han cosmology or the Song uses of number.

The essence of correlative thinking was the making of correlations. A correlation was a resonance between two things of the same category. It was a specific mode of causality that took on different concrete forms.[30] In the Han, any two things of the same *qi* came to resonate and made one correlation. In the Song, resonances occurred along more regulated movements of *qi*. That *qi* circulated in the world and made one thing respond to another provided the foundation for the belief that *qi* propelled all transformations in the world—a belief held by Han and Song thinkers.

Compared with the Han, however, Song thinkers spoke less of concrete resonances, such as one between the sun and the deer. This was not because of a decline in correlative thinking, but was the result of a more organized way of using it. Song scholars contained myriad ad hoc correlations within new systems featuring more definite structures, which constituted more effective instruments of organization. Examples include not only technical cases like Shen's use of Course-*Qi* but also the philosophies of Zhou Dunyi, Shao Yong, Zhang Zai, and Cheng Yi.

The reorganization of *qi* enabled Song thinkers to more accurately monitor the use of correlations to suit their specific purposes. For instance, they generally refrained from correlating policy making with meteorological occurrences (especially natural disasters) to ensure that the imperial monarch (and his bureaucracy) enjoyed a greater degree of autonomy.[31] But in some realms of statecraft, such as making ceremonial music, Song literati still made extensive references to correlations between cosmic forces and human affairs.[32] The regulation and redistribution of correlations did not render correlative thinking obsolete in any sense. Correlative thinking remained equally real to Song thinkers, despite the Han version falling out of popularity.

The foregoing excursion into the broad issues helps reveal the implications of "individuation." First, it shows why a revolutionary renunciation of number was not possible for a Song person. The ontology dictated that number was not a theory that could be easily discarded. Second, it demonstrates that individuation, although not as revolutionary as renunciation, was difficult and required an elaborate coming to terms with existing stipulations. Third and most importantly, it shows that a negotiation of epistemological commitment was possible between a net and knots, either in a political context or in Course-*Qi* numerology. A number scheme, though holding normative power, did not command an exclusive, fixating epistemological commitment. Participants in Song politics were allowed to bracket some of the epistemological implications of correlative thinking (such as resonances between court policies and natural disasters) and remodel the scheme in accordance with their own experience. Similarly, Shen Gua took the liberty to seek new epistemic guides beyond the received Course-*Qi* system and revise existing rules on the basis of sensory knowledge.

## Separating Things from Representation

One ramification of Shen's effort to individuate "things" was his further thinking on the relation between "things" and their representations. Two methodological points are worth clarification. First, rigorously speaking, the separation between sensory "things" and representational schemes was no longer individuation but an escalated version of it, because it indicated an ontological demarcation between experience and theory. Indeed, in the following contexts, Shen started to contemplate an ontological schism related to but distinct from his handling of the Course-*Qi* system. Second, a note should be added to my discussion of representation. Shen did not invent a concept that specifically designated "representation." The awareness of the distinction between a "thing" and its representation was an epistemological assumption underlying his praxis, a commitment he demonstrated through action. The significance of this assumption should not be dwarfed by the lack of normative articulation, because it exerted an ineluctable influence in shaping his concrete decisions.

Not surprisingly, Shen demonstrated this insight in astronomical work. His initial series of discussions regarding representation focused on measurement—a kind of mathematical representation.[33] To Shen, the biggest "thing" was the firmament, and the procedures of mathematical measurement, along with the instruments which reified them, were postulated entities for representational purposes.[34] This view, however, was not a consensus Shen shared with lay people or even professional computists. The issue prompted his repeated efforts to articulate the difference between the actual and the representational.

Once, for instance, a literatus colleague expressed doubt about the correct use of the armillary sphere.[35] Since the horizon ring in the sphere was supposed to be the ground, this fellow official asked, shouldn't the sphere be placed precisely on the ground rather than on a platform over thirty feet high? In Shen's response, he pointed out that the problem lay in the blurring of the distinction between "actual number" (*shishu* 實數) and "scale number" (*zhunshu* 準數).[36] The actual number was the real distance between a heavenly body and a particular location in space, a sensory "thing," whereas the apparent distance was what people read from measurement, a proportional reduction of the former. As Shen explained, the scale of this particular armillary sphere approximated the following ratio:

1 *fen* (0.31 centimeters/0.12 inches) = thousands of *li*
(hundreds of kilometers/miles)

As an "actual number," the height of the platform measured a few *zhang* (1 *zhang* approximating 3.17 meters/10.39 feet), but as soon as it was scaled down according to this ratio, the "scale number" would become too minute to be taken into consideration. It was a mistake to evaluate a numerical value without paying attention to its nature, "actual" or "scaled." Therefore, Shen concluded, "how can a few *zhang* affect the measurement of the grand Heaven and Earth?"[37] He further reasoned that if a discrepancy occurred within the scaled sphere—in this case, the space encircled within the armillaries—its impact should be proportionally scaled up, thus becoming no longer negligible. Raising or lowering the brace by one *fen*, for instance, would result in a difference of thousands of *li* in calculating actual distances.

Shen's main point was to show his confused colleague an ontological distinction: the universe was a "thing," an entity in human experience, and the armillary sphere provided a mathematical representation of it. The armillary sphere was supposed to be a model of the universe. With the numerical marks engraved on the armillaries, the instrument helped map the qualitative structure of the cosmos into a numerically based one. As soon as one switched his eyes from the stars in the sky to the engravings on the rings, he traveled from the domain of sensory objects to that of mathematical representation.

Another inquiry from a different colleague invoked a similar confusion. This time, the so-called twenty-eight lunar mansions, the Chinese model of organizing constellations, was called into question. The Chinese recognized twenty-eight groupings of stars along the celestial equator, which constituted a measurement system assisting an observer in locating a heavenly body/celestial phenomenon. Unlike the evenly divided Western zodiac, this system did not have equal parts.[38] As observed in Shen's times, a small mansion was merely a few degrees in breadth, and a big one spread over thirty degrees. His colleague questioned this disparity, with the assumption that a measurement system always dispensed equal and consistent divisions.[39]

Shen responded by again invoking the distinction between a "thing" (stars) and measurement. He explained how the lunar mansion system came into existence by first emphasizing that the firmament and its measurement system were two separate entities. The latter—a construct—came into existence due to astronomers' intention to subject the former—a "thing"—to calculation. The first step was thus to generate artificial measurement on the basis of natural phenomena:

> Celestial phenomena originally had no degrees. Calendrical computists had nowhere to assign numbers, so they stipulated that one day's mean solar travel was over 365 degrees.
>
> 天事本無度，推曆者無以寓其數，乃以日所行分天為三百六十五度有奇。[40]

Then, these astronomers felt the need to find more reference points in the sky to establish their new measurement system, so they set out to locate twenty-eight stars as determinatives. Hence the second step, from numbers back to "things":

> Since [the astronomers] had divided [mean solar travel into degrees], there had to be things to assist memorization so they could look and then measure. Therefore, [they] memorized according to stars which fell right on [certain] degrees.
>
> 既分之，必有物記之，然後可窺而數，於是以當度之星記之。[41]

Each "determinative star," along with the natural grouping of stars in its vicinity, formulated a lunar mansion. According to Shen's understanding, these twenty-eight stars were as many as the early astronomers could find to fit their measurement model. Therefore, it was incorrect to say that astronomers did not intend the mansions to be even; the stars, being "things," were naturally uneven to begin with.[42] It was a straightforward issue if one did not confuse the firmament with the measurement system built on it.

Another example that demonstrated Shen's acute awareness of representation concerned the "lunar roads." In principle, the Yellow Road (ecliptic), the Red Road (equator), and the nine lunar roads were models of the apparent movements of the sun and the moon, hence representations. Shen addressed all of them as "imposed names" (*qiangming* 強名), rather than "what really existed" (*shiyou* 實有), putting forth a dichotomy comparable to that between representation and the "thing" it stood for.[43] The Yellow Road and Red Road were relatively straightforward models; the lunar roads, however, were more convoluted in appearance. To accommodate the moon's complex positions relative to the sun and its swing in velocity, earlier astronomers had divided the apparent lunar travel into different sections, marking each segment with a unique color. For instance, the section south of the Yellow Road was "Vermilion Road," and the section north, "Black Road." Shen found it alarmingly problematic that many computists in his times naively assumed that the moon indeed had nine different orbits. Once again, they failed to grasp the difference between the phenomenal "thing" (the lunar apparent movement) and its representational model.

Shen's long-term engagement with mathematical representation in astronomical work enabled him to develop a balanced stance over this issue: he embraced its utility without losing sight of potential issues. Using the armillary sphere as an example, he praised the benefit of using representation on the theoretical level:

> When degrees are in the sky, [humans] make an astronomical instrument so that degrees come to be in the instrument. When degrees are in the instrument, the sun, the moon, and the Five Planets all gather in the instrument without being interfered by the sky anymore. When the sky no longer intervenes, it is not difficult to know what is in the sky.
>
> 度在天者也，為之璣衡，則度在器；度在器，則日月五星可以搏乎器中，而天無所豫也。天無所豫，則在天者不為難知也。[44]

The first "degrees" refer to the locations of heavenly bodies; after the transfer from "the sky" into "the instrument," they became readings on the armillaries—the positions of the sun, moon, and planets in the measurement model. This was the transformation from the realm of "things" (stars) to that of representation (degrees). The world "no longer interfered by the sky" was the representational world, which claimed a new ontological status and thus sundered all direct links with "things." Once the representational system achieved its autonomy, "it is not difficult to know what is in the sky" because all entities and distances could be measured and calculated in numerical terms. Mathematical representation enabled astronomers to calibrate the vast universe in finite numbers at a manageable scale.

Shen's endorsement of the utility of representation did not distract him from constant watchfulness for its postulated nature; in a way, it prompted him to look out for erroneous models. He once made a bold panoramic statement: that a calendar was by nature representation, and it was a flawed representational model. In his words, "calendars were the product of speculations" (*li yi chu hu yi ye* 曆亦出乎臆也).[45] The invocation of the word "speculations" revealed Shen's alertness toward the constructed nature of calendars and the predilection to err. In his critical eyes, a calendar could never fully represent the phenomenal world, because again it captured the "coarse" and yet often missed the "intricate."[46]

Shen carefully investigated inaccuracies in calendars. First, erroneous predictions abounded in history. For instance, in the late 1060s Venus conjoined with Mars in the lunar mansion Zhen, an important celestial event that as many as eleven calendrical systems failed to predict. Beyond ad hoc failures, a calendrical system lacked the capacity to accommodate certain complexities. Although a calendar could pinpoint the location at which a celestial phenomenon happened by assigning a numerical degree to it, the system was unable to reveal other essential details, such as the celestial event's position relative to the ecliptic.[47] Based on these considerations, Shen posed a pointed critique of contemporaneous calendrical experts for being too reliant on established astronomical models, too convinced of the veracity of their content, and too ignorant of the artificial nature of these constructs. The theoretical awareness of representation turned him into an outspoken advocate of astronomical observation: after all, only through observation would humans be able to reconnect with the phenomenal world.

Shen's discussion of the thing/representation distinction in the context of calendars affords an interesting comparison with his previous thinking on "things" versus number. To a modern audience, both a mathematical astronomical model and a number system are constructs, "world pictures" representational in nature; furthermore, astronomy is a more accurate ("scientific") representation than numerology. To Shen's peers, number and calendar were "self-so" (*ziran* 自然) and a calendar was part of number.[48] Yet to Shen, number (in its three basics) was reality and calendar an artifice; calendrical systems were infinitely insufficient approximation of the deep reality.

In affirming the value of distinctive "things," Shen encountered various challenges in the world he inhabited. He needed to negotiate a certain degree of individuation of "things" from larger orders so he could productively use hearing and seeing, which he deemed a valuable source of knowledge. He also had to keep a discerning eye on representations, which might or might not facilitate human understanding of the objects themselves. These epistemological insights surfaced not only in his highly technical work, such as calendrical studies, but also in a broad realm of statecraft as his career evolved. In the next two chapters, I consider Shen's insights in the context of the grand New Policies reform.

CHAPTER 5

# Reforming the World (1071–1075)

In 1055 Shen Gua was still struggling in the countryside as a clerk; in 1065, he was a novice associate working between the stacks of the Imperial Libraries; and in 1075 he became a new political star as vice head of state finance. Despite that Shen had achieved success in a series of quantitative projects from the 1060s through 1075, as I discussed in chapter 3, an important political event directly propelled his rapid rise: the so-called New Policies reform. In this reform, Shen applied his adeptness with "things" to a variety of fields, emerging steadily as a piecemeal engineer indispensable to the most ambitious reform initiative of his time.

This chapter starts with an introduction of the New Policies and continues by analyzing the political alliance between Shen and Wang Anshi, the reform leader. In the following, I introduce Shen's concrete work in four major reform areas, including ritual reconstruction, hydraulic finance, rural relief, and security management. In the last section, I discuss his military work on the Song–Liao border.

## Overview of the New Policies

The year 1068 witnessed the emergence of Wang Anshi as the cynosure of political change. Having secured Emperor Shenzong's endorsement, Wang launched a reform unprecedented in scale and impact. Known as the New Policies, the reform reached almost every aspect of Song life—the economy, bureaucracy, military, legislation, and culture—and brought about a sea change. With aims to enrich the state and strengthen the army, Wang thrust the machinery of state into various areas previously free of governmental influence and turned the state into the predominant actor in education, land use, trade, and poor relief.[1]

To understand Shen's participation in the reform, a brief introduction to the concrete initiatives is necessary. The following acts were both policies central to Wang's entire reform agenda as well as areas where Shen intensively participated. The Green Sprouts Policy (*qingmiao fa* 青苗法) provided the blueprint for a rural credit system. To keep small farmers solvent and lessen their dependence on "engrossing" (*jianbing* 兼併) landlords, the state extended loans to agricultural cultivators in the spring and retrieved payment in two installments in the summer and fall.[2] It was enacted primarily as an economic welfare measure.[3]

Another significant policy, the so-called Hired Service Policy (*muyi fa* 募役法), concerned the recruitment of service personnel for local government (*zhiyi* 職役).[4] Prior to the reform, the government of a locality organized the population into a property-based, ranked hierarchy and requisitioned servicemen from the upper-grade houses by rotation. Wang's new policy aimed at replacing requisitioned recruits with paid volunteers. Previously, only part of the population, usually rural landowners and well-to-do farmers, were liable for service rotation. The new system extracted a cash tax from all households to cover the costs of hired laborers. As a result, privileged social sectors (for instance, official households) once exempt from service now had to contribute by paying taxes.[5] Green Sprouts and Hired Labor represented the state's attempt to redistribute economic resources to foster a higher degree of social equality.

The Mutual Security Policy (*baojia fa* 保甲法) allowed the state to organize the population into a family-based security network. The policy grouped households into communities of different sizes, each standing as a security unit guarded by armed local residents. The system primarily served to tighten local control and maintain order, and it later became a means by which Wang reformed the military establishment.[6]

Finally, the Policy of Farming and Hydraulics (*nongtian shuili fa* 農田水利法) augmented the state's direct participation in the construction of agricultural infrastructure for enhancing economic productivity.[7]

All four policies clearly demonstrated the central tenet of Wang's reform: an expansive, activist state apparatus to boost the economic and military strength of the empire.

## Rapport with Wang Anshi

In their younger days, Shen Gua and Wang Anshi maintained a personal friendship, not close but enduring.[8] Their acquaintance went back to the 1050s, when Shen and his brother went to Wang to seek an epitaph for their late father (see chapter 1). In the late 1060s they struck up a more substantial connection in the compound of the Imperial Libraries, where they served as associates. In 1068, Emperor Shenzong appointed Wang as Hanlin Academician and summoned him to the court to prepare for the New Policies.[9] From that moment onward, the friendship between Wang and Shen became a political partnership.

Wang recruited Shen into his reform camp at the turn of the 1070s, presenting Shen with a great political opportunity, and in the next few years Shen received quick promotions. In 1071, Wang appointed Shen secretariat supervisor of the Office of Justice (*jianzheng zhongshu xingfang gongshi* 檢正中書刑房公事).[10] In this position Shen served as one of the executives in the subcouncil for the Compilation of Secretariat Regulations (*bianxiu zhongshu tiaoli si* 編修中書條例司), a key bureau that supervised the entire reform through monitoring bureaucratic organization. All executives in this office were presumably the most trusted and crucial consultants to grand councilors. During this time, they directly reported to Wang and institutionally reinforced his power in the decision-making process.[11] Starting in 1073, Wang entrusted Shen with comprehensive executive power in monitoring multiple initiatives of the New Policies, especially those concerning fiscal affairs. Although their close cooperation ended abruptly in 1074, the power Shen enjoyed was substantial. As observed by his begrudging foe, Cai Que 蔡確 (1037–1093), Shen "participated in each and every step of planning, whether significant or trivial, in the New Policies."[12] To a large extent, Shen's fast rise in the early 1070s was a consequence of the reform as well as Wang's relentless promotion of this grand campaign.

To implement a plan as vast as the New Policies, Wang was eager to recruit like-minded associates, and Shen proved an appealing ally in a number of ways. He came from a social background Wang found familiar and trustworthy. The recent generations of Shen and Wang families had followed similar life trajectories: both resided in the south, and both had a relatively long history affiliated with the officialdom. Both men fared well on the examinations and came to serve in the capital as the new generation of southerners.[13] Wang was known for being partial to southerners in general, and Shen was a natural candidate for his consideration.[14]

Given their similar upbringings, Wang and Shen evinced a common commitment to action. In the officialdom, action consisted of myriad daily transactions, where Wang envisioned a literatus par excellence able to immerse his intellectual self. Wang was no doubt one of the most aggressive advocates of active learning, which provided the intellectual foundation for his famous political slogan "activist rule" (*dayouwei* 大有為, literally, "making a huge difference").[15] Wang specifically disapproved of scholars who "relegated affairs of ritual, music, jurisdiction, and administration to the bureaucracy and regarded them as none of their business."[16] He emphatically argued that "undertakings" (*shi* 事), that is, action, was the vital evaluative criterion of a literatus. He thus proposed to include undertakings as the final test for screening personnel; in his own words, "to know [a person's] talent, test him on his speech; having observed his speech and behaviors, test him with undertakings. The so-called scrutiny is to test one with undertakings."[17]

Shen extolled Wang's emphasis on action by reiterating that the current age was "the grand times of activist rule" (*dayouwei zhi shengji* 大有為之盛際).[18] He emphasized that action was vital to following the *dao* of sages in statecraft, opposing the separation between words and deeds. For instance, in a letter to Zhang Qizhi 張器之 (ca. eleventh century), Shen stated:

The *dao* of early [sage] kings indeed resides in the Five Classics. Since the Han Dynasty started to recruit personnel based on words, the Five Classics became a skill of scholars. As words and deeds separated, the *dao* of sages started to become obscure.

先王之道，誠在於五經。自漢始以言舉人，而五經為學者一藝。言行分立，而聖人之道始晦。[19]

Shen's interest in "things" naturally drew his attention to the various practical contexts where "things" nestled and operated. In his opinion, concrete affairs regarding the running of the state, even "the minor chores such as legal punishment, administrative meetings, and affairs regarding rice and salt" (*zhufa qihui miyan zhi xiwu* 誅罰期會米鹽之細務) were urgently relevant business, because "the failure to accomplish even a single item of these things would cause damage" (*yi shi bu zhi ze zhi you suo fei* 一事不至則知有所廢).[20] He joined Wang's camp as one of those most adept in "undertakings."

Wang and Shen also shared other important assumptions. For instance, when discussing the dichotomy of "righteousness" and "profit," both defended profit seeking as an amoral—not immoral—act. Although they became interested in this issue for different reasons, their philosophical conclusions converged. As I discussed in chapter 1, Shen expurgated "profit" mainly to validate his personal ambition, especially in the early years, when he was busily seeking career advancement, whereas Wang was motivated by a grander agenda: to defend his reform initiative. The overall purpose of the New Policies—enrich the state and bolster the army—was plainly profit seeking, and the attention Wang invested in fiscal reforms made this explicit.[21] Wang defended himself by asserting that "profit" for the common good was essentially different from selfish interests. Like Shen, he advocated collapsing the distinction between "righteousness" and "profit," striving to highlight a new vision whereby "profit is the culmination of righteousness."[22]

## *Piecemeal Fixes of the Reform*

In practice, the ideological complementarity between Wang and Shen quickly developed into a convenient pattern of cooperation. Wang, a system builder, provided vision and sketched a total-view plan in ambitious strokes, while Shen, a piecemeal engineer, came to the task of troubleshooting whenever the system faltered. For a while Shen served as one of the most efficient cogs in Wang's machinery.

Shen's utility as a troubleshooter became apparent when he cooperated with Wang in reforming the Southern Suburb Ritual, the triennial state rite performed at the winter solstice to worship the High Lord of

August Heaven (*haotian shangdi* 昊天上帝), a deity who embodied Heaven. It remained a paramount political event throughout the dynasty because of its symbolic reinforcement of the imperial legitimacy.[23] Given its ideological significance, the ritual tended to be a sumptuous event incurring lavish expenses.[24] Wang proposed to reduce the budget of the event; perhaps he secretly entertained the thought of containing the imperial monarch through coopting the most important court ritual. Although not part of the aforementioned major policies, this project carried high political stakes.

Wang's first attempt to recruit a project supervisor resulted in a minor crisis. His choice was Li Ding 李定 (fl. 1070s), surprisingly, a little-known minor official. At the time of his promotion, Li served as an investigating censor (*jiancha yushi* 監察御史) whose job was to gather public response to state policies; before that, he was a low-ranking local administrative assistant. Li probably won Wang's favor by filtering farmers' complaints on the Green Sprouts Policy. The policy was already under attack from a wide circle of officials, and Wang was desperate to fend off criticism. Greatly pleased to have support from Li, Wang immediately appointed him to supervise the important Southern Suburb Ritual project. To dissenters, however, Li was nothing more than an opportunist. They protested and pressured the emperor into canceling the appointment.[25] The controversy roiled and threatened to end the ritual reform.

Amid the contention, Wang urgently summoned Shen to save the ritual reform, and to Wang's delight, Shen promptly achieved that goal. In 1072, less than a year after his appointment, Shen submitted a new ritual program featuring a 10,000-string reduction in budget.[26] In addition, he instituted a wide gamut of changes concerning locations, order of procedures, and many other details featured in the ritual.[27] Based on his reorganization and revisions, Shen composed a 110-chapter treatise, *The Code of the Southern Suburb Ritual* (*Nanjiao li shi* 南郊禮式), which was later published under Wang Anshi's name.[28] The text was the first attempt in the Song to codify this significant ritual on a comprehensive scale.

The success of the ritual project solidified Wang's trust in Shen. Unlike many other partisans Wang favored because of political considerations, Shen displayed a true talent in concrete skills. By proving to be a reliable assistant in helping Wang turn his grand ideas into reality, Shen earned a series of further opportunities.

In 1073, Wang called on Shen to help in another major crisis, this time concerning dams and dikes in the south. The Policy of Farming and Hydraulics had gone awry. The act intended to fund construction of agricultural, mostly hydraulic projects. But in the south, where waterways played a vital role in economic activity, the scale of Wang's plan was enormous and the challenges substantial. This time, Wang summoned Shen for the specific purpose of monitoring the finances of the hydraulic reform.

Prior to Shen's involvement, a similar crisis had engulfed the project's previous director, Jia Dan 郟亶 (1038–1103), who found himself steeped in controversy shortly after the launch of the initiative.[29] Only recently promoted by Wang from the lower ranks, Jia joined the campaign with great enthusiasm, planning a sweeping line of new constructions in the Liangzhe area. Jia envisioned a five-year project requiring 4,000,000 man-days.[30] The massive cost raised concerns, however, and Jia's method of procuring funds specifically invited heated criticism.[31] His original reckoning was straightforward: the reform would draw on the poor for labor and on the rich for funds, a practice established in Song precedents.[32]

Given the unprecedented scale of the project, the old practice failed to meet the demands. The financial burden proved overwhelming for rich and poor, and the social consequences were grave. The extraction of funds on behalf of the state led to an inevitable redistribution of public and private interests, and the political authorities felt free to sacrifice certain groups or regions to the relative advantage of others. Within a year, people's resistance quickly swelled to the extent that the confidence of the emperor and Wang waned, and Jia was dismissed.[33]

As soon as Shen stepped in, he decided to curtail the scope of the project. Wang gave his acquiescence to possible concessions, as he indicated to the emperor that one reason to recommend Shen was his "prudent" (*jinmi* 謹密) personality.[34] Shen implemented two modifications. First, he trimmed Jia's ambitious plan, reducing an all-out campaign to regional experiments. Shen designated two prefectures, Suzhou and Xiuzhou, as locations for trials. In these areas, lakes and waterways were already draining, leaving land ready for reclamation, and local labor was relatively abundant. Shen expected these pilot regions to provide successful models for other areas. Moreover, he expected the initial success to convince local farmers to join the labor force on a voluntary basis.[35]

Having reduced the size of the project, Shen switched the source of funding. He suspended the previous levies and pressed for the government's investment. He reasoned that since the state had initiated such massive construction, it was responsible for funding the project. The new proposal issued governmental loans to cover the costs and hired farmers as wage labor.[36]

Shen's revision of the project showed ingenuity in adapting Wang's policy to changing particular circumstances. The next year, a severe drought hit the Liangzhe area, striking the prefectures of Changzhou and Runzhou especially hard. Peasants forced from their homes fled to neighboring districts, and Shen suggested that the state hire these migrants as workers in the hydraulic projects.[37] The suggestion worked well, and Wang was particularly pleased.

After averting the hydraulic crisis in the south, Shen further convinced Wang of his talent in financial engineering. In the same year (1073), Wang assigned Shen to join his campaign addressing agrarian inequalities. A key act that served this cause was the Green Sprouts Policy, a state-operated rural loan system.

Green Sprouts was initially a response to the old, faltering system of rural relief, the Ever-Normal Granary System (*changping cang* 常平倉).[38] Starting in 992, a network of state-run granaries was built across the empire. These granaries purchased rice from small farmers at a satisfactory price (2 to 10 percent above the market rate) in the harvest season and sold it at a discount of one-third off prevailing prices in the spring. The state also used the granary stocks to sell or loan grain to low-income households.[39] By this means, the state was able to ease stress on the poor caused by food scarcity and enhance their ability to fulfill their social dues. After decades-long operation, however, the system started to crumble as the granaries hoarded grain and cash, keeping them out of circulation. Multiple reasons were behind the failure. For one, the number of existing granaries was relatively small and their geographical distribution was limited to major prefects and counties. For another, complicated and time-consuming operational procedures anteceded failures to deliver resources to farmers in time.[40]

Green Sprouts sought to improve the charitable function of the old granary system with a more progressive monetization initiative. The policy required the granaries to sell grain reserves and use the income as a source of state loans. In place of great landowners and money lenders, the state assumed the lending function and extended loans to small farmers. The reform served multiple goals Wang desired: force the rural population into a market economy, curb the power of rapacious landowners, and strengthen the social welfare function.

By the time Shen came to the rural relief project in 1073, years had passed since the Green Sprouts proposal, and Wang had been promoting the new system with uncompromised conviction. He demanded a complete takeover, with all reserves in the old granaries monetized and the capital transferred to the Green Sprouts fund. Some local officials followed his order and emptied the granaries without retaining any backup resources; many who showed reservations were forced out of the government.[41]

Unsurprisingly, the Green Sprouts act became the center of controversy throughout the reform years. Wang's aggressive implementation of it generated numerous cases of impeachment and scandal, including the instance of Li Ding. Even Emperor Shenzong was unable to soften Wang's resolution. Not until 1074, when Wang tentatively stepped down from his councilorship, did the emperor seize an opportunity to mitigate Wang's radical approach. Shenzong ordered local officials to distribute half the granary reserves as poor relief and half as loans.[42] The emperor's choice of a middle road perhaps represented a general agreement among officials who were troubled by the absurd contentions. His decision reveals that in spite of Wang's headlong push, not all grain stocks had been monetized and some granaries still functioned in a modest way.[43]

In light of this background, Shen's way of participating in rural relief was intriguing. As a close partner of Wang, he arrived in the Liangzhe Circuit with a proposal to extend the government granary system (rather than eliminating it). He witnessed how, despite high crop yields in the area, farmers still ran into cash shortages. In addition, the ongoing famine made life difficult for everyone in the region. Shen's solution was to set up government granaries in a couple of prefectures.[44] These granaries would purchase grain from farmers at harvest and resell it to them at below-market prices in the seeding season, providing immediate relief to the local population.[45]

Shen's move was not to promote the new policy but to rehabilitate the old system. Curiously, his decision did not provoke Wang's ire, and the entire project seems to have been implemented as planned. One plausible explanation is that given the severity of the natural disaster, the only convenient means for alleviating the famine was to resurrect food stocks and dispense them immediately. Acting in the capacity of troubleshooter, Shen decided to opt for a solution of immediate efficacy. Perhaps Wang could not find Shen blameworthy in the face of a persistent crisis.

Another reason for Wang's acceptance of Shen may have been Shen's adaptive—rather than duplicative—use of the old system. Shen had incorporated another reform act of Wang's into his rehabilitation of Green Sprouts. While organizing the granaries, Shen divided the population into the so-called mutual security units, the fruits of the Mutual Security reform. In Wang's design, every ten households formed a small guard, every five of these small guards formed a large guard, and every five large guards made a superior guard.[46] Wang expected these mutual security units to tighten social control and thus forestall banditry.[47]

While Shen was in the process of resuscitating the granary system, he made sure the arrangement of granaries strictly corresponded to the mutual security units; thus the grain distribution patterns closely meshed with the new social fabric.[48] One major benefit of this system was that it prevented local farmers from using false identities or misinformation to secure resources illegitimately, which made the charitable function of the granaries more equitable.[49] This corrected a major defect that had invited attacks on the granary system. In addition, the granaries enhanced the Mutual Security reform by creating a new incentive for local participation. From the perspective of a farmer, the imposed security units now took on positive meanings associated with community welfare.

Despite his revision of Green Sprouts, Shen had convinced Wang of his resolute support for the Mutual Security initiative, which became his next focus. The nationwide promotion of this policy consumed more time than did any other act, and Shen made major contributions during the process.[50] In 1074, Wang appointed Shen to supervise the implementation of Mutual Security in the Hedong and Hebei circuits in the northwest.[51] At this time, it had been four years since Wang's first announcement of his plan.[52] From 1070 to 1073, Wang had focused on testing and ameliorating his proposal, first in the pan-Kaifeng area, and then in a selection of key circuits. After 1073 he started to promulgate the reform nation-wide.[53] Wang recruited Shen to join the reform at the cusp of this region-to-nation transition, and he put Shen in Hedong and Hebei, pilot circuits for early experiments.

In addition to tightening local control, Wang envisioned another important goal, also his ultimate aspiration regarding mutual security: to turn the local militia into a real peasant army and replace the notoriously corrupt mercenaries. To this end, he contemplated two measures: annex the mutual security units to other established local militia, for example, the so-called righteous braves (*yiyong* 義勇), and establish effective military drill, a matter Wang had pushed for since the beginning of the reform despite Shenzong's misgivings.[54]

As soon as Shen arrived in Hebei and Hedong in 1074, he worked diligently to turn Wang's plans into concrete management procedures. Shen made specific arrangements to assist merging different local militia, suggesting that mutual security soldiers should join the righteous braves in following a unified drill schedule. In the past, the righteous braves attended a month-long session of "drill and inspection" (*jiaoyue* 教閱) in the prefect, while mutual security men observed a shorter session at the county level.[55] In Shen's proposal, the two systems would converge, with both groups gathering in the prefect for thirty days of annual training.[56]

Shen also devised a number of measures to enhance the efficacy of the drill system. First, he called to stabilize and augment the source of funding on national and local levels. Not only should the capital deliver the regular provisions in time, Shen suggested, but the prefect government should also contribute when central government sponsorship was insufficient.[57] Second, Shen aimed to improve the supervisory mechanism by involving more government officials as mentors. He asked judicial commissioners (*tidian xingyu* 提點刑獄) to lead the "drill and inspection" and required a recruitment commissioner at the Military Commission (*anfu si xuanbing guan* 安撫司選兵官) as well as an official from the county government to supervise the drill of righteous braves.[58] In other words, the training system for local militia came under the control of officials who held positions in the central government (including Shen) as well as those who served at all local levels. Thus, local militia became more tightly connected with the centralized officialdom. In an attempt to

systematize the drill curriculum, Shen called for establishing clear criteria for selecting soldiers and appointing instructors to the drill faculty.

All the measures he proposed were solid stepping stones toward
Wang's ultimate plan to transform armed farmers into professional fight-
ers. His effort to integrate the mutual security units with existing forces
received imperial reinforcement a couple of years later. In 1076, Emperor
Shenzong issued an edict dictating that all local militia, including the
mutual security armies and the righteous braves, would follow the same
drill schedule in the same localities in the Hebei region.[59] Shen's attempt
to enforce the connection between the mutual security militia and the
officialdom developed further: in 1075, the mutual security units officially
became a subordinate component of the Ministry of Military (*bingbu*
兵部), under whose direction better-organized selection and drill procedures were put into practice in 1076.[60] Shen's policies contributed in multiple ways to the systematic annexation of the mutual security units by the
regular military in 1079.[61]

Shen also adapted the Mutual Security Policy in a variety of ways to
better suit the local needs of Hebei, specifically the Dingzhou area (Dingx-
ian in modern Hebei). Dingzhou was a strategic town on the Song–Liao
border, a stronghold against the Khitan Liao's cavalries. The population
was relatively sparse and the number of towns few.[62] Shen figured that a
rigorous application of the mutual security system would help tighten the
social fabric and strengthen the defensive function. Some officials suggested that certain groups of locals could be settled under Mount Xi
(a ridge protecting Song soil from the Liao) and act as a deterrent to Khitan intruders. Shen disagreed. To disperse the population like that, he
argued, was not only disrespectful to the people but a poor strategy that
would embolden enemies by offering them easy opportunities to kill the
innocent. Instead, Shen proposed to implement the mutual security structure so that locals could fight with greater morale and better organization.
By his estimate, the mutual security arrangement could mobilize tens of
thousands of soldiers from this area, a considerable addition to the current forces.[63]

To ensure a thorough application of the mutual security system in
Dingzhou, Shen introduced a facilitating tool, the "system of wards and
markets" (*fangshi fa* 坊市法), an arrangement of urban residences into

regimented units.[64] This system was a throwback to the Tang Dynasty, when the model for organizing an urban population was to segregate social groups in walled wards and restrict their activities by curfew.[65] Although the regimented paradigm had gradually given way to the emergence of open urban space in major Song cities (such as Kaifeng), the ward walls remained in Song people's memory as a means of organization and might have inspired Shen in Dingzhou.[66] One reason for Shen's reenactment of the ward system was to protect against the Khitan cavalries, who frequently harassed Song border residents and took some of them back to the Liao. When Song victims managed to flee captivity and return home, border guards, unable to tell whether they were Khitan spies, refused to allow them in.[67] The new system reinforced population registers and put security checks at ward entrances, which brought peace and order to neighborhoods.[68] More importantly, the stringent division of residential wards provided a clear structure for further organization of mutual security units, and good order among the local population promised enhanced defensive power.

## Securing the Border

While concentrating on the Mutual Security Policy in Hebei, Shen also paid attention to border security in general and played an active role in developing new tactics.[69] Based on a geographical survey he conducted of the region, Shen proposed as many as thirty-one defense strategies. For instance, worried that the Khitans might use wood from Mount Xi to attack with fire, he suggested setting up a fire prevention apparatus.[70] He carefully compared the two major models of defense—arboreal and hydraulic—and argued in favor of the latter, because wet and boggy ground most effectively impeded approaching cavalry.[71] To strengthen hydraulic defenses, Shen proposed turning local waterways into continuous water obstacles and developing new rice paddies. He was concerned with potential loopholes in an arboreal defense, which relied on planting elms and willows on the border to block Liao troops. He argued that the Song's major technology of strength was archery, and a tree wall would protect the Liao against arrows from the Song. Also, to make siege weapons such as scaling ladders, Khitans had to acquire wood, and the Song trees might provide the source.[72]

In addition, Shen urged Song officials to carefully gauge the strengths and weaknesses of both sides before devising war strategies. He argued forcefully that the Song army should focus on their own special unit—bows and crossbows—instead of imitating the Khitans' use of mounted troops. This was a soft pushback at Wang Anshi's Household Horse Breeding Policy (*baoma fa* 保馬法), an act that enlisted local households on the border to breed military horses and help expand the cavalry.[73] In Shen's opinion, archers were effective in countering cavalry, a technological advantage the Song should assert and amplify. Shen found it highly problematic to abandon one's own strength and pursue that of others.[74] Evidence suggests that he held a long-term interest in archery and made the argument above on the foundation of a good understanding of the technique. Later in *Brush Talks*, he wrote extensively on the technical details of archery as he reminisced about his career, for instance, characteristics of a quality crossbow, particular models of bow he observed, and archery skills inspired by mathematical calculations.[75]

A few years into working in the Hebei region, Shen managed to turn his thoughts into more permanent forms, such as "things" and writings. First he managed to make a number of artifacts that facilitated his work on the border. In the eighth month of 1074, Emperor Shenzong appointed Shen to lead the Directorate for Armaments (*junqi jian* 軍器監), a position in which he devoted efforts to designing and manufacturing weapons. Together with a colleague, Zhang Dun 章惇 (1035–1105), Shen made a model chariot based on two classical references, the *Rituals of the Zhou* and the *Classic of Odes* (*Shijing* 詩經). Shen and Zhang provided a meticulous textualization of the model, with each component of the chariot identified with a designation and measured in all dimensions. In addition to making the object, they specified five ways of using the chariot. Shenzong personally examined the final product with satisfaction.[76]

Another artifact Shen made was a relief map of the Song–Liao border, which grew out of the geographical survey he conducted while preparing to make security plans for this region. Shen represented the topographical features of the territory with a cartographical model made from wheat paste and shredded wood fiber. To preserve it better in cold weather and make it travel-friendly, Shen made an alternative copy with lightweight wax. Impressed by Shen's work, Emperor Shenzong summoned ministers to examine the map and commissioned the border circuits to make wooden duplicates for their own reference.[77] This is one of the earliest known relief maps in Chinese history, the one I mentioned at the opening of this book.[78]

In addition to making artifacts, Shen composed two important treatises based on his fieldwork in the war zone. The emperor commissioned him to study the so-called Nine-Army Formation (*Jiujun zhenfa* 九軍陣法), a lost infantry tactic allegedly created by the Tang general Li Jing 李靖 (571–649), and articulate his findings in a text.[79] In the same year, Shen coauthored an important military work with fellow official Lü Heqing 呂和卿 (fl. 1070s), a treatise titled *Principles of Fort Building* (*Xiucheng fashi tiaoyue* 修城法式條約), endorsed by Emperor Shenzong as the official guidelines for fortress construction.[80]

Shen's success in the borderland turned out to be highly rewarding. Not only was Wang Anshi satisfied with his service in promoting the Mutual Security Policy, the emperor was pleased with Shen's innovative approaches toward border issues. His appointment as emissary to the Liao negotiation was a measure of the emperor's personal acknowledgment of Shen's achievements in this area (see chapter 3).

The first five years working for the New Policies constituted the best times in Shen's career. In his various projects, Shen dealt with his favorite, "things," in his characteristically hands-on way. While Wang Anshi was devoted to building a new system structured with broad ideological goals, Shen focused on ensuring their implementation in concrete praxis. The small, particular achievements Shen accumulated enabled him to claim an indispensable role in Wang's campaign and, furthermore, to begin asserting a more independent role in imperial politics. In the following chapter, I introduce Shen's attempts to exert his independent influence and the dramatic consequences that ensued.

# CHAPTER 6

## *Buffeted by the World (1075–1085)*

From 1075 through 1077, Shen Gua's life was on a rollercoaster. In 1075 Emperor Shenzong appointed him the vice head of state finance; at the end of 1076, Shenzong further promoted Shen to supervise the national economy as the fiscal grand councilor; eight months later, in 1077, the monarch exiled Shen from the capital.

By the time of his demotion and exile, Shen was a broken man. Beyond the collapse of his career, he had come under siege apparently because of his moral character. Wang Anshi, previously his mentor and close colleague, subjected him to relentless moral censure. Shen's reputation as a capricious opportunist began to gain currency among his peers. During the second phase of his service in the New Policies, Shen turned his piecemeal fixes of the system into consistent middle-road policies, which, for ideological and personal reasons, precipitated his dubious reputation and political downfall.

In this chapter, I introduce the factional conflict during the New Policies as a general background and follow with an account of the dissolution of the Wang–Shen cooperation. The next three sections provide analyses of Shen's middle-road polices in Hired Service, the salt trade, and currency reform. The chapter concludes with a discussion of Shen's political reputation and how his policies turned into a moral issue.

## *The Reform as a Maelstrom*

The most important context of Shen's sudden rise and fall was the polarized ideological atmosphere that prevailed during the New Policies period. Intense factional infighting ended careers of many bureaucrats, and Shen was one of them.

Factional conflict may have been a constant during the entire eleventh century, but it reached its zenith in the 1070s.[1] The launch of the New Polices in the late 1060s led to three decades of ideological confrontation between two groups: reform advocates led by Wang Anshi and antireformists represented by Sima Guang. The 1070s saw the reform come into full swing and factional tension reach an unprecedented intensity. Wang Anshi's experience evidenced the ferocity of the storm: in spite of his iron will to sustain the reform, he was forced to resign twice, first in 1074 and again in 1076, due to torrents of opposition directed at him by the antireformists.

In a nutshell, the ideological conflict between the two factions centered on one issue: the role of the state. Wang envisioned a state claiming all-encompassing power and universal regulatory responsibilities. His new policies thus focused on "creation," that is, the continued building of institutions along the lines of expanding regulatory objectives. Sima Guang, in contrast, strongly opposed enlarging the state apparatus because of his firm belief in the sufficiency of current governmental structures, which he saw as a heritage from the sage kings. Thus he emphasized "restoration," that is, maintaining the existing system and adjust it only when problems arise.[2]

Disagreement over the state's function provoked repeated debates on the conflict between public power and private interests. In the context of finance, should the former command the latter? Wang responded affirmatively to this question. His economic philosophy assigned the highest priority to the state's regulatory power and developmental initiative.[3] In Wang's agenda, the active state fulfilled two primary roles: it participated in the economy through direct monopolies and indirect commercial taxation, and it committed state power to boosting economic activities.[4] Sima stood in polar opposition to the notion of an expansive state and urged the government to leave wealth in private hands. In his opinion, the state should more cautiously manage its revenue, and all agents, public and private, should return to their dutiful roles.[5]

The enduring contestations between Wang and Sima were certainly not a question of disagreement in academic debates only. Behind the shields of ideological differences, the two camps clashed over the balance of power and partisan interests. The absurdly intense infighting came largely from Wang's relentless and oppressive style in dealing with divergent opinions. Before his first resignation in 1074, Wang was the single unquestioned head of government and had the emperor's full approbation. Within a brief period of time, he managed to dominate personnel recruitment, control the flow of information, and create whole new institutions to suit the reform's needs. The multiple cases of personnel controversies I discussed in the previous chapter attested to his strong wish to promote partisans and purge dissidents. Eventually he hoped to achieve "ideological unanimity," which he deemed as essential for a full measure of success.[6] To that end, Wang showed little mercy in repressing disagreement. His temerity in enforcing uniformity, however, only led to its opposite: protest and bickering quickly escalated in spite of his best efforts to muffle them. Soon enough, Wang found himself stuck in a vicious circle: the more intensely he suppressed dissidents, the stronger the counterattacks he had to face.

With a growing sense of insecurity, Wang eventually extended purges into his own coalition. After his first resignation, subfactions started to form within the reform cadre. His successor, Lü Huiqing 呂惠卿 (1032–1111), former protégé and assistant, quickly turned against him and began eliminating people loyal to Wang. Lü's betrayal stirred a new spate of factional struggles and irritated the emperor. In 1075, Shenzong reinstated Wang as part of his plan to oust Lü.[7] Wang faced new prospects sure to provoke his anxiety: not only had he lost unchallenged dominion in the court, he also had to handle competition for power within his own faction. His reaction to any slight discordance of opinion among his followers became excessively intense, and he watched many bureaucrats change overnight from partisans into foes. Shen was not the only reform partisan exiled by his comrades, nor was his sudden fall an exceptional tragedy. After all, stability was a luxury in this age. Within or without Wang's circle, the ceaseless friction left hardly any high official in peace.

The moral attack on Shen exemplifies another characteristic of the reform period: a rhetorical strategy prevailing in almost all factional quarrels. The strategy followed a straightforward thesis: a dissenting opinion, normally by a political foe, was condemnable because it reflected a moral deficiency in the speaker. When one judged the validity of a policy or opinion, the motive and character of the proponent rather than the idea itself became the primary consideration. The reasoning was essentially circular: a contrasting opinion betrayed wrongful motivation and dubious morals; therefore, anyone holding such a view must be morally weak.

The central concept of this moralistic rhetoric was the dichotomy between the "superior man" (*junzi*) and the "petty man" (*xiaoren* 小人).[8] A faction member invariably saw himself as a "superior man" and addressed an opponent as a "petty man," leaving little space for compromise. Sima Guang summarized the demarcation as follows:

> The incompatibility between a superior man and a petty man can be aptly compared to the impossibility of placing ice and hot coal in the same container. Therefore, if the superior man attains power, he would reject the petty man. If the petty man attains power, he would banish the superior man. This is a natural pattern.[9]

The fundamental distinction between a superior man and a petty man, also the source of the (in)validity of their ideas, was one of motivation. Possible motives were defined along two dyads: "public good" versus "self-interest" and "righteousness" versus "profit," in a black-and-white manner.[10] From a superior man's point of view, a petty man always and only sought self-interest or profit, hence the unreliability of his opinions. Only a superior man would consistently guard the public good and assert righteousness, which provided a substantial grounding for (any of) their policies. The three binaries constituted a paradigm of identification/judgment that drew heavily on morality for vocabulary yet certainly went beyond moral issues in implication. In such a historical context, moral accusations were closely associated with and often directly caused by political antagonism. In Shen's case, it is reasonable to analyze the moral aspersions against him within broader conditions.

Indeed, the ideological divide and intense political atmosphere during the reform set the general background for Shen's rollercoaster experience, and his interactions with Wang Anshi provided the immediate antecedents of this dramatic turn of events.

## *Parting Ways with Wang Anshi*

Starting in 1074, the pleasantly simple leader–follower relationship between Wang and Shen dissolved; in its place emerged a rivalry imbued with rancor. From 1074 through 1077, the career trajectories of these men moved in contrasting directions. Wang was twice driven out of the position of grand councilor, reluctantly coming to terms with a waning career. Shen, by contrast, had been spiraling upward. His promotion to the top was primarily due to the emperor's need to fill the vacuum caused by Wang's resignations. This reversal of power surely entailed a change in Wang and Shen's mutual perceptions, and it almost unavoidably provoked Wang's animosity.

In 1074 Wang started to lose his centralized grip on power backed by the emperor's unreserved support. That year a prolonged drought led to famine in the north, and thousands of refugees moved to the vicinity of the capital. Despite Wang's best efforts to isolate the reform from the crisis, the emperor felt pressed to investigate the social distress by soliciting different points of view on it. Because of the emperor's personal encouragement, the long repressed strife among officials developed into a flurry of memorials, some of which revealed strikingly gruesome consequences of the reform. At last, neither Wang nor the emperor could stand up to the pressure. In the fourth month of 1074, Wang submitted his resignation, and Emperor Shenzong accepted it.[11]

Yet in many ways Wang's retreat exacerbated the situation. To replace him, the emperor appointed Lü Huiqing, who, as noted already, accelerated the disintegration of the reform camp and caused further strains at the court. In less than a year, the emperor decided to reinstate Wang and launch a campaign to erase Lü and the influence of his clique.[12]

Shen's ascendance to a top power position occurred in the middle of the unfolding of the emperor's restoration plot. The monarch saw in Shen someone able to address both the damage Lü had caused and the power vacuum Wang had created. He first appointed Shen provisional head of state finance (*quan fa qian sansi shi* 權發遣三司使, "Provisional Acting Commissioner of the Three Fiscal Agencies") in the tenth month of 1075 as a replacement for a Lü partisan.[13] This happened shortly after Wang's temporary return. A year later, Wang resigned for a second time and the emperor replaced him with Wu Chong 吳充 (1021–1080) and Wang Gui 王珪 (1019–1085), neither of whom expressed clear support for the New Policies.[14] This decision was a shocking revelation to Wang: the emperor had publicly abandoned his full support for the reform. Thus Wang's decade-long exertion to further the New Policies came to an end. Two months later, Shen was named the official head of state finance (*Quan sansi shi* 權三司使, "Acting Commissioner of the Three Fiscal Agencies"), clearly not as a Wang partisan but as a balancing force between waning activism and resurgent antireformism. Over time Shen had grown out of the shadow of Wang's dominance, and the emperor's increasing trust in Shen greatly accelerated his parting from Wang.

In a world fragmented among polarized factions, the middle way was a rough road to travel, and Shen met with slander and hostility along the way. Lü Huiqing was swift to discern the monarch's intention to use Shen against him. He observed the close bond Shen had developed with the emperor, and grudgingly acknowledged that Shen almost always had the emperor's ear;[15] with no attempt to conceal his preference, the emperor refuted and chastised Lü for being unreasonably jealous.[16]

Ironically, the most ferocious attack on Shen came from his former mentor, Wang Anshi. Just a year earlier, Wang had recommended Shen to the emperor for hydraulic projects in Liangzhe with unreserved enthusiasm. Now, when the emperor considered appointing Shen to the Ministry of Military, Wang called Shen an "awful flatterer" (*kongren* 孔壬)[17] and accused him of "furtively sabotaging the New Polices" (*yinju huai xinfa* 陰沮壞新法).[18] Emperor Shenzong suspended the promotion, but he took Wang's acrimony lightly. The monarch admitted that Shen was not a "fine man" (*jiashi* 佳士) but insisted that his talent should not be easily dismissed.[19] Despite Wang's objection, he eventually promoted Shen to head of state finance. Wang and Shen officially went separate ways.

## Adjusting Hired Service

As Shen assumed the office of the fiscal grand councilor, he inherited all the problems and ill-feeling the New Policies had thus far accrued: economic strains, repressed anger, and convoluted power structures. Shen understood the mission the emperor had entrusted to him, and he immediately initiated revisionist plans. Under his directive, a series of changes were introduced into such realms as Hired Service, the salt voucher system, and currency management. Shen intended to effect a gradual movement away from the radical activist ideas Wang espoused.

Shen started with the Hired Service reform in the Liangzhe area. In 1074 he had assisted in promoting this policy, which attempted to replace the mandatory rotation of service with wage labor and ensure that the population shared service obligations on a more equal basis. But nation-wide implementation of this policy had imposed excessive taxation on a wide range of people, including the poor. Shen's new plan aimed to mitigate the suffering the policy had caused. He proposed to reduce taxation in general, and on that basis redistribute the tax/service burdens to better match local conditions.

Shen's plan was simple: instead of adhering to the previous tax system that divided the population into twenty-five grades, he regrouped people into three large coalitions by their status prior to the reform and made new rules.[20] First, those with no previous obligation for service—official households, households with only one adult male or no male, and religious establishments—continued to submit a cash tax, a continuation of the Hired Service Policy. Second, people summoned for a heavy service load, usually farmers, would receive cash compensation for their service. Third, people with a light, reasonable load would serve with no further taxation obligations. Shen relieved the latter two groups from paying cash taxes by offering them an option to return to the old requisition system. An estimated 28,000 households in the Liangzhe area could receive tax exemptions.[21]

In the meantime, Shen arranged for the local government to run on a smaller budget. He calculated that revenue generated from the government-owned factories (*fang* 坊), market gatherings (*chang* 場), and ferry tolls (*hedu* 河渡) would cover the salaries of one category of servicemen—those in guard duty—who incurred the greatest expense. Another group, service people in office duty, could be paid with taxes submitted by the first category of taxpayers (official households, etc.) plus the remainder of the government revenue.[22] In this plan, fewer households would have to pay, and Shen hoped that "the resources among the population would even out spontaneously" (*minli zijun* 民力自均).[23] Although in effect he restored the old system to enable the revisions, Shen considered his plan a nod to the original egalitarian spirit of the Hired Service Policy, and he insisted that his revision was an amelioration, not a regression.[24] In a confident mood, he suggested to the new conservative-leaning grand councilor Wu Chong that the change be implemented nationwide.[25]

In theory, Shen's revision was reasonably justified, but its implications became problematic given the ideological tenor of the times: it could be easily read as a betrayal of Wang. Shen's eager proffering of this policy to the new leadership after Wang's resignation cast suspicion on his real motive. This incident became a notorious "change of mind" that evidenced Shen's "sabotage" of the New Policies.

## Restoring the Salt Vouchers

In addition to Hired Service, Shen revised another institution vital to the economy—the state monopoly on salt. He dealt specifically with the so-called system of salt vouchers (*yanchao* 鹽鈔), a key instrument the state used to control the salt trade.[26] In rectifying the voucher system, Shen was able to improve and augment the state's indirect control mechanism while reducing the New Policies' direct intervention in the salt trade.

The Song government drew on agriculture and the monopoly of certain commodities as primary sources of revenue, and salt was the most important of these commodities.[27] During Emperor Shenzong's reign, income from salt reached 12 million strings, over half of total state monopoly revenues (22 million strings).[28] The state monopoly consisted of two basic mechanisms: government distribution and merchant distribution. In the former system, the government dispensed salt directly through state-run shops, which materialized the state's strongly regulatory involvement.

In the merchant distribution system, the state kept a tight grip on its merchant distributors by requiring them to conduct business in a number of government-designated models. The salt voucher system was one of the regulatory models.[29]

The use of salt vouchers traces back to the late tenth century, a time when the Song Dynasty was at war with Tangut Xia (1038–1227), a neighboring state in the northwest. Due to a shortage of supplies on the warfront, the Song government sought cooperation from merchants, who shipped food to the border and received state-issued vouchers as rewards. The merchants could later redeem the vouchers in the capital for cash or purchase salt (in addition to spices, tea, and a few other designated products) in certain areas appointed by the state. Afterward, they had freedom to resell these commodities to customers and reap the profits.

Soon after its launch, the system started to show cracks, and it required a major adjustment in the eleventh century. To maximize profits, merchants tended to inflate the price of the supplies they shipped. In addition, the massive transportation fees became a burden imposed on the population. In 1044, an official named Fan Xiang 范祥 (d. 1060) instituted a new rule to mitigate these problems. The rule granted merchants the flexibility of not having to first ship supplies to the border before receiving vouchers. Merchants could streamline their itineraries and avoid unnecessary costs on the road. Fan's scheme proved successful and continued in the next few decades.[30]

The advent of the New Policies period disturbed equilibrium in the salt trade. Not surprisingly, Wang Anshi saw the salt monopoly as a rich source of revenue, and he decided to minimize the portion of merchant distribution. The reformists banned merchant trade in some regions and gradually increased the scope of restriction. As a consequence, merchants suspended trips to the border and a supply crisis in these areas ensued.

A more serious crisis arose from the salt vouchers themselves. Even as they gradually lost value as merchants' activities were repressed, the government kept printing vouchers in large quantities. The surplus vouchers were an expedient response to a shortage of cash, but the government did not foresee the repercussions of overprinting. When the state started to use salt vouchers in lieu of normal currency in unregulated ways, the problem worsened. Excessive issue of salt vouchers stimulated inflation, and the whole system became seriously corrupted.[31]

The inflation crisis spurred another round of heated debate between Wang partisans and reform adversaries. The critics charged that Wang had yet again shattered an efficient old system. Zhang Jingwen 張景溫 (fl. 1070s), Wang's representative in the salt affairs, ferociously defended the ban on merchant trade and suggested extending it. Wanting an independent opinion for a final judgment, Shenzong summoned Pi Gongbi 皮公弼 (d. 1079), a salt commissioner active in Shaanxi, northwestern China. Pi was highly critical of Wang and pushed for a total reversal: restore merchant distribution, buy up devalued vouchers, and actively prevent further inflation. The acute disagreement once again threw the monarch into bafflement. He left the decision to the new fiscal grand councilor, Shen Gua.[32]

Shen's course of action turned out to be intriguing for his contemporaries and historians. He first declined Pi's proposal, but as soon as Wang stepped down in 1077, Shen reversed his decision.[33] Shen and Pi joined in an effort to resuscitate the old system. In addition to retrieving old vouchers, they called for extra cash payments from merchants who had purchased salt with old vouchers to cover the difference in value between old and new vouchers. These policies aimed at clearing out inflated old vouchers and keeping merchants in the system.[34]

An extensive document Shen later composed proved that his change of mind was by no means capricious. He laid out detailed plans in writing along with a rationale for his revision. He proposed four major modifications. First, excessive voucher printing should be suspended immediately, and he recommended setting a fixed annual quota at 2 million strings of coins. Second, in certain regions, the central government should maintain a fixed rate rather than allowing arbitrary regional differences in the price of salt. This proposal specifically concerned the type of salt called Xie salt, one of the most widely used kinds of salt in the eleventh century.[35] The distribution area of Xie salt consisted of three major regions: eastern, western, and southern.[36] In Shen's times, the government deliberately kept the salt price in the western region lower to repress private trade from the other side of the Song–Xia border. This policy created a serious imbalance between the eastern and western areas. Cheaper salt from the west was extensively consumed in the east; at the same time a glut of salt produced in the east sat idly in reservoirs. Shen suggested revoking this policy.

In addition, he proposed that the authority to issue vouchers be strictly limited to the central finance commission, the Three Fiscal Agencies (*sansi* 三司). He specifically pointed out that such authority should be taken away from "external bureaus" (*waisi* 外司), including the Xie Salt Bureau (*Xie yan si* 解鹽司), a circuit-level office augmenting military funding by regulating the Xie salt trade.[37] Although the Xie Salt Bureau could assist in dispensing vouchers, Shen suggested, that the Three Fiscal Agencies should exercise the exclusive power to print vouchers because they had clear knowledge of state salt reserves and could determine the required number of vouchers. This measure to recentralize the voucher system meant to forestall inflation.

Last, Shen was concerned with the damage various bureaus had jointly inflicted on the circulation of vouchers. Many salt vouchers, he believed, had been taken out of the salt trade and improperly hoarded among the people.[38] He thus suggested that the state stick to a unified price in all salt-related business conducted by the government. For instance, in addition to the distribution models, various bureaus in the Song state also "exchanged salt" (*yu yan* 鬻鹽) with their subjects. In times of need, the government granted salt to an individual/household as a loan, and the recipient repaid in silk.[39] Even though this bartering process involved no currency, Shen insisted that salt be traded at the same value as in the voucher system, and that the bureaus refrain from arbitrarily manipulating the price. That way, people would have no incentive to hoard salt vouchers.

Shen identified three major loopholes caused by institutional misconduct in handling salt vouchers.[40] The departments (*sheng* 省) and courts (*si* 寺), agencies at the central government, often used salt vouchers to trade for other goods at random values they determined. The Fiscal Commissions (*zhuanyun si* 轉運司) in various circuits sometimes traded salt for land contracts, circumventing the whole voucher system. The Chamberlain for the Palace Revenues (*taifu* 太府), the agency responsible for managing nongrain revenues at the central government, traded vouchers for cash with prefects for the purpose of cutting transportation costs. To remove stagnant salt vouchers and restore a healthy circulation cycle, Shen argued that these problems must be immediately rectified.[41]

In addition to his detailed understanding of the voucher system, other evidence from Shen's earlier experience revealed his commitment to a less activist policy in the salt trade. Prior to 1077, on multiple occasions he had expressed support for the expansion of private distribution. In 1075, for instance, Emperor Shenzong summoned Shen to discuss the frequent occurrences of salt smuggling in Chengdufu in southwestern China, a major locality for the private production of well salt. The emperor was considering whether to shut down the wells drilled by private salters and transport state-made salt to sell in the area. Shen immediately disagreed and argued that since the state allowed franchised salt wells, it should also tolerate private trade.[42] Eventually he talked the emperor out of the idea. The same year, Shen also dissuaded the head of state finance from abolishing merchant activities in two regions in the north. He went to great lengths to defend the private sector of the system, even referencing the founding monarch of the Northern Song. According to Shen, the first emperor explicitly stipulated that salt trades organized by merchants should be allowed in Hebei (one of the northern circuits).[43] Given his public record of support for private business, Shen's advocation for Pi should be no surprise.

Shen's handling of salt vouchers illustrates a dilemma he faced during those years: how to proceed without directly challenging Wang. As Wang stumbled through ups and downs, Shen observed and waited for "interims" to institute his own policies. From Shen's perspective, his actions might seem a cautious act of self-protection, but to Wang, Shen's obvious opportunism capitalized on his misfortune. Shen's eager cooperation with the new leadership—blatant sycophancy in the eyes of many—further exacerbated Wang's contempt. Shen might have truly believed that his actions were just and not simply designed to gain political advantage, but the situation nevertheless cast an ugly light on his motives and rendered his intellectual reasons nearly irrelevant.

## Easing the Cash Famine

Another area where Shen initiated major revisions was currency. As mentioned in the case of salt vouchers, a cash shortage loomed large in the background of all the smaller crises. Recognized as "cash famine" (*qian huang* 錢荒), the cash shortage was a relatively new economic issue that first occurred in the Tang Dynasty. The subsequent Song, especially the New Policies era, witnessed ardent efforts at monetization, which led to a severe paucity of cash in circulation. As currency circulating in economic activities became increasingly scarce, the purchasing power of a peasant's produce declined, and deflation ensued.

The causes of the shortage were manifold.[44] Generally speaking, the "famine" was a relative insufficiency against the background of rapid development. The Song mints generated six billion copper coins a year in late eleventh century, a level of production unrivaled by any other Chinese dynasties.[45] However, the increase failed to keep up with even more rapidly rising needs.[46] From antireformist officials in the eleventh century to modern scholars, many believe that Wang Anshi's forceful monetization initiative rendered the gap more egregious. The New Policies aggressively promoted money as a form of taxation, downplaying the significance of other exchange media. Since the classical period, Chinese farmers had used certain commodities, for example, grain and cloth, to pay taxes. The reform policies, such as Green Sprouts and Hired Labor, coerced peasants into selling their produce before obtaining cash to pay their debts and service taxes. In so doing, the state extracted an unusually high amount of currency from the local economy, and it shifted a disproportionate part of the revenue to cover the cost of its bloated bureaucracy and army.[47] As Wang aggressively expanded his reform apparatus, the government budget grew and the deflationary situation worsened. By the time of Wang's retirement and Shen's ascension, the cash crisis had provoked an intense factional debate.[48]

In his analysis of the problem, Shen carefully walked a middle road. He listed five causes and solutions without placing outright blame on the New Policies. First he suggested restoring a ban on the use of copper, a sanction Wang had lifted.[49] The soaring price of raw copper had encouraged people to melt coins to make daily life implements (such as a copper utensil).[50] Suppressing consumption of copper would stabilize its price and prevent coin mutilation. Shen's second suggestion concerned his previous effort to reform salt vouchers. He proposed stabilizing the value of the vouchers to rebuild people's confidence in using them as an alternative exchange medium.[51] Third, he suggested further efforts to seek other possible money objects and invoked precedents:

> In the ancient times, gold, silver, pearls, turtle shells, and scallop shells were all things substituting for currency; [people] did not rely on [copper] coins only.
>
> 古為幣之物，金銀珠玉龜貝皆是也，而不專賴於錢。[52]

Fourth, Shen urged the state to release cash stagnant in reserves, such as at local Ever-Normal Granaries.[53] Here Shen made a deliberate effort to circumvent the New Policies, which, as antireformists believed, was mainly responsible for cash hoarding. He pointed to the old granary system before the Green Sprouts reform swept in. Last, Shen warned about the outflow of Song copper coins to other countries, such as the Liao and Japan.[54]

Although some of Shen's points, such as the first and last, were conventional arguments also seen in the writings of his contemporaries, the other three proposals constituted a persuasive theme that probed the phenomenon of cash famine in depth. All three points boiled down to an emphasis on circulation, or, as Shen put it, "the benefits of currency lie in circulation" (*qian liyu liu jie* 錢利於流借).[55] He explicated the idea through a simple hypothesis: in a town with 100,000 strings of coins available, if all this money were hoarded in one household and kept out of circulation, it would remain 100,000 strings even after a century; if used in the market and allowed to circulate among all households in the town, however, the money would soon produce profits and reach a million strings. Shen reverted to the current situation: in an Ever-Normal Granary of a small town, the cash depository could amount to as much as 10,000 strings; if all these funds were to circulate freely, the cash famine would soon be ended.

Shen considered money crucial to the economy as a convenient exchange medium rather than a stagnant entity with a fixed intrinsic value. Thus the utility of currency did not depend on any specific material form: well-backed salt vouchers constituted an alternative (similar to other promissory notes, such as paper currency), and expensive metals other than copper would also be viable as money. This was how Shen's three points came to cohere.

Shen's stress on circulation touched on a key causal factor leading to the cash famine: the aggressive promotion of monetization primarily driven by the state's directives was not yet fully integrated into economic life. The circulation of currency to a certain extent stood as a system on its own, and its marriage to productive activities in the Song society was not yet seamless. Farmers and landowners would hoard currency as a commodity of intrinsic value for the purpose of strengthening financial security (as in the case of salt vouchers). The state, too, would sometimes keep currency out of circulation to build stronger reserves instead of pumping it back into the local economy. The cash famine was not a simple, absolute deficiency, and its solution required a systemic change in economic dynamics. Shen was not the only (much less the first) person to understand the nature of currency as a vehicle of exchange, but his multipronged effort to address circulation amid the cash famine demonstrated an insight beyond conventional wisdom.[56]

## Crystallizing His Reputation

It is not clear to what extent Shen managed to carry out his currency reform in practice, as political calamities descended on him later in 1077. Despite his timid efforts to downplay the connotation of "betrayal" in his revisions, censures such as "awful flatterer" kept coming from Wang and others. Shen's reputation as an opportunist gradually became consolidated by the late 1070s, as his political career was simultaneously drawing to an end.

At first, Emperor Shenzong resisted these attacks and insisted on using Shen; the situation continued until 1077 when the final, lethal blow was struck. The General Censor, Cai Que, submitted a long impeachment against Shen. At that moment, Cai was the new leader of the reformists after Wang's second resignation.[57] Calling Shen "treacherous" (*popi* 頗僻) and "capricious" (*fanfu* 翻覆), he enumerated major betrayals Shen had committed: the flip-flop on the Hired Service Policy and fawning on the new antireform leadership, among others. Cai added accusations against Shen's morality, claiming that although Shen enjoyed tremendous imperial grace, he "had done nothing useful" (*yi wu suo bu* 一無所補) in his position, and he only tried to please the subordinates while being a toady for the superiors.[58] Shen responded to this critique with calm and humility, admitting that it spoke some truth. Touched by his honesty, the emperor let the case rest.

Cai tried to enforce the impeachment with an attack from a different angle. This time he pointed out that Shen had stealthily slipped his proposal for changes to the grand councilor (Wu Chong) rather than to the emperor, implying that Shen, who had betrayed the New Policies, would continue to betray the monarch. Cai's fresh criticism hit a sensitive nerve. The emperor exiled Shen to Xuanzhou (in southeastern modern Anhui) and announced the decision by echoing the recurring accusation of "capriciousness":

> I promoted official Shen Gua all the way to serve as my assistant. You (Shen) have participated in all my creations of policies and institutions. However, at the beginning, [you] did not carefully reckon gains and losses; in the end, [you] resorted to frivolous rhetoric. [You] tailor [your ideas] just to suit mine. You act capriciously and speak inconsistently. What hopes do I have when my most trusted subordinate acts like this![59]

Curiously, many clues suggest that Shenzong's decision to banish Shen was a reluctant one; the invocation of caprice might have been more of a strategic response to pressure than a decision on true beliefs. In the following two years, the emperor tried twice to bring Shen back to the capital with promotions, but both efforts were thwarted by remonstrance officials.[60]

Eventually the emperor found a new opportunity for Shen: the warfront between the Song and Tangut Xia. During the whole reform period, the Tangut Empire had presented a constant military threat on the Song border, and Shenzong wanted to eradicate the menace.[61] To his chagrin, the idea of tackling the Xia had received little support even from the most activist officials. In 1081, a coup d'état occurred at the Xia court and threw the entire regime into disarray. This incident rekindled Shenzong's hope. He quickly appointed Shen as the general supervisor of a new military maneuver.[62]

The opportunity boosted Shen's hope for rehabilitation. Although his official job description was to manage logistics, he worked diligently to

expand his utility in a variety of areas. These two years became a con-
densed version of his previous multifarious career. He applied his mathematical skills to devising the most efficient transportation plans.[63] He
spent abundant time surveying the area, analyzing topographic conditions, and building new fortifications.[64] He used his previous experience
in organizing local militia to strengthen this relatively new military base,
mobilizing local residents to join a defense alliance.[65] He even found a
way to use his musical expertise, boosting the troops' morale by composing patriotic hymns and teaching the men to sing.[66] Although assuming
a civilian supervisory function, Shen even volunteered to lead battles
whenever he deemed it necessary.

Shen worked with frenzied diligence in the vast borderland, but his
efforts failed to resurrect his career. In 1082, a siege cosupervised by Shen
and Xu Xi 徐禧 (1035–1082) turned into a fiasco and cost the lives of tens
of thousands of Song soldiers, including Xu himself. This incident—
known as the Yongle debacle—permanently shattered the emperor's
ambition to conquer the Xia.[67] Although it might be reasonable to blame
poor coordination between Shen and Xu for the catastrophe, the emperor
held Shen fully responsible. Out of fury he stripped Shen of all political
ranks and detained him in custody in a Buddhist monastery in Suizhou
(in modern Hubei Province) for three years.[68] Shen not only failed to rescue
his reputation and restore his besmirched career, he sank even further.

The language of morality in the eleventh century was so deeply
intertwined with political and ideological meanings that it never simply
reflected one's personal qualities. Shen provides a vivid case of how personal opportunism and centrist political beliefs can conspire to create a
reputation as a petty man. From his early struggles with criticism for
profit-seeking propensities as a clerk to the abrupt termination of his
career for caprice, contemporaneous narratives in the Song consistently
portrayed Shen as a man with moral deficiencies. This discourse peaked
with an incident involving Su Shi, Shen's widely respected peer. In 1079,
Su was framed and imprisoned for slighting the imperial authority in a
series of sarcastic poems. The incident became known as the Crow Terrace Poetry Case (*Wutai shi'an* 烏臺詩案).[69] Shen was believed to have
turned in one of those subversive poems, which he had sought from Su
as a sign of friendship. It remains a matter of debate in modern scholar-

ship whether this story is fictional.[70] Yet once again, the anecdote fits so seamlessly into Shen's profile that its veracity becomes nearly irrelevant.

Shen Gua's swift ascent in the political world arose from his ability to make revisions, and he plummeted for the same reason. Despite all the political factors affecting the course of events, from the start of their cooperation, Wang and Shen clearly held distinctive approaches to reform: Wang aimed at building a system, and Shen focused on piecemeal revisions. When the political alliance was strong, the two could cooperate effectively, but when political conflict arose, trust became impossible. Beyond the political vagaries in the age of factional struggles, the split between Wang and Shen spoke to deeper intellectual differences about the nature of learning. In the following chapter, I subject these differences to an epistemological analysis.

# CHAPTER 7

## *In the System (and Then Out)*

During his decade-long service in the top echelons of the officialdom, Shen Gua had multiple revelations, the most important being the deep and eventually uncompromised difference between him and Wang Anshi. Beyond its political and ideological consequences, this difference bore critical epistemological implications that connected the concrete doings of Wang and Shen in the New Policies era to key issues about learning. In this chapter, I introduce the epistemological schism between the men by anatomizing the concept of system.

I start with a discussion of systems and end with an analysis of Shen's withdrawal from them. The first five sections define a system and its major epistemological implications. To follow, I introduce Wang Anshi's system and a few other systems as comparisons. The last section returns to Shen and introduces his departure from system building.

## *System*

In the most general sense, a system is a total view. It is a collection of elements meaningfully organized to the effect that the entire picture is more than the sum of particular parts. In the current study, the elements of a Song system consisted in "things," the broad concept encompassing objects, affairs, and sometimes humans (see chapter 2). The ultimate meaning of the organization of elements lay in unity. For a Song thinker, having a total view was to be able to contain and unify infinite particulars in the phenomenal world within a definite order.

In my usage, a system is not a generic "philosophical system," which describes a coherent set of ideas or, for that matter, any coherent line of thinking. Empiricism, for instance, is a system in modern philosophical discourse. A Song system was a philosophical system in which the propositional content specifically designated a total view and intended to unify the particulars within that view. An empiricist, modern or medieval, would not provide a total view like this, hence my designation of Shen's empiricism as a "nonsystem" in chapter 9.

The current discussion of systems is not meant to be a comprehensive recapitulation of eleventh-century intellectual history with a new conceptual outfit. Instead, my narrative of system building aims to tease out the epistemological dimension of Song thought. The primary function of a total-view system was undoubtedly to illuminate orders in the world, and a significant purpose of ordering was to guide knowing. My discussion of systems focuses on how the Song thinkers envisioned epistemic guides and how they organized their epistemic praxis accordingly. Thus, before delving into particular systems, I introduce "modeling," the mainstream way of knowing practitioners of learning followed.

## Modeling

Modeling was a way of knowing. It was the epistemic praxis that a practitioner of learning engaged within the framework of pursuing the *dao*. It designated how a *dao*-seeker was supposed to generate new beliefs/actions in particular instances. It presumably surpassed a sensory engagement with an individuated "thing" ("knowing from hearing and seeing") and regarded instead the "well-placed-ness" of the "thing" in larger orders. Some Song thinkers, such as Zhang Zai and Cheng Yi, addressed this type of knowing as "knowing from virtuous nature," which may lead to a mistaken assumption that it referred to moral knowledge (narrowly construed). In fact, this type of knowing regarded a range of orders of relations ("models"), of which morality was just one kind, hence my choice of a more generic term—"modeling."

A thorough understanding of modeling requires elucidating two concepts. One is "order of relations" (alternatively, "deep orders"), which I call the content of modeling.[1] To model was to be able to place a "thing" in a relation and then properly deal with it. The relation at issue was normative in nature and provided the knower with epistemic guidance. This whole process should be understood with the caveat that one did not have to be motivated by a conscious perception regarding a relation (content fully articulable) whenever responding to a "thing." In fact, one could reach the new belief/action spontaneously "in an instant of brute clarity."[2] This point takes us to the next concept key of modeling: the heart-mind.

The heart-mind shouldered the heavy lifting of what I call the psychology of modeling. *Xin*, the Chinese word for "heart," which also assumed the function of mind, was the epistemic apparatus that enabled modeling and acted as the counterpart of the senses in sensory knowing. Among other functions, the heart-mind helped one in spontaneously discerning order without having to engage in any deliberative activity. The stipulation of the heart-mind's special function was an ancient heritage, a consensus held by most eleventh-century thinkers; an effective philosophical design that fully capacitated this function, however, did not emerge until in Cheng Yi's thinking (see chapter 10). In accordance with this development, I organize my analysis of modeling into two sections: I focus on order of relations—the content of modeling—in the current chapter, where I introduce the common basics of system building, and I elaborate on the heart-mind psychology in chapter 10, where I discuss a more advanced stage of system building in Cheng Yi's design.

## Orders of Relations

Orders of relations claimed central significance in modeling because they served as epistemic guides. Such a stipulation was cause and effect of Song literati's efforts to envision the ineffable *dao* in articulable forms. The epistemic belief that one should know through larger orders was premised on the cosmological belief that the cosmos—the multiplicity encompassing "things," humans, and all their activities in one vast manifold—moved with a structured dynamism. The latter proposition calls back to the generative scheme linking the *dao* with "things," a point I made in chapter 2 and briefly recapitulate here:

*dao* → the numinous → intermediate stages (e.g., number and figure) → ten thousand things

This formula was an effort to parse the structured dynamism of the *dao* into discernible orders; each of the intermediate stages stood for a large order, a realm in which distinctive relations resided. In chapter 2 I cited the example of Shao Yong's work, in which he highlighted number and figure as two intermediate orders. As I demonstrate shortly, orders resided not only in numerical ratios but also in more concrete realms such as morality and governance. Compared to the "trivial" (*mo*) particular "things" in their characteristically chaotic state, these orders appealed to Song thinkers as more "fundamental" (*ben*) in terms of illuminating patterns, big or small, in the enveloping flow of the *dao*.

Turning to Wang Anshi, the central figure in this chapter, I explore the range of orders of relations in his conception. Wang presented his understanding of larger orders in a narrative meant to instruct followers in pursuit of the *dao*. According to Wang, since the *dao* had generated various orders before it disseminated into particular "things," one should approach the *dao* in exactly the reverse order of the generative sequence. In his own words, "the (generative) sequence of the *dao* develops from fineness (*jing* 精) into extreme coarseness (*cu* 粗), and the sequence of learning the *dao* starts with coarseness and arrives at extreme fineness."[3]

Thus one had to advance through multiple stages to eventually reach the *dao* and engage a certain order of relations at each stage. Wang laid out each condition in detail. First, a learner took his departure in a world filled with things "worth reflecting on and employing" (*ke si ke wei* 可思可為), and he should explore the "functions" (*yong* 用) of these things for the purpose of "settling his self" (*anshen* 安身). To settle oneself in the world was to find one's proper place, and as soon as that happened, one must "enhance his virtues" (*chongde* 崇德). Afterward, the learner was able to "apply himself to all under Heaven" (*zhiyong yu tianxia* 致用於天下). The finest service he could provide to the world was to "maintain order and yet forget not disorder, and settle in peace and yet forget not danger" (*zhi bu wang luan, an bu wang wei* 治不忘亂，安不忘危).

Once the learner "completed his (worldly) career" (*shiye bei* 事業備), he would be ready to enter the next stage, "exhausting the numinous" (*qiongshen* 窮神). Amid the numinous, he would know "the intricate and the manifest" and "the soft and the firm" (*zhi wei zhi zhang, zhi rou zhi gang* 知微知彰，知柔知剛), which would connect him to the *dao*.

Upon reaching the *dao*, the learner achieved an unobstructed awareness of interconnected unity and was able to navigate among the ten thousand things with spontaneous sagacity. He no longer needed to "reflect," because "all worries returned (and vanished) into unity" (*bai lü zhi gui hu yi* 百慮之歸乎一). That was precisely how one "obtained all things under Heaven without reflection." Instead of relying on mental or behavioral exertions, one now simply swam in the flow of things "with no reflection or action" (*wu si wu wei* 無思無為), remaining "dormant and motionless" (*jiran budong* 寂然不動).[4]

From this account, it is clear that Wang acknowledged multiple orders, which ranged from small to all-encompassing, and from the concrete to the ineffable. In the sequence he stipulated, the first order of relations—also the most concrete and smallest in scale—was the function of a thing. Function connected an object or an activity to a human purpose, thus constituting a relation. For instance, the function of a chair is to be sat on, in which case the chair serves the purpose humans see in it.

The next two larger orders in Wang's account were concepts well known to students of Chinese history. The first, concerning "virtues," was the moral order. Morality was arguably the most popular subject among the Song literati. In their vocabulary, the moral order was closely associated with "ritual propriety" (*li* 禮), which referred to ethics in general. In this context, people contemplated a "thing" (likely a process rather than an object) within an interhuman relationship.

To follow, the order in which one "applied himself to all under Heaven" and concerned himself with "order" and "disorder" was that of governance, which bore immediate relevance to the literati's participation in statecraft. In Wang's rendition, this order primarily consisted of institutions, or *fa* 法, within which people examined and managed "things" within institutional relations.

From the order of governance, Wang leaped into the numinous, the clairvoyant stage immediately before the consummation of unity, the *dao*. The numinous and the *dao* were even larger orders, albeit with no articulable structures. By making the leap, Wang glossed over a couple of models frequently invoked by his peers, such as number, the order of relations Shao Yong favored. Nevertheless, in describing the numinous with dyads such as "the intricate and the manifest" and "the soft and the firm," Wang was alluding to number, for these terms were saliently associated with the *Change*.[5] Although perhaps disinterested in deeply engaging number, Wang was well aware that "things" had their proper places in accordance with numerical relations.

Of special note, although Wang presented these orders as separate and self-contained realms, he did not mean—because it was not the case in reality or in his perception—that these orders were mutually exclusive and only appeared in one's experience in the rigorous hierarchical sequence. The sequential arrangement obviously meant to match the generative scheme of the *dao*; it served as a didactic tool to remind followers of the graduated degrees of fundamentality and exhort them to go higher (and thus closer to the *dao*). In real life, an individual most likely encountered orders in a random manner and experienced epiphanies much more freely.

Also, there was no uniform way to generalize about the mutual relationships between these orders. Since they embodied different degrees of fundamentality, the orders were certainly not ontological equivalents. For instance, orders prior to the "birth" of the ten thousand things (in the generative scheme) assumed priority over those after; number was thus more fundamental (and in a sense, more cosmological) than ritual priority. Mutual relationships between orders became even more convoluted when some Song thinkers fused orders into different combinations (often with contesting understandings of the ranking of fundamentality). Functionally speaking, these orders were epistemic equivalents because they all stood for structures that guided people's understanding of the world. When Wang listed the orders in a continuous gamut, he pointed to their functional equality in an epistemological sense.

## Orders of Relations as Epistemic Guides

The orders of relations served not only as way stations of the *dao*-seeking process, but as epistemic guides for a *dao*-seeker in particular instances of knowledge acquisition. The two functions became conflated owing to the nature of the *dao*: from the viewpoint of an individual, to arrive at the *dao* meant that he could generate an appropriate belief/action in any situation. The process to approach the *dao* thus entailed a long-term endeavor to make as many good decisions as possible, in all contexts in one's life. Just as Wang suggested, one should constantly "gauge an undertaking with the *dao*" (*yi dao kui shi* 以道揆事).[6] Given that the *dao* featured no articulable content and could not provide explicit guidance, Wang's injunction meant that when performing any concrete task, one should always align his new belief/action with a correspondent order, wherein he found his epistemic guidance.

It is imperative to examine the nature of orders of relations as part of the discussion of their epistemic role. First, any of the aforementioned larger orders—either as an entire set or as individual relations—had a real, experiential existence in the phenomenal world. This was the case for the functions of physical objects, ethics, and institutional relations, as well as the more abstract-looking number. For the convenience of further discussion, let me cite a specific relation from each larger order (all except the number example are from Wang's writings):

I. "Function": the function of a crossbow—as in its relation with human users—was for defense.[7]

II. Ritual propriety: the relation between a caring father and a filial son, who mourned for his father for three years after his passing.[8]

III. Number: a yinyang relation between a yin turtle and a yang deer (see chapter 4).

IV. Institutions: a hierarchical relation wherein "the worthy rules the unworthy, and the honored rules the lowly."[9]

These relations were not abstract laws associated with a suprahuman lawgiver or a realm transcending the experiential world. For a Song person, modeling was thus a process in which he actualized existing orders (such as the aforementioned) in his personal experience dealing with "things." Contrary to a Cartesian counterpart, the Song person saw his action not as fitting the "thing" into an order as a conceptual framework, in essence a representation of reality. He did not believe an order of reality was a product of subjectivity, either his own or any superior other's.

Moreover, the experiential character of these relations should be understood with the caveat that they were not—in a rigorous sense— empirically extrapolated patterns, which derived their existence from experiential repeatability. All foregoing relations were correlated with, and to a great extent anticipated, empirical successes, experiential comfort, or a general feeling of "naturalness." From the perspective of a Song thinker, however, it was definitely not the case that he could extrapolate any relation that happened to occur to him to be a relation based on the criteria of empirical success, comfort, or general appropriateness. For Song literati, there was a bounded (though not fixed) selection of orders of relations, an issue I later clarify with the metaphor of repertoire.

Beyond its ontological nature, another aspect of an order of relations— what Donald Munro calls a "fact-value fusion"—actuated its capacity to guide knowing.[10] Any relation in the foregoing examples had both a descriptive dimension, that it was so and so, and an evaluative dimension, that it ought to be so and so. A crossbow, when related to humans, was a weapon of defense and also ought to be a weapon of defense. A deer was yang, and a turtle was yin; a deer ought be in the yang position when juxtaposed with a turtle. The normative implication became even more salient in orders concerning morality and statecraft. A father–son relation was characterized with the son's filial piety, and meanwhile ought to be so. A worthy person sat in the position of leadership, and any institution should place the worthy in leading positions.

Therefore, a person about to generate a proper belief/action regarding a certain situation should acknowledge the normative status of a correspondent relation and take it as a guide to his response. When a person picks up a crossbow and draws it against his enemies, for instance, the action results from his understanding of the crossbow–human relation bound in the crossbow's function. A doctor contemplating powdered deer antler as a potential ingredient in a medical recipe should be aware of its yang quality (and may mix in powdered turtle shells as a yin ingredient to counter any excessive yang effect). When a father dies, his son should mourn in accordance with the father–son relation; in the particular example I mentioned, Wang reached the belief that the son should adhere to the classical prescription and stay in mourning for three years.[11] When appointing a leader to a group, the recruitment authority should evaluate the candidates in terms of the worthy/unworthy binary. Wang in this specific case concluded that the current remonstrance officials (in his opinion, not truly worthy) should not be assigned the type of overarching supervisory power they assumed.[12] All these relations were viewed as fundamental, "self-so" patterns/structures of the dynamism enveloping the entire cosmos, hence their normative status in the enterprise of knowledge production.

How would a knower access these orders and relations, given that he had to depend on their guidance? This is a question that requires us to complement the philosophical discussion above with a historical analysis. The access to orders of relations was both experiential and textual, and the two aspects were intimately interlocked. On one hand, in theory, a grasp of orders could be the result of long-term, reflexive engagement with the world, and thus deeply experiential in character.

On the other hand, in practice, people relied on testimonies for knowledge of orders. Song literati were not looking at the world as a blank slate where they were at liberty to create or discover orders all on their own. All foregoing larger orders, as my previous discussion implies, were bounded realms where certain content had already been established as the foundation of further developments. Up to this time, historical testimonies in their long-term accumulation had formed a discourse on the orders in the world; these accounts constituted an indispensable (if not dominant) resource for Song thinkers' further observations. The historical testimonies mainly resided in old texts, including the Classics, on which basis contemporary epistemic authorities (a concept I explain in the next section) continued to contribute updated testimonial accounts.

Summing up, to explore the world was essentially an experiential matter, for the ultimate purpose was to know the world better and engage it more productively. It was not, nevertheless, purely an experiential task in praxis, for literati did learn about orders in propositional forms (with a readiness to actualize and experience them). The experiential and textual aspects of ordering the world were inextricably connected and mutually enforcing; to Song thinkers, they served the same purpose.

To clarify the relationship between orders of relations and testimonies, I employ the concept of repertoire as a framing device. Every Song thinker who drew on the written culture for knowledge of orders worked with the assistance of a repertoire of textual testimonies. The repertoire contained many historical basics—antecedently valued knowledge passed down by the authority of tradition—and yet the repertoire was not exclusively historical, as the contemporary epistemic authorities kept producing new testimonies on the basis of old ones. The repertoire was dynamic and ever-growing.

The accumulated textual tradition provided abundant historical testimonies. To readers familiar with Chinese history, it should already be clear that the orders on Wang Anshi's list had ancient roots. "Function" was the dichotomous counterpart of "body" (*ti* 體); the body-function dyad originated in the ancient classic *Change*.[13] Virtue and ritual propriety were concepts with the longest possible pedigrees, tracing back to ancient sage kings.[14] The discussion of institutions was also a prominent part of the cultural discourse since the *Rituals of Zhou*.[15] Number, as I introduced in chapter 4, was a prevalent theme in classical and nonclassical old texts. All the categories contained a rich accumulation of accounts on specific relations, providing eleventh-century thinkers with references of elaborate kinds. In explicit and implicit ways, these historical testimonies served as authoritative references that conditioned the ways Song thinkers observed the world and articulated their new insights on orders.

The significance of historical testimonies, however, does not mean that the talk of order in the Song remained strictly identical with received literature. While Song literati inherited the basics in historical testimonies, they felt comfortable improvising variations on them. For instance, in the realm of number, Shao Yong created a system on the foundation of yinyang, the Five Processes, and hexagrams with elaborate new formulas. In terms of ritual propriety, as the Song literati spoke of the five basic relations received from Mengzi—between father–son, ruler–minister, elder brother–younger brother, husband–wife, and friend–friend—they continued to enrich the discourse of relationality with more detailed proposals.[16] Regarding institutions, Wang Anshi was able to name an ancient origin for each of his institutional reforms, yet the institutional apparatus he eventually created surpassed any ancient precedent in complexity.[17] The contemporary talk of order, especially as formulated by key intellectual figures, became new testimonies.

The overall significance of textual testimonies was both cause and effect of the close connection between the Song literati's world-ordering ambition and the paramount attention they paid to written culture. Given that the repertoire was supposed to include all discourses on order in a long span of history, it reserved expansive and expanding space to encompass all relevant content in the whole written culture in both historical and contemporaneous dimensions. In this sense, relying on the cultural repertoire was synonymous with using the written culture as a repertoire.[18] Modeling, after all, was essentially the epistemic praxis of participants of the written culture, who had privileged access to historical testimonies and held an entitlement to produce contemporary references.

After exploring the forms in which orders of relations existed as a source of knowledge, let me push the central epistemological question further: at the instant of generating a new belief/action, how exactly did an individual receive the epistemic guidance from orders of relations? The answer is related to the psychology of modeling and again involves multiple options related to the experiential-textual complexity in accessing orders. One could certainly follow a fully articulated rule as part of a testimony, thus taking the textual route. For instance, a son could follow the classical prescription of a filial son's duty and stay in mourning for his deceased father for three years. A more favored procedure was to let the normative force of an order prompt one to a belief/action without having to rely on any deliberative procedure. The son in this scenario could be prompted by the father–son relation to a spontaneous decision, be it mourning for three years or uttering three cries at the funeral (and leaving).[19] This spontaneous process attested to the knower's true experiential engagement with the relation as a fundamental structure of the world. I further explain the second procedure in chapter 10 with the example of Cheng Yi.

## Epistemic Authorities

For multiple reasons, modeling as a way of pursuing knowledge required the presence of epistemic authorities. An epistemic authority is a person who serves as an authoritative guide for another person's beliefs.[20] In the Song intellectual world, an epistemic authority could be historical (an ancient sage) or modern (a contemporary leading scholar). In modeling there was a general reason for his existence: a common *dao*-seeker (not a sage) had difficulty claiming perfect epistemic self-reliance. Most people were still in the middle of seeking the *dao*; responding to every situation in perfect accordance with the *dao* remained a constant aspiration, a "matter of concern," rather than a "matter of fact" that could be unfailingly accomplished.[21]

The more specific raison d'être behind epistemic authorities concerned the access to orders of relations. An eleventh-century literatus often relied on testimonies for knowledge of relationality, either in lifelong learning or at a particular moment of generating new knowledge. These testimonies came from epistemic authorities: the ancient sages provided many in old texts, and the contemporary authorities continued to produce modern interpretations.

To put it simply, an epistemic authority assumed guiding power by supplying testimonies of orders in the world. The subtle implications of his authoritative role were at least twofold. First, an epistemic authority was more of a broker than a source of knowledge in the epistemological sense; second, he was thus not supposed to directly command others' beliefs. Although following epistemic authorities was an indispensable action in learning, one did not have to depend on them for fully specified ideas. The second implication was both cause and effect of the stipulation of the psychology of modeling: to follow a fully articulated rule (made by an authority) was less preferable than following one's heart-mind.

On a separate but related note, the second point was also a key premise held by Song literati in their veneration of the ancient sages as the ultimate epistemic authorities. An understanding of the soft power of epistemic authorities (compared to the coercive role assumed by political authorities) is crucial for an accurate grasp of Song antiquarianism, which, among other things, was *not* dogmatic fixation on tradition.

Epistemic authorities did not exist only in ancient times; contemporary literati were also active in pursuing such a role. For instance, all system builders were epistemic authorities. Now I return to systems and their builders.

## *Wang Anshi's System*

A system was a comprehensive program for ordering and knowing. It was a system builder's organized thinking on orders and his systematic instructions on how to appropriately engage the world in the prospect of achieving the *dao*. Some eleventh-century literati built their total views on one focal order of relations, and Wang Anshi was perhaps the most prominent example among them.

Wang's system centered on the institutional order, or *fa* (institutions/ regulations/policies). In most contexts, the direct translations of *fa* should be either "policies" (as in "New Policies") or "regulations," both terms rooted in the institutional world.[22] Wang's assemblage of *fa* was essentially a system of governance, in which "things" were related to one another along the institutional lines of a government.

Let me first introduce the philosophical orientation of the system and then its content. Wang chose a highly concrete order of relations situated close to the beginning of the *dao*-learning process. He connected the *dao* with *fa* as follows:

> When the *dao* is in the affairs of government (*dao zhi zai zhengshi* 道之在 政事), noble and lowly have their proper place, last and first have their proper order, many and few have their proper number, and slow and fast have their proper time. Their institution and deployment depends on *fa*, but putting them into practice depends on the person.[23]

Wang clearly generated this view from the perspective of a ruler, as he put the emperor at the center of the system: "[He who] occupies the position of the Son of Heaven, makes the *dao* all under Heaven thoroughly illuminated and thoroughly complete."[24]

Of special note was the scope of Wang's system. The choice of the institutional order did not mean that he was concerned with a narrowly construed political order only. Just like other system builders, his vision resided in the larger *dao* (as he explicitly stated in an expression such as "when the *dao* is in the affairs of government"). In Wang's belief, a sound system of governance would ensure proper placement of "things" around the world and thus constituted an effective approximation of the *dao*. In other words, a good institutional apparatus (the realization of the institutional order) set the grounds for all other orders to come to actualization. Thus, Wang envisioned a universal power of *fa* in the practical sense, in contrast to the metaphysical type of universality embodied in number or *li*, orders a modern audience regards as abstract entities.

The content of Wang's system—the New Policies—attested to the kind of practical universality Wang intended. During the reform Wang stipulated a body of *fa* and each was a specific prescription regarding a concrete governmental issue. Chapters 5 and 6 describe the key economic policies, such as the Green Sprouts, Mutual Security, and Hired Labor. The reform, however, was not just meant to reorganize economic and political affairs, issues conventionally associated with governmental functionality; it was also aimed at transforming morality and customs among the population through educational reforms and economic leveraging.[25] All the acts came together to define a fully specified system of governance with the following expectations: a flourishing economy characterized by equality and productivity, a strong military, an extensive education system, a robust moral culture, and an extensive state apparatus functioning as part of the order as well as the engine sustaining the order—in one word, an approximation of the *dao*.

A salient feature of Wang's system was the full specificity of its content, and a high degree of normativity ensued. In his world, a governmental policy of his design stood for what ought to be done, assumed aggressive regulatory authority, and remained entitled to political enforcement. Wang's *fa* was thus so normative that it bordered on being coercive, which became an issue for governing and knowing, a problem I address in the next section.

In sum, Wang's system was a fully specified total view of governance. It became the most influential system in his times due to its marriage with political power. The immediate influence Wang's system exerted on the day-to-day transactions of statecraft also easily diverts modern scholars' attention away from its origin in the scheme of pursuing a larger, cosmic *dao*. By collating his thinking on system building into a complete narrative in this chapter, I show that just like many other *dao*-seekers, Wang had always kept the cosmic *dao* in his view while building his system, and the lack of cosmological/metaphysical flavor was mainly due to his particular choice of focal order.

Now I turn to the issue of epistemic authorities, which, in Wang's example, had a dual significance. On one hand, in building the system Wang surely aspired to become a contemporary epistemic authority entitled to guide people's actions on a comprehensive scale. With his system closely wedded to political power, however, his role transmuted into a political authority and thus sanctioned his aggressive tendency to prescribe and coerce.

On the other hand, Wang acknowledged the guidance he received from ancient epistemic authorities—the sages—specifically in the sense of learning about the order of relations he paid most attention to. In Wang's perception, the ancient sages had also devoted most of their efforts to building versions of *fa*, namely, "policies of the sages" (*shengren zhi fa* 聖人之法).[26] Wang often depicted ancient sages as occupied as he was in devising effective institutions. For instance, in the "Ten-Thousand-Words Memorial," he spoke of how sage kings had designed educational institutions in antiquity:

> The kings and dukes in antiquity instituted schools from the capital to counties; [at the schools] they appointed instructors and screened them rigorously. All things concerning ritual, music, punishment, and administration at the imperial court were taught at the schools. What literati observed and studied consisted in words and virtues of former kings as well as their conceptions in ruling the world; the talents [of the literati] thus could apply to [ordering] the state and the world.[27]

The passage shows Wang identifying multiple points of inspiration from the ancient kings' institutional work. First, they extensively built schools at both national and local levels; second, they appointed well-qualified teachers. Third and most important, the school curriculum focused on issues pertaining directly to statecraft, which fostered the cultivation of effective bureaucrats. Later in his own educational reform, Wang included all three points as items on his agenda.[28]

The way Wang learned from sages provides a typical example of what I have called the soft power of epistemic authorities. While finding the sages' policies ultimately inspirational, Wang argued that he followed the "conceptions" (*yi* 意) the sages held in creating their institutional rules and explicitly opposed duplicating the rules themselves. In the foregoing example, Wang advocated following the three general points underlying the ancient schooling system rather than replicating the exact praxis. In his reasoning, the vast difference between antiquity and his times made simple replication unproductive. In his own words, "[the times] of the two emperors and three sage kings is more than a thousand years away from now. . . . The changes and the circumstances were not the same; the methods to apply were also different."[29]

Wang then proceeded to explain "conceptions" and why they remained timeless:

> However, there was not necessarily any difference in their conceptions in organizing the world and the state, nor in what they made the most fundamental and what they granted priority.[30]

The implication was that a key aspect of such "conceptions" was a sense of "origin" and "branches," that is, acumen to discern order—the relational patterns underlying specific institutional arrangements. Thus, Wang saw the ancient sages as kindred spirits who thoroughly explored *fa* as an order of relations. He developed his own formulas on the foundation of the basic patterns he discerned from the sages' doings.

Obviously, Wang's reliance on the ancient epistemic authorities was flexible. Although it is reasonable to argue that such flexibility facilitated Wang's use of antiquity for his own goals, his reference to the sages was not simply an expedient rationalization. In fact, the soft guiding power of epistemic authorities sanctioned space for followers to improvise in accordance with their own predilections. Like Wang, many other system builders sought testimonies on orders from sages and then presented the ancient wisdom with selective foci of their own choosing.[31] The sages in Wang's perception provided especially useful lessons on *fa*, and those in Shao Yong's were particularly devoted to number.[32] The pattern recurred in many other contemporaneous attempts at system building.

## Other Systems in Comparison

Here I offer glimpses of a few other systems for two purposes: to make comparisons with Wang Anshi and to weave the discussions of individual systems into a coherent narrative. The eleventh century witnessed two major ways of system building, which eventually turned into two stages of a long-lasting developmental process. Wang Anshi and Shao Yong represented the first stage; each of them built a system with one focal order of relations. Some others, such as Cheng Yi, developed a new metaphysical scheme to incorporate all orders of relations, representing the second stage of system building. In comparing these two stages, I highlight the concept of "unity," the ultimate goal for any system, and discuss the different philosophical designs Song system builders proposed to substantiate it.

As with Wang, Shao Yong chose one focal order to build his system: number.[33] Shao deemed numerical relations the most effective model to properly place "things" and thus constructed a whole edifice on it.[34] Similar with the Course-*Qi* system I discussed in chapter 4, Shao's system aimed to "order the world" (*jingshi* 經世) with numerical formulas. Drawing on the *Change* as the main source, he extensively used trigrams and quaternary taxonomies to organize all phenomena (including time) in the world.[35]

Thus, Wang's and Shao's systems were comparable in the sense that both were highly normative in providing epistemic guidance. In their respective ways, the systems dispensed fully articulated prescriptions.[36] Wang provided followers with guidance in the form of institutional specifics; Shao offered elaborate diagrams and calculation methods. The mechanisms through which the systems generated normativity were not identical. Unlike institutional relations, numerical patterns were not a literal delineation of concrete praxis, featuring what modern minds view as a certain level of abstraction (with the caveat that it was not a regular abstractum; see chapter 4). As a result, Wang's prescriptions featured more concrete empirical content, whereas Shao's system was specific in terms of providing elaborate structures. From the perspective of a follower, nevertheless, the effects were similar, as both schemes left limited space for improvisation.

Another commonality lay in Wang's and Shao's visions of unity. Indeed, they were successful in bringing out a sense of unity against the chaos of the particulars, and such unity depended on uniformity. In each system, the unifier was the focal order of the system builder's choosing; homogeneous relations—institutional or numerical—constituted an overarching structure and allotted each particular "thing" a definite place. Whereas Shao asserted that particular "things" became unified due to their alignment with the numerical patterns of $qi$'s movement, Wang understood uniformity even more literally: he built a concrete system of governance, fully specifying all details he deemed necessary and thus asserting placement for everything.

Uniformity was no doubt an effective expression of unity, yet it posed potential problems. System builders like Wang and Shao faced a dilemma: their choices of focal order were at odds with the multitudinous dimensions of pursuing the *dao*. Neither Shao nor Wang meant to exclude other orders as necessary guides for knowing the world. Both of them explicitly endorsed the multistage sequence of pursuing the *dao*, and both invoked other order concepts in their writings. For instance, while adamantly rooted in *fa*, Wang also referred to *li*, *qi*, and numerical ratios.[37] It was nevertheless unclear how his *fa*-centered system could accommodate other orders and coordinate with the complex *dao*-seeking journey.

In addition, Wang's and Shao's systems featured excessive normativity that potentially abused their privileges as epistemic authorities and, more importantly, contradicted the psychology of modeling. The systems exhorted adherents to follow instructions fully articulated in words or numbers, leaving them little room for reflection, much less spontaneous reactions springing from their hearts-minds (see chapter 10 for details). The problem was especially salient for Wang. With its fully specified content, Wang's system commanded belief from followers and, when wedded to political power, further coerced obedience. Although Wang himself learned from ancient epistemic authorities in the appropriately flexible manner, he did not pass this privilege on to his adherents. Such was the epistemological dimension of his political dogmatism. As his opponent Su Shi sharply observed, "Wang's writings are not necessarily bad; the problem is that he wants to make everyone the same as he."[38]

The second stage of system building featured innovative attempts to address both problems. Su Shi and the Cheng brothers (mostly Cheng Yi)—the key thinkers I choose to represent this development—were familiar with stage-one systems and known for their active critiques of them. Indeed, Su and Cheng were prominent protesters of Wang's New Policies and formed their own dissident factions.[39] Su designed a system with no exclusive fixation on any particular order of relations. He spoke copiously of the structuring function of ritual propriety and institutional

forms; he invoked "constant patterns" (*changli* 常理) and numerical relations.[40] Su practically included all commonly seen orders of relations without committing to any of them as his particularly favored way station
between the *dao* and "things."

What held Su's thinking together as a system was a stout commitment to unity, which, according to him, was an "intuitive" one.[41] The *dao*
was one, a unified source of everything concrete in existence. A learner
thus should strive to work "upstream" toward the source.[42] In the course
of "swimming," Su encouraged fellow *dao*-seekers to immerse themselves
in all possible *dao*-conducive activities and learn through constant exposure to all orders.

While sharing a number of similarities with Su, the Cheng brothers
introduced more innovation. They coined the concept *li* (Coherence) to
encompass all possible orders between the *dao* and "things," thereby proposing a new kind of unity detached from uniformity.[43] Among the myriad
relations by which one might have an epiphany on the *dao*-seeking journey,
a "Coherence" could be a numerical ratio, a father–son relationship, or a
crossbow with a user-friendly design. The Chengs addressed the issue of
excessive normativity by developing further arguments on the heart-mind
in coordination with the new metaphysical claim regarding Coherence.
They consolidated Su's "intuitive" proposal with a concrete philosophical
plan, a contribution that merits its own unit of discussion in chapter 10.

In sum, the epistemic praxis of literati in the eleventh century
revolved around the exploration of orders of relations. The systems—
philosophical products extrapolated from such epistemic praxis—were
total views that organized "things" into meaningful orders. Although
Wang's system gained the most salient political influence, his many peers
competed in the same intellectual arena, proposing a variety of conceptualizations of order with different foci and priorities. These endeavors
composed what I call the mainstream of system building.


## Dropping out of the System

The foregoing introduction to system building gives Shen Gua's deviance
from systems the rich background it deserves. Shen dropped out of the

system in two senses: in his departure from Wang's New Policies and, more generally, in his drift away from modeling as a way of knowing. Here I offer an analysis of the first point—the epistemological schism between Wang and Shen. In the next two chapters, I explore the second point, a much more elaborate undertaking that merits its own discussion.

An epistemological difference took root at the very beginning of the cooperation between Wang and Shen, and ironically, it was intended to facilitate a productive partnership. In the New Policies, Wang, the system builder, recruited Shen, the piecemeal engineer, to manage the hands-on details of his broad schemes. Wang had a clear expectation of his followers vis-à-vis the grand system of *fa*: they were supposed to implement his fully specified formulas with rigor. In his words, they were supposed to "guard" it:

> Material wealth brings together the multitude under Heaven; *fa* organizes the material wealth under Heaven; bureaucrats guard the *fa* under Heaven. If bureaucrats are not qualified, *fa* may not be well guarded even if it exists; if *fa* is not sound, fiscal management may falter even if material wealth abounds.[44]

Shen came to the guardian role with a keen awareness of Wang's intentions. Early in their cooperation, Shen demonstrated ardor for fulfilling the responsibility of enforcing Wang's policies. On a number of occasions, he applauded Wang's *fa* with enthusiastic prose. Shen claimed in an essay that Wang's employment of *fa* complied with one of the two key ruling traditions descended from ancient sages (presumably Legalism).[45] In his typical way, Shen transposed thinking into doing: he built his reverence for *fa* into the clepsydra he designed at the observatory. On the wall of the constant-level receiver, Shen lodged a decorative knot shaped as a "literatus" (*shi* 士), presumably a male human figure.[46] The constant-level receiver functioned to regulate the water levels in the reservoir and the receiver, maintaining the equilibrium of the entire system (see chapter 3). Shen saw this stabilizer as an analogue of literati, for they were in charge of "generating *fa*" (*shengfa* 生法) and thus responsible for upholding the equilibrium of the state.[47] Shen announced his endorsement of *fa* in bronze as well as in words.

Over time, Shen's approach to fulfilling the guardian role started to deviate from Wang's expectations. In the cases discussed in chapters 5 and 6, instead of rigorously following Wang's prescriptions, Shen implemented them with revisions. He broke the monotony of the Green Sprouts with a partial restoration of the old granary system. He downsized the hydraulic project in Liangzhe and altered the structure of financing. He reorganized the Hired Service Policy and put a brake on the expansion of the state monopoly of salt by resuscitating the merchant distribution mechanism.

Despite the political, partisan readings of Shen's actions I introduced in chapter 6, it is not difficult to see that he sustained an epistemological consistency in revising Wang's system. First, he had an unwavering goal. In each instance, Shen aimed for empirical success in some form. In the Green Sprouts case, he wished to enable immediate famine relief for local villagers. In the hydraulic project, he hoped to rescue the initiative from complete collapse. In adjusting the Hired Labor Act, he aspired to reduce taxation, and in the salt monopoly he expected to work out a solution for the supply shortage.

The steady emphasis on empirical success led to a consistent epistemological stance, one that kept a distance from grand ideological schemes and focused on concrete details, most of which came from hearing and seeing. Shen constantly revised the system in response to the experiential consequences it incurred. In evaluating outcomes and devising changes, he focused attention on examining particular, local conditions. For example, in replanning the hydraulic financing, he surveyed the conditions in all relevant regions covered by the project before selecting localities for pilot experiments. In adjusting the Hired Labor Act, he inspected the financial status and needs of all relevant social groups in the region before arranging for fairer redistribution of taxation. Many of Shen's decisions were efforts to break up the uniformity of Wang's *fa* and render them more applicable in specific contexts.

Shen dropped out of the system at a point when Wang was no longer willing to accept Shen's piecemeal revisions, especially as the leader-follower power structure between them was unraveling. In chapter 6 I described how Wang regarded their split as a result of Shen's political calculation—a blatant ideological betrayal. From Shen's point of view, his revisions were still intended to complement the system (as had always been the case), and his parting of ways with Wang was a contingent result of power struggles. Notably, Shen never admitted any intention to turn against the New Policies, and he continued to identify with the reformist ideology until the end of his life. This stance is clearly evident in his writings after retirement: in his last book *Brush Talks*, Shen consistently presented New Policies and Wang Anshi in positive terms and earnestly wished for an amicable relationship with Wang.[48]

Despite Shen's unwillingness to express his differences with Wang in political terms, he was well aware of the epistemological schism and quietly stood by his choice. Indeed, Shen was not a system builder. His withdrawal from Wang's system implied a deeper, tacit resistance to system building. He announced no systematic scheme known to his peers. While he worked extensively on all fronts of the reform, he made no attempt to bring unity out of his wide range of experience. Quietly and resolutely, he turned away from making his own total view. In the extensive body of historical sources on his sayings and doings during the New Polices, rarely did Shen feature in a discussion on unity or any grand principles generated thereof. His accounts of projects and his rationales were filled with information on concrete conditions and procedures. Judging from the extant sources, Shen made no systematic effort to interpret the words of ancient sages, including his favorite, Mengzi. In making and rationalizing his decisions, he showed less interest in relating to epistemic authorities and in repeating received testimonies.

The uneven distribution of the current chapter is an honest representation of Shen Gua's situation: set against an extraordinarily complex mainstream, he remained in discursive silence. The imbalance is not entirely surprising. In an age of systems, having no system meant having no proper intellectual voice. After dropping out of Wang's system, where could Shen go? For a while he was perhaps too exhausted to think further, for he was still struggling for survival in the wake of political strife. It took him another decade to present his final answer in words.

# CHAPTER 8

# *Brushtalking the World (1085–1095)*

The last ten years of Shen Gua's life yielded sweet fruits of emancipation. This decade first witnessed the relaxation of his political shackles, followed by the liberation of his intellectual voice. During this period Shen enjoyed the life of a recluse and completed his most important work, *Brush Talks from Dream Brook*, released as his last intellectual statement.

This chapter begins with an analysis of Shen's eremitic life with specific attention to the new cultural meanings he sought in retirement. Against this background, the following section introduces *Brush Talks* as Shen's last vow, which he crafted with devotion and rigor. The third section analyzes the textual designs of the book, and the last part further clarifies the meanings of these designs by distinguishing *Brush Talks* from a conventional Song encyclopedia.

## *Return to Dream Brook*

Shen's final freedom emanated slowly from the unraveling of his confinement. Since losing his last battle on the northwestern border in 1082, he had been held in solitary custody in a Buddhist monastery called Fayun Temple (*Fayun si* 法雲寺) in Suizhou. For three years he did not see a single familiar face, not even his own family.[1] His detainment eventually ended after Emperor Shenzong's death in 1085. The new regent, Empress

Xuanren (1032–1093), released Shen and transferred him to Xiuzhou, a prefecture in the lower Yangzi Delta.

Thrilled to return to his home region, Shen found the familiar views intoxicating. He rejoiced in the smallest pleasures encountered on the road, writing stanzas to exult in mischievous flowers brushing across his face.[2] He responded to invitations from local authorities to visit new schools in the region and wrote in grandiose prose about the meaning of education.[3] His world suddenly softened and expanded into bright vistas.

The blissful times in Xiuzhou afforded Shen time to complete a project he had started a decade earlier and pursued only intermittently while floundering in political quicksand: a compilation of maps for the grand Song empire.[4] Titled *Atlas of Prefectures and Counties All under Heaven* (*Tianxia zhouxian tu* 天下州縣圖), the atlas consisted of twenty maps: one big map of the state, one small map of the state, and eighteen maps of the circuits.[5] Shen applied a number of innovative techniques. Instead of measuring distances along established routes, he provided linear distances, each between two geographical points measured as the crow flied.[6] He also used the so-called six principles of mapping (*zhitu liuti* 制圖六體), a system requiring the measurement of six key variables.[7] In addition, he used twenty-four directions instead of the conventional four or eight to enhance the accuracy of the maps. Pleased with the final product, the empress granted Shen further freedom of travel and restored his status as a civil servant in 1088.[8]

Shen celebrated his regained freedom with another move. He and his family relocated to Runzhou (close to modern Zhenjiang in Jiangsu Province), where he had purchased an estate early in 1077 but had never had a chance to inhabit it. When he first visited the property in 1086, Shen was amazed to find it looking exactly like a landscape he saw in a dream in his thirties: below a hill a creek frolicked, and above it, flowers extended into a coat of iridescent silk.[9] The estate was quiet and secluded, located in the wilderness outside of Runzhou. In 1089, as soon as Shen obtained the empress's permission to choose a residence at his own liberty, he moved his whole family there and happily settled. In the last breath of life, he became a free man again.

Although the estate may have been modest, Shen adored it to the utmost.[10] He installed elaborate landscaping plans and adorned each component of the garden with a poetic name. He named the creek running through the estate "Dream Brook," commemorating its presence in his youthful dream. The hill above the brook he called "Mound of Hundred Blooms" (*Baihua dui* 百花堆). On the top of the mound, a hut named "Hall of an Elder at Brookside" (*An lao tang* 岸老堂) overlooked the landscape.[11] Also on the hill, the "Hall of Whistling Wind" (*Xiaoxiao tang* 蕭蕭堂) stood amid a swath of bamboo.[12] On the estate one might hear nothing other than a deer softly treading through ancient woods. Some visitors found the place too quiet and forlorn and turned away with a frown.[13] Shen, however, was delighted in the solitude, as he found pleasant company in the natural surroundings.[14]

In Shen's own words, his settlement by Dream Brook was a "return" (*gui* 歸) he had been longing for. Given the proximity between Runzhou and his birthplace, Shen's move was a homecoming of sorts. More importantly, to "return" was a recluse's ideal, a new intellectual persuasion embraced by a man freed from the leviathan's grip.[15] While he had spent the majority of his life striving to "advance" (*jin* 進) his career, the idea to "retreat" (*tui*) and eventually "return" had remained constantly in Shen's rhetoric as well as in his mind. The mention of retirement from public life was a necessary moral claim of detachment, and the thought of realizing it must have been therapeutic (to say the least) to one who had endured the scathing politics of the eleventh century. Throughout his years in service Shen had written a number of verses to express his willingness to "return." He titled a poem "Plans for Return" ("Gui ji" 歸計) and yet ended by lamenting that "it seems that the plans for return are delayed again."[16] In another poem he voiced the need to "long for a return" (*si gui* 思歸) because "the grey hair is getting numerous and the body aging, and yet the home mountains are only accessible in pictures."[17]

Shen's vision of a return involved three indispensable elements: the solace of religion, the tradition of a mountain retreat, and the creativity of active leisure. Interestingly, an essay he wrote early in his life presented these very elements, attesting to his long-term contemplation of what he valued. While visiting his elder brother, Shen Pi, in Ningguo (where Pi served as registrar, see chapter 1) and sojourning with him to prepare for the examinations, Shen came to a mountain on which a Buddhist monastery was perched high. Enchanted by the view, he penned the following thoughts:

> For people these days, it has to come to [difficult situations] such that they encounter absurdities and conflicts, or they find their own talents unfit for

the times and yet have no undertakings such as those of the fields, the woods, or crafts to return to, would they slow down and recede. They remain free and proud among woods and rivers, leisurely observing and contemplating on the ancient worthies for the rest of their lives. I have not yet become so resolute in making my mind to advance or to retreat, but I will take the opportunities to render my self useless and lodge on external things, to step high and look afar, to dash in the clear waves of meandering springs, and to take a shade under the exuberant shadows of opulent trees.

今之人必至於乖謬齟齬，材智不合於時，去無田疇山林百工之事以歸
其身，而後逶遲偃蹇，肆傲於山林水石之間，悠然暇觀，思古人而終身
焉。雖然於進退之決，予未能如彼其果也，要無所用其身而寓之外物，
登高而望遠，激流泉之清波，翳茂樹之繁蔭，則予將有遇焉。[18]

Many years later, after personally experiencing all the "absurdities and conflicts," Shen was perhaps amazed at how well he had accomplished his young dream of return at such a place as the Dream Brook Garden. First, like the image of the monastery sitting high on the cliff, the Buddhist religion hung a beacon that became increasingly luminous in Shen's final stage of life. Shen had always been interested in Buddhism. In 1080, while in exile in Xuanzhou, he wrote an inscription by invitation for a Buddhist monastery called Xingguo Temple as far away as the Yunzhou prefecture (in modern Jiangxi). In an apparent testimonial to Shen's reputation as an ardent supporter of Buddhism, a monk traveled all the way on foot to seek his writings.[19] Shen wrote a number of verses to commemorate his visits to Buddhist establishments, and in *Brush Talks* he discussed objects and practices associated with Buddhism a number of times.[20] His attachment to the religion visibly strengthened after 1077, the beginning of his political troubles, and lasted well into the end of his life.

In his final years, Shen expressed his appreciation of Buddhism in ways he had not had the chance or the gumption to use in the political world. He repeatedly referred to Buddhism as the *dao* and to the Buddha as a sage. In the inscription he prepared for the Xingguo Temple, he spoke of the meaning for a literatus to follow the Buddhist *dao*:

Not only does the [Buddhist] *dao* move people, but also, many of those who learn about the dharma are frequently able to detach themselves from the concerns of gains, losses, and advantages.

不獨其道有以動人，而學其法者，多能自處於得喪勢利之外。[21]

In an inscription written for a monastery in Xuanzhou, Shen opened
the narrative with the celebration of the Buddha's rise. He addressed the
Buddha as "a great sage" (*da shengren* 大聖人) and quoted his sagacious
teaching, that "it is all illusory (*wang* 妄); the idea that it is illusory is
also illusory."[22] He proceeded to expound on the Buddha's *dao*:

> The *dao* is nondual;[23] being nondual is being nonsingular. Precisely because
> of [the truth of] nonduality and nonsingularity, names have nothing to
> attach to, not to mention words.[24]

While in confinement at the Fayun Monastery in Suizhou, Shen was invited to write an inscription for a new bell installed at the temple. He
agreed and posed a laudatory rhetorical question, "doesn't one know that
what the heavenly bell embodies (*yu* 寓) is the *dao*?"[25]

Following the Buddhist teaching, Shen lamented the suffering in the
world, because "the sentient beings have lost contact with their essences
and chase after fluctuating consciousness."[26] Fortunately, the Buddha,
"the sage, using his amazing power of knowledge," undid all woeful
attachments and brightened the path for the suffering.[27] As he finally
settled in the Dream Brook Garden, Shen described Chan Buddhism,
in addition to his zither, chess, ink, vermilion (pigment for painting),
tea, singing, chatting, and wine, as his most valued "nine guests" (*jiu
ke* 九客).[28]

The second element of a return—the serenity of a mountain life—
also naturally materialized by Dream Brook. With its characteristic re-
clusiveness, the estate afforded Shen a utopian realm in which he could
avoid social woes and fulfill the ideal of a recluse in a "free and proud"
spirit. In his fantasy, he imagined socializing with hermits in the past,
specifically with three of his favorite ancient poets: Tao Yuanming, Bo
Juyi 白居易 (772–846), and Li Yue 李約 (ca. 788–806).[29] Tao was in-
arguably the most famous recluse up to Shen's times. By taking up the
practice of a recluse, Shen found a new, respectable way of life after his
reluctant detachment from public service and added a fresh facet to the
literati identity he had spent a lifetime molding.

In addition to fulfilling the cultural ideal, Shen experienced new revelations regarding humans and nature while residing in the wilderness.
In the essay he wrote for the Hall of Whistling Wind, he ruminated on

the relationship between "humans" (*ren* 人) and "things," with the latter designating nonhuman living creatures. He opened the passage by describing a peaceful scene in which he played the zither in the garden, accompanied by birds chirping above and fish gliding beneath. But when he moved toward them, the birds would dash in flight and the fish dart away. Amused, Shen realized that these animals mistook him for part of the vegetation until he moved. He turned to think about the historical relationship between humans and nature, lamenting that "things (other living creatures) vis-à-vis humans, they have been like Zou and Chu (two states at war) for long."[30] At the beginning of human history, beasts and birds inhabited all parts of the Earth, while people lived in caves and "depended on [resources] spared by birds and beasts."[31] Later, humans "chased away tigers and rhinoceroses, dispelled dragons and snakes," "removed or cultivated the forests as they saw fit," placed "capture nets in hills and extensive fishing nets in rivers and lakes," and "exhausted their ideas to trap and fight."[32] Obviously, a schism had opened between what humans desired and what animals needed. In Shen's words, "what humans abandon is what fish and birds seek as well as what fish and birds enjoy."[33] He suggested alleviating the confrontation by regulating human aggression, as humans would be able to maintain "a deep (presumably stable and sustainable) existence if they rest where things (other creatures) do not seek."[34]

The third element, the "undertakings of the fields, the woods, and crafts," constituted Shen's major activity while he was in the mood of retreat. He conducted a variety of hands-on creative activities and turned his thoughts about them into writing. He wrote a book titled *Records of Forgetting and Recollecting* (*Wanghuai lu* 忘懷錄), which consisted of a mélange of tips for living a mountain life.[35] The topics were diverse, including the furniture and equipment he used for leisure (such as a comfortable recliner), gardening, cooking recipes, and animals he observed in the wild.[36] He gathered what he knew about tea and compiled a treatise called *On Tea* (*Cha lun* 茶論).[37] In addition, Shen found his long-term interest in medicine still vibrantly alive, so he compiled a ten-chapter *Efficacious Prescriptions* (*Liangfang* 良方), which contained medical recipes he had personally found to be effective.[38]

Shen's return to Dream Brook would have been perfect except for the misfortune of family trouble. Although he had maintained a congenial

relationship with his father-in-law Zhang Chu for decades, Shen failed
to get along with his second wife, Née Zhang, and the tension between
them seemed to escalate in his later years. In one altercation, she ripped
off his beard, taking bits of flesh with it. Née Zhang's violence was a pain
Shen tried to alleviate through meditation in the woods. A few years
into their life on the new estate, Zhang died suddenly. Shen was unable
to recover from the trauma she inflicted, and he sank even deeper into
mental instability.[39] After an unsuccessful attempt at suicide and a few
more years of struggle, he died quietly by Dream Brook at the age of
sixty-five.

## *His Last Vow*

Although Shen was clearly appreciative of the opportunity to return to
religion, nature, and handicrafts after surviving political turmoil, he
did not surrender all of his ambitions. During his hermit's repose, he
managed to recuperate from exhaustion and quietly directed some ener-
gies into composing his most famous work, *Brush Talks from Dream
Brook*. This book constituted his last intellectual vow, a delayed answer
to previous ambiguities and conflicts in his career, and most importantly, a textual feat to make sense of his doing and thinking vis-à-vis
systems in his times. The vow remained tempered in tone, as it grew out
of a peaceful, healing state of mind, but it was not to be underestimated
for rigor and strength, which Shen endeavored to demonstrate with carefully curated content.

Shen prefaced the text with a terse and modest note. He first presented it as a product of solitude by titling it "Brush Talks," for in the
wilderness by Dream Brook no one was around to chime in with his
thought, and writing felt like a silent dialogue with his pen and inkstone.[40]
The content of the text, as he said, consisted of recollections of previous
conversations with guests. Shen cautiously toned down its political connotations, identifying sensitive issues he wished to avoid:

[I] have not presumed to make private records of sagely (imperial) decisions, state policies, or any matters close to the imperial court and state

departments. As for things regarding the praise and condemnation of fellow literati today, [I completely avoid them], as not only do I wish to avoid speaking ill of people, but I also have no desire to mention the good [in them].

聖謨國政，及事近宮省，皆不敢私紀。至於繫當日士大夫毀譽者，雖善亦不欲書，非止不言人惡而已。[41]

In so claiming, Shen denied the book's connection with contemporaneous politics. Neither official activities concerning policy making nor moral judgments insinuating factional confrontation were his concern any longer.

He described the content of the text as follows:

What I have recorded is nothing but [things] in the mountains and among woods, as well as some candid casual talk, none of which bears any relevance to others' gain and loss. [This book] covers everything including even street talk. [I] have also included hearsay, some of which may be partial or erroneous.

所錄唯山間木蔭，率意談噱，不繫人之利害者；下至閭巷之言，靡所不有。亦有得於傳聞者，其間不能無缺謬。[42]

He ended the preface with a characteristically humble conclusion:

[The content of this text] would be too shabby to serve as my "words." But it is acceptable since I do not harbor any intention of making "words."

以之為言，則甚卑，以予為無意於言，可也。[43]

"Words," or, more accurately, "memorable words," referred to the discursive mark every learned man since antiquity aspired to leave in history. *Yan* alluded to *li yan* 立言, one of the "three kinds of cultural immortality" (*san buxiu* 三不朽) in the classical discourse.[44] Literally "establishing words," the phrase meant that one could leave a lasting legacy by propounding ideas (often in textual forms). Shen nevertheless vocally disavowed the ambition to present *Brush Talks* in such a fashion.

Shen's humble prologue should be taken with a grain of salt. In a number of ways, *Brush Talks* easily protruded through the self-effacing facade to reveal its true vigor. First, the text included information far ex-ceeding "things in the mountains and among the woods," a topic he had already engaged in the *Records of Forgetting and Recollecting*. Although *Brush Talks* also featured content connected with mountain life, such as gardening tips and observations of wild animals, it did not recycle any information from the *Records of Forgetting and Recollecting*. Given that Shen composed the two texts during the same years, it is hard to deny his intention to distinguish them as separate projects. *Brush Talks* was not a more extensive sequel in which he continued to ruminate on his life in seclusion.

Second, Shen's claim to steer clear of contemporary politics was gen-uine, yet it should not be taken as a total renouncement. Instead, a good part of *Brush Talks* consisted of his recollections of his service in the officialdom, including a lot of information on fiscal, legal, and military policies; daily transactions in the bureaucratic machinery; and the ac-tivities of fellow literati in a variety of official capacities.[45] What Shen wished to evade—as he successfully managed to do—was the factional, controversial aspects of statecraft. He rendered few judgments that might have potentially offended authorities or induced factional contro-versies. He rarely spoke of the monarchs who reigned during his years of active service. He mentioned Wang Anshi a number of times, mostly addressing Wang's nonpolitical engagements (such as poetry and per-sonal life) and casting him in an absolutely positive light.[46] When he reminisced about the New Policies, Shen recalled the many technical details of his own participation, unavoidably in the spirit of an advocate, yet he never articulated his penchant in forthright terms. Moreover, he remained practically taciturn on anything or anyone with strong antire-formist ties. *Brush Talks* was Shen's best effort to keep factional politics at arm's length.

In addition, "casual talk" cannot define *Brush Talks*. Shen's presen-tation of many topics (such as the fiscal policies) featured very technical details that were by no means the result of casual chats or random recol-lections. Furthermore, the possibly partial or erroneous "hearsay" Shen apologized for occupied a small portion of the text, as the majority of the content derived from his personal experience. His specific mention of this source of knowledge revealed his reluctant endorsement, which spoke to a caution against casualness rather than a confirmation of it.

The relationship between *Brush Talks* and Shen's previous written work revealed his serious devotion to his last work: the book covered the entire spectrum of topics in Shen's earlier writings, yet showed little overlap in content with them. Before *Brush Talks*, Shen was already a prolific author. He enjoyed a reputation as someone who "liked to write" and boasted a list of works as long as thirty-five titles, mostly single-subject monographs.[47] In his early years, Shen composed a few sets of classical commentaries. During the three decades of his official career he accumulated a stack of bureaucratic writings. Among them, six treatises documented the calendrical reform he conducted, including a seven-chapter record of the calendar itself, a couple of proceedings on the reform procedures, and two separate accounts on key technologies.[48] Shen also composed two accounts of his adventurous mission to the Liao: one narrative, a lengthy memorial on the negotiations; and one cartographical, maps of the Liao accompanied by verbal explanations of local particulars.[49] He wrote about a host of miscellaneous obligations he had shouldered during the reform (1072–1077), ranging from setting stylistic rules for edicts to micromanaging food budgets.[50]

Shen copiously engaged topics outside work, too. He once composed a pamphlet of stories on ghosts, spirits, and other paranormal occurrences and titled it *Records on Clear Nights* (*Qingye lu* 清夜錄).[51] In a long rhymed account he appraised the styles of renowned painters.[52] His lifetime interest in medicine generated a number of pharmacological texts, including the early *Prescriptions of the Numinous Garden* (*Lingyuan fang* 靈苑方) and the later *Efficacious Prescriptions*.[53]

The impressively oeuvre was a legacy Shen wanted to continue and augment by bringing *Brush Talks* to completion. Except for a small number of medical recipes recycled verbatim from *Efficacious Prescriptions*, *Brush Talks* primarily featured new content spanning the breadth of his entire oeuvre. Shen made his pursuit of originality explicit: in a number of places in *Brush Talks* he stated that some items were meant to complement his previous work as new additions and that he would like to avoid repeating the old.[54] In sum, Shen's last book was a new project rather than a nostalgic summation of old work; the attention he invested rendered it anything but a pack of "casual" remarks.

There is, however, a point beyond the self-effacement of Shen's denial of *Brush Talks* as serious "words." Compared to the works of his peers,

*Brush Talks* stood out as a peculiar last vow. Writing a book as one's final statement had become a trend among literati of Shen's generation. Heightened factional struggles rendered political longevity a precarious luxury even among the most powerful, and the end of an official's career often coincided with the silencing of his political voice. Because so many statesmen spent retirement in exile or forced silence, they turned to writing books to preserve their ideas. These last books often embodied, on one hand, deep insecurity in the face of personal and political mortality and, on the other, a burning zeal to proffer one's finest thoughts.

A number of famous literati, such as Wang Anshi and Su Shi, concluded their oeuvres with book-length treatises. Wang produced *Explications of Characters* (*Zi shuo* 字說) four years before his death.[55] Disguised as a lexicon, this book offered his last ideological pronouncement: a view of the world through the lens of Buddhism. Having abandoned his fixation on institutions, Wang proposed to seek "conceptions of sages" from Buddhists instead of from ancient kings. His new vision held that "In the Three Dynasties sages mostly emerged from literati, and after the Han they appeared among Buddhists."[56] During his final years in exile in the southern extreme of the empire, Su Shi completed a commentary on the *Classic of Documents*, a textual feat meant to "illuminate the extinct learning of high antiquity" and advance his interpretations of the sages' conceptions (in many places unsurprisingly at odds with Wang Anshi).[57]

Not coincidentally, in their last works these famous system builders remained fixated on reinforcing systems, albeit in mutated ways. Despite the sharp turn toward Buddhism, Wang stood firm in his belief that the vicissitudes in the world could be organized into a unifying order. Su, the dissident who protested Wang's excessive normativity, decided to leave his own prescriptions in the marginalia of the Classics, combating norms with norms. The spirit of systems resounded in the last books with increasing intensity and prompted further utterance of memorable "words" literati wished to pass along as their legacies.

Against this background, Shen surely had a reason to distinguish *Brush Talks* from the more conventional "words," for he was not a system builder and had no interest in branding himself as one. His disavowal in the preface was more declaration than surrender: *Brush Talks* was certainly not shoddy craftsmanship, and it was unlike any other book-length vow that delineated a system.

## *Textual Designs of* Brush Talks

To discern the intellectual messages embodied in *Brush Talks*, an examination of its thoughtful textual designs serves as an initial step. Three major features—inclusive coverage, item-by-item layout, and thick description—marked Shen's effort to implement his agenda in textual forms.[58]

First, *Brush Talks* covered all kinds of knowledge that came under Shen's command, and he implied no hierarchical distinction between one kind and another. As I introduced previously, *Brush Talks* encompassed all the topics in his earlier writings and yet transcended their respective genres. The last book boasted a new, encyclopedic framework featuring seventeen divisions of content, each of them a taxonomic category.[59] The divisions ran as follows: "Precedents," "Identification and Verification," "Musical Harmonics," "Figure and Number," "Human Affairs," "Policies and Administrative Issues," "Tactics," "Belles Lettres," "Calligraphy and Painting," "Particular Skills," "Implements," "Numinous Marvels," "Strange Occurrences," "Errors [and Corrections]," "Wit and Sarcasm," "Miscellaneous," and "Medicine."[60] Many of these designations are literal and self-evident: "Musical Harmonics," "Calligraphy and Painting," "Belles Lettres," and "Medicine" were precisely sections regarding the named fields of knowledge. "Figure and Number" referred to number-related studies, such as the calendrical and numerological. "Numinous Marvels" and "Strange Occurrences" recorded phenomena associated with paranormal forces. "Precedents," "Identification and Verification," and "Errors [and Corrections]," as I analyze in chapter 9, were sections demonstrating certain methodologies in diverse inquiries.

Under each taxonomic heading, Shen assembled a total of 507 short, self-contained jottings. The sizes of the divisions varied, but not to an extent indicative of an overt preference for one topic over another. *Brush Talks* in its entirety presented a rich, encompassing stream of information evenly punctuated by classificatory efforts (see table 1).

In addition to a broad coverage, the text featured an item-by-item structure.[61] Each item asserted independence and completeness for the content it contained, and no item claimed systematic connection with another. *Brush Talks* was thus an assemblage of "nuggets of knowledge."[62]

This design highlighted the autonomy of each nugget, confirming its singular value and discouraging the formation of larger connections likely to lead to systems.

The classificatory system coupled with the item-by-item arrangement evinced a strong sense of egalitarianism, which downplayed possible social connotations of knowledge in different fields. While belles

Table 1
Structure of the Modern Standard Edition of *Brush Talks*

| | |
|---|---|
| "Precedents" (*gushi* 故事)<br>   Chapter 1, items 1–30<br>   Chapter 2, items 31–41 | "Calligraphy and Painting"<br>(*shuhua* 書畫)<br>   Chapter 17, items 277–97 |
| "Identification and Verification"<br>(*bianzheng* 辯證)<br>   Chapter 3, items 42–70<br>   Chapter 4, items 71–81 | "Particular Skills" (*jiyi* 技藝)<br>   Chapter 18, items 298–318<br><br>"Implements" (*qiyong* 器用)<br>   Chapter 19, items 319–37 |
| "Musical Harmonics" (*yuelü* 樂律)<br>   Chapter 5, items 82–110<br>   Chapter 6, items 111–5 | "Numinous Marvels" (*shenqi* 神奇)<br>   Chapter 20, items 338–56 |
| "Figure and Number" (*xiangshu* 象數)<br>   Chapter 7, items 116–42<br>   Chapter 8, items 143–50 | "Strange Occurrences" (*yishi* 異事)<br>   Chapter 21, items 357–87 |
| "Human Affairs" (*renshi* 人事)<br>   Chapter 9, items 151–82<br>   Chapter 10, items 183–88 | "Errors [and Corrections]"<br>(*miuwu* 謬誤)<br>   Chapter 22, items 388–400 |
| "Policies and Administrative Issues"<br>(*guanzheng* 官政)<br>   Chapter 11, items 189–212<br>   Chapter 12, items 213–23 | "Wit and Sarcasm" (*jixue* 譏謔)<br>   Chapter 23, items 401–19 |
| "Tactics" (*quanzhi* 權智)<br>   Chapter 13, items 224–44 | "Miscellaneous" (*zazhi* 雜志)<br>   Chapter 24, items 420–49<br>   Chapter 25, items 450–79 |
| "Belles Lettres" (*yiwen* 藝文)<br>   Chapter 14, items 245–62<br>   Chapter 15, items 263–73<br>   Chapter 16, items 274–76 | "Medicine" (*yaoyi* 藥議)<br>   Chapter 26, items 480–507 |

SOURCE: Hu Edition of MXBT (see appendix 1)

lettres, bureaucratic precedents, and policy discussions were proper subjects frequently featured in elite writings, Shen juxtaposed medical recipes and fantastic tales with them as well as machinery and toolmaking. Resembling the technical outlook of a reference book, *Brush Talks* listed each category as one equal component of a classificatory scheme without indicating any sociocultural differentiation. The sense of equality further materialized at the level of items: one tip on making a durable archery bow enjoyed the same status as another on how to properly rhyme when writing poetry.[63] In the universe of *Brush Talks*, one bit of particular knowledge was equal with another.

The third important formal feature in *Brush Talks* was the prominence of description. In the majority of items, Shen simply described a particular object or procedure. His richly detailed descriptions often demonstrated a primary interest in the essential qualities and attributes of a "thing." Occasionally he went beyond description and ventured into causation, but he showed a tendency to adhere to local and immediate causes in descriptive terms.[64]

A good example of Shen's thick description is his account on movable-type printing, including information on the basic setting of this technology and hands-on details regarding procedures. Shen wrote from the perspective of a practitioner ready to operate the apparatus. Following his description, the reader learned how to properly place the types so that the surface of the block remained even, how to choose materials so that the types could be conveniently reused, and how to prepare types so that common characters in a Chinese text could be adequately covered.[65] Shen's mastery of detailed description accounted for this item becoming the most frequently cited historical source for China's first use of movable-type printing.

Shen's preoccupation with description became even more evident in "Policies and Administrative Issues." This section takes the reader on a brisk tour through all the specific capacities Shen had served in during his years in office: hydraulic construction (items 207, 210, and 213), legislation and jurisdiction (items 196, 198, 199, 201, 202, 209, and 214), tea business (items 189, 190, 219, 220, and 221), rice trade (items 192 and 222), taxation (items 194, 195, and 215), salt monopoly (items 211 and 212), and military defense (items 191 and 200). Most items focused on describing technical details of these subjects. Items 220 and 221, for instance, introduced aspects of tea monopoly in numerical terms accurate to the decimal point, filling pages with elaborate strings of numbers.[66] Quantitative expression allowed Shen to fulfill his descriptive mission.

In sum, the three major characteristics—inclusiveness, item-by-item style, and detailed description—afforded Shen a structure to contain a rich range of information in an orderly fashion. Organization, nevertheless, was a means rather than an end. A clear understanding of Shen's intellectual agenda requires a comparison between *Brush Talks* and other texts with similar organizational patterns, a task I take up in the next section.

## Not Simply a Collectanea

To readers familiar with the Song textual world, the structural features of *Brush Talks* almost immediately associate it with a popular genre: *leishu* 類書, "collectanea of categorized knowledge" (or, conventionally put, encyclopedias).[67] Shen was certainly aware of contemporaneous encyclopedias. Among the seventeen categories he used, more than a few of them were taxonomic designations commonly seen in Song collectanea. In a *leishu*, the textual organization was similar, too: each classificatory heading grouped an assemblage of items diverse in content.

Broad coverage was another feature that rendered *Brush Talks* comparable to a collectanea. Like a modern encyclopedia, a Song collectanea aimed at exhibiting a comprehensive host of subjects so that readers might easily retrieve basic information on a topic. Take the *Imperial Collectanea of the Taiping [Xingguo] Era* (*Taiping yulan* 太平御覽, hereafter *Taiping Collectanea*), for example. This gigantic compilation was a late tenth-century product, with 55 categories and 5,474 subcategories. Purportedly, it provided an inventory of the known universe: from the firmament to the human world, from the imperial house to the bureaucracy, from the military to rituals, from diseases to practical instruments, from clothing to food, and from animals to plants, among many other things.[68]

To conflate *Brush Talks* with collectanea, however, encourages a gross misunderstanding of Shen's intention. Most obviously, a collectanea was a compilation, whereas *Brush Talks* was an auctorial text. This distinction has a number of important implications. First and foremost, a collectanea was often a collective work and thus lacked a distinctive intellectual voice associated with an individual. The *Taiping Collectanea*, for instance, was commissioned by the Song state and involved a board of editors.[69] Second, a collectanea was by definition a reference book, a repertoire of information awaiting potential future use. The two factors produced the intellectual neutrality typical of an encyclopedia.[70]

As a consequence of the absence of an intellectual voice, the form of presenting information in a collectanea also differed from that of *Brush Talks*. Take the *Taiping Collectanea* for example: it is best described as a storehouse of excerpts, and, drawing on the same metaphor, an assemblage of "nuggets of texts" (instead of "nuggets of knowledge"). The compilers primarily culled short segments of text from certain sources and brought them together under topical headings. While placing them in a sequence, the compilers kept the original appearance of these textual nuggets unmolested, maintaining their many diverse forms. For example, forty-five items appeared under the heading of "fog" (*wu* 霧). Some items provided normative definitions of fog, such as the one from *Approaching Refinement* (*Erya* 爾雅); many others were anecdotes or historical tales involving the occurrence of fog, and a few items were poetic verses featuring fog.[71]

As a storehouse of excerpts, the *Taiping Collectanea* proved useful and convenient to common literati as a typical reference book. Scholars could easily spot helpful references—either a definition, a quotation, or a citable story—while seeking sources for their own writings.[72] As in any modern reference book, a collectanea entry offers basic information and raw material—a departure point for further pursuits.

In contrast with a derivative reference book, *Brush Talks* featured a fiercely original single authorship. Instead of excerpting from existing sources, Shen crafted every item on his own, taking ownership of both the content and the writing. As I have previously discussed, he did not even reuse his own earlier writings. Moreover, the inclusiveness of *Brush Talks* was by no means comprehensiveness for the sake of comprehensiveness.[73] Shen covered all areas he had participated in or witnessed, and only these areas, because he intended to highlight his personal perspective. The breadth of the text matched the author's life experience. For instance, the selection of categories on cultural pursuits in *Brush Talks* immediately spoke to Shen's interests: "Musical Harmonics," "Figure and

Number," "Belles Lettres," "Painting and Calligraphy," "Implements," and "Medicine" were all fields he had engaged in during different stages of his life. The inclusion of these categories no doubt had to do with the general significance these fields held for literati and, more importantly, with the special status they once enjoyed in Shen's own experience and the numerous original insights they had once generated and continued to promise. Thus *Brush Talks* bore only a superficial resemblance to an encyclopedia.

In sum, *Brush Talks* was the last work Shen completed by Dream Brook, where he relished quiet time to recuperate from political traumas and to bring clarity to his thinking. The text was impressively broad in coverage, yet unlike a collectanea, it spoke in a distinctive intellectual voice, which was Shen's last, solemn intellectual vow.

How could an assemblage of miscellaneous knowledge constitute a coherent intellectual message? Indeed, *Brush Talks* lacks a clear significance that modern readers can easily grasp, mainly because Shen dispensed no formal theory or propositional principles. Although the author might have worked hard to build his distinctive voice, how could a reader use *Brush Talks* differently from a conventional reference book? In the next chapter I respond to this question with an analysis of the nonsystem Shen was presenting in this book.

# CHAPTER 9

# Building a Nonsystem

In *Brush Talks*, Shen Gua searched for a conscious intellectual voice, which in many ways was meant to validate his epistemological choice vis-à-vis systems in his times. One question he had to confront when pursuing such a goal was whether challenging a system would necessarily lead to chaos. If not, what kind of coherence could possibly arise from old systems crumbling? Shen responded with a nonsystem.

This chapter starts by revisiting knowing from hearing and seeing, an important category that registered Shen's epistemological predilections and served as the departure point of his nonsystem. I follow by introducing the genre *biji* 筆記 (notebook), a key cultural resource Shen appropriated for building his nonsystem. The next two sections analyze the meaning of reliability and how methods for reliability constituted the core of the nonsystem. I go on to compare the nonsystem to systems and survey the epistemic community the nonsystem was supposed to engage. The chapter concludes by summarizing the philosophical characteristics of the nonsystem under the rubric of empiricism.

## Brush Talks *and "Knowing from Hearing and Seeing"*

To examine how *Brush Talks* responded to the dominant discourse on systems, we need to probe deeper into background issues Shen intended to address, starting with "knowing from hearing and seeing." Although Shen himself did not invoke hearing and seeing to characterize *Brush Talks*, a historical discussion of the text can hardly avoid this concept. For one thing, since Song times, readers and handlers attached that label to *Brush Talks*.[1] For another, the text indeed included rich content on Shen's sensory encounters with the world, the look of a meteor, the mechanism of erosion in mountain formation, and the color of petroleum, all resulted from effective observations through hearing and seeing.

Shen's disuse of the phrase may or may not indicate deliberate consideration, but one complication is worth mentioning when we examine "hearing and seeing" as a text label. As I have shown in chapters 2 and 4, knowing from hearing and seeing was a philosophical argument that designated treatment of a "thing" as an individuated entity. In contrast to modeling, it focused on the sensory qualities of a "thing" and eluded its placement in deep orders, thus it was often slighted as the inferior type of knowing.

When used to label a textual practice, "hearing and seeing" became a broader term. It acquired a new orientation with which one could—should one wish—overcome the negative association the term bore as the inferior cousin to modeling. In this context, the phrase highlighted the presence of an individual knowledge seeker with epistemic autonomy. "Seeing" and "hearing" referred to acts of observation with an emphasis on the observer's personal presence rather than his visual and auditory capacities (which might or might not be involved). For instance, when a person claimed to learn something from hearing, he could mean that he obtained the information by hearing from a friend, stressing that he learned it from a social occasion—a conversation with this friend—in which he personally participated. His hearing modality was certainly involved in the exchange, but the central point was not that auditory perception served as a source of knowledge. Learning something from one's own hearing and seeing could mean that he knew it from reading on his own. In these scenarios—social transaction and book reading—the knowledge seeker exercised certain critical skills to obtain knowledge. These critical skills were sometimes related to hearing and seeing in the narrow sense and sometimes not necessarily so.

The wider horizon of hearing and seeing resulted from the textual practice by Shen and others who advocated a broad assertion of epistemic autonomy by exercising one's perception and other critical skills. The two-tier epistemology remained a relevant issue, but whether it assumed central significance depended on an individual thinker's epistemological stance. A person could solely indulge in sensory inquiries, should his personal interest direct him so, or he might as well venture into discussions of deep orders, using critical skills befitting the broader hearing and seeing. In *Brush Talks* Shen coordinated his extensive observations of the world with contemplation on deep orders he found relevant, as he had consistently done in his epistemic praxis. He did not, and certainly was not obligated to, confine himself to the philosophical denigration of sensory knowing. In the following discussion, I explicate the new breadth of hearing and seeing Shen was committed to under the rubric of reliability.

## Brush Talks *as "Biji"*

In relation to knowing from hearing and seeing, Shen had one convenient cultural resource ready for use. In the literary praxis of Song literati, one specific kind of text, *biji*, was a particularly flexible genre that accommodated diverse ways of presenting information. Literally "notebooks" or "written notes," *biji* was a highly inclusive label representing a large body of texts produced over a long period of time: roughly from the fifth century to the end of the nineteenth.[2] Due to its wide-ranging content and lack of stable structural traits, it has remained a nebulous category that approximated a genre.[3]

A number of common features can be identified across the majority of notebooks produced in the Song. First, they often adopted an item-by-item style and stipulated no further rules for the size, structure, or mutual relations of these items, for which the "nuggets of knowledge" format in *Brush Talks* provides a good example. Second, the possible subjects of a *biji* were highly diverse.[4] Some were intended to supplement official histories with personal records, and others cited paranormal occurrences and entertaining gossip; in addition, many *biji* concerned one or multiple subjects of specific knowledge, such as a cultural hobby or a material object.[5]

Third, as a deviance from mainstream genres (such as poetry, official essays, and classical commentaries), a *biji* could invoke either positive or negative associations, depending on the context. Northern Song officials often chose to compose a notebook after retirement; the undertaking partly served to entertain oneself and implied a diversion from serious commitments. For this reason, some *biji* authors, including Shen, felt the need to apologize for the petty nature of their work despite any serious intellectual purposes they might have harbored.[6] In some other contexts, the deviance of notebook literature bore a positive implication due to its capacity to afford an alternative voice to official records (especially histories). I elaborate this point next.

Among *biji* authors, Shen was able to find some kindred spirits partial to "knowing from hearing and seeing" for a variety of reasons. The flexibility in content and form sanctioned by the *biji* genre accommodated the impulses of some Song literati to see, hear, and write on their own with ample liberty. With its "deviant" implication, *biji* provided a convenient venue, sometimes even an empowering setting, for literati to assert their epistemic autonomy beyond the confines of official discourse. Thus, to jot about one's hearing and seeing and compose notebooks became compatible choices egging each other on.[7]

A more accurate understanding of Shen's participation in notebook production requires a closer look at classifications within the genre. Song *biji* authors started to develop certain assessment criteria, one of which specifically divided notebook literature into two types, one less credible and the other resembling serious scholarship. Zhang Bangji 張邦基 (fl. 1130), for instance, articulated this division in terms of subject matter. In the postscript of *Random Records from the Mozhuang Estate* (*Mozhuang manlu* 墨莊漫錄), he called the first kind "clairvoyant and mysterious, reckless and strange" and named examples such as *Records of Murkiness and Oddities* (*Xuan guai lu* 玄怪錄), a Tang collection of fantastic tales.[8] In contrast, the second type included those that "had been extensively referenced by chroniclers of later ages," and they were "worth reading and transmitting."[9] *Brush Talks*, according to Zhang, belonged to the second group.[10]

In Zhang's view, the first type of *biji* deserved less credibility because it covered topics inappropriate for official historiography. Notwithstanding the inaccuracy of this generalization, the binary categorization of the notebook literature already enjoyed considerable currency prior to Zhang's times.[11] In the eleventh century, famous *biji* writers such as Fan Zhen 范鎮 (1007–1088) and Ouyang Xiu expressed similar opinions when discussing their own work.[12] Their contemporary Shen should also be aware of the evaluative framework as he joined the notebook community. Shen would be pleased with Zhang's classification of *Brush Talks* as the more credible type of *biji*, as he looked to the genre for its flexibility to accommodate new intellectual leads rather than relaxation of scholarly rigor.

## Reliability

While *biji* provided a suitable platform for hearing, seeing, and the exercise of critical skills, a key issue Shen had to address concerned the whole point of these engagements. Why was a commitment to hearing, seeing, and critical thinking a commendable merit? In *Brush Talks* Shen provided his answer: reliability. In contrast to building a system centered on unity, this text demonstrated that one could deal with "things" in a variety of nonsystematic ways and achieve reliable results. *Brush Talks* was a book of reliability in two senses: it presented concrete cases of reliable knowledge, and it demonstrated the means to achieve such reliability.

Shen's fierce pursuit of originality in *Brush Talks* provides the first evidence of his emphasis on reliability. Throughout the text, he posed himself as a responsible knower (or a dutiful observer of other responsible knowers) whose beliefs constantly merited some kind of positive evaluation. In his words, these beliefs were "trustworthy" (*xin* 信), "verified" (*yan* 驗), and "necessarily so" (*biran* 必然), among others.[13] In modern parlance, we might say well-grounded, provable, reasonable, and so on. One epistemological concept that encompasses this family of characterizations is reliability. Reliable knowledge is credible, grounded knowledge, and reliability indicates truth-conduciveness without rigorously subscribing to certainty or logical infallibility. The term accurately captures Shen's interest in good knowledge and disinterest in total or absolute truth.[14]

Reliability in *Brush Talks* was best characterized as a practical commitment rather than a propositional theme. As shown already, it had no

exclusive bond with any particular word, and featured no unified terminology; instead, it brought together a range of concepts exercising the
same function of qualitative distinction. This distinction set a responsible knower and his reliable beliefs apart from many possible epistemic
vices. Shen's intention was to demonstrate a range of practical cases ex-
ecuted by such a knower, rather than posit a theory of reliability in propositional terms.

In fact, in most items Shen used action rather than language to
denote reliability, and in these cases he demonstrated his commitment
by relying on descriptive rather than normative textualization. The following item provides a typical example. In this entry Shen was contemplating the apparent movements of the five planets and their methods
of calculation. He observed that calendrical experts in the past most
frequently made computational errors in sections where the planets
were in retrograde or came to a standstill. He announced this observation as follows:

> I have examined ephemerides from the past through the present, and [I
> have found that] most (calculation) errors occur when the planets are in
> retrograde or in a standstill.
>
> 予嘗考古今曆法五星行度，唯留逆之際最多差。[15]

Shen highlighted his presence, described his working method, and
presented his belief in a matter-of-fact and assertive tone. A comprehensive survey of previous calendars was a sound course of action that
likely produced a reliable belief regarding past errors, so Shen presented this belief with confidence. Such was his way of pronouncing
reliability.

The most convincing evidence of his focus on reliability lay in Shen's
extensive efforts to ground the seemingly most "unreliable" knowledge—
things perceivable and yet of a fantastic nature. In *Brush Talks* he devoted
two sections to the fantastic and mysterious, respectively titled "Numinous Marvels" and "Strange Occurrences." Most accounts in these categories are plainly hard to believe: a Buddha-shaped turnip, dragon eggs,
a clairvoyant who foresaw his date of death, wood patterns resembling
writing, and so on.

Shen's pursuit of reliable knowledge in the fantastic realm is notably
singular. Jotting about clairvoyant phenomena was often associated with

the less credible type of notebook literature; thus, most authors who participated in the subgenre were unlikely to consider reliability as a high priority (at least not in the eyes of their readers). Against this background, Shen's approach to the fantastic was counterintuitive: he found ways to demonstrate the reliability of ostensibly unreliable knowledge and presented his records as confirmatory notes rather than short fictions.[16]

The means Shen found most apt to ground reliability in the fantastic world was to offer personal witness. In every item, he explicitly demonstrated that a trustworthy source—himself or a knower he deemed responsible—had witnessed the uncanny. For instance, he recounted that while working in Runzhou he had personally seen the Buddha-shaped turnip at the residence of a colleague named Li Binke 李賓客 (fl. eleventh century), offering exact details regarding who and where.[17] The enormous eggs people regarded as those of dragons, as uncanny as it sounded, had been "repeatedly seen" (*lü jian zhi* 屢見之) by Shen.[18] The clairvoyant who successfully predicted his day of demise was Shen's distant relative, and Shen had been "made privy to the details" (*zhi zhi shen xiang* 知之甚詳) of his deeds.[19] In these cases, he consistently worked to show his readers that the uncanny things he recorded existed in perceptual experience and thus merited serious attention. In seeking reliable knowledge from the least plausible domain, Shen convincingly demonstrated the extent of his commitment to reliability.

Another clue illuminating Shen's attachment to reliability was his sensitivity to ungrounded knowledge, vividly manifested in his habitual inclination to correct others' errors. Both "Identification and Verification" and "Errors [and Corrections]" grew out of his reaction to the unreliable beliefs he had observed in other knowers. For example, he criticized the famous Tang text *Miscellaneous Morsels from Youyang* (*Youyang zazu* 酉陽雜俎) for being "largely absurd," with the information on flora and fauna in its record specifically erroneous.[20] The reason Shen gave was straightforward: the author "simply documented things in foreign lands without having any grounds."[21] The blatant dismissal of reliability was what provoked him to speak against this text.

In another item Shen criticized the famous classicist Zheng Xuan 鄭玄 (127–200) for identifying an object based on incomplete evidence and guesswork. One passage in the *Book of Documents* narrated that when King Wen of Zhou (d. mid-eleventh century BCE) was imprisoned by his rival King Zhou of Shang (d. ca. 1046 BCE), King Wen's devoted follower tried to bribe King Zhou with "a big shell resembling *che-qu* 車渠" for the release of his master. Zheng Xuan interpreted *qu* as "net" and *che* as "chariot," so the bribery was understood to be a shell as large as a protective net cast over a chariot. Shen called this absurd, because he knew of a kind of shell whose name was precisely *chequ*, normally seen in the South Sea. Thus, the gift should be a South Sea shell of a normal size, which had nothing to do with a chariot cover. To acquire and incorporate information concerning this specific type of shell was indispensable in forming a reliable judgment, hence Shen's critique of Zheng.

While Shen's pursuit of reliability was mostly evident in cases where he addressed serious knowledge, it also appeared in sections where he seemed to write more leisurely, such as "Human Affairs." Modern scholars tend to address items in this category as gossip and anecdote, and I have no doubt that Shen put the material together with a more entertaining mood. There is no sign, however, that his commitment to the reliable quality of his accounts bowed to the desire to entertain. In fact, the angle of reliability may help us decipher some elusive items in this section. For instance, the shortest item in "Human Affairs" (as well as in the entire *Brush Talks*) contained only thirteen Chinese characters and had a central message hard to pinpoint. The laconic account stated:

Wang Ziye never ate animal flesh for his whole lifetime, and he was fine with that.

王子野生平不茹葷腥，居之甚安。[22]

Wang Ziye (Wang Zhi 王質, 1001–1045) was a descendant of a wealthy family. History depicted him as a virtuous man who was "self-disciplined and active in charity," in contrast to his brothers, who led sumptuous lives typical of rich men.[23] Shen's account, though somewhat chatty in tone, was certainly not "without productive value or factual reliability."[24] Indeed, it was meant as a piece of evidence that corroborated Wang's historical reputation. In all fairness, the very brief account was more useful (in terms of reliability) than witty.

Thus, *Brush Talks* was a book of reliable knowledge. How did Shen achieve this reliability? The nonsystem, which consisted of a variety of his methods, provided the answer.

## *The Nonsystem and Reliability*

*Brush Talks* presented knowledge that had passed Shen's tests of reliability, and the methods he employed to achieve reliability constituted the nonsystem he sought to build. Notably, Shen presented this nonsystem in descriptive rather than normative terms. *Brush Talks* did not advance a formal theory of knowledge; instead, it provided a descriptive demonstration of how reliable knowledge was achieved.[25]

Shen exhibited an attention to methods aiming for reliability when he stipulated the structure of *Brush Talks*: whereas some sections such as "Musical Harmonics" and "Medicine" were plain taxonomic categories each representing one class of "things," others, such as "Precedents," "Identification and Verification," and "Errors [and Corrections]," represented methods and processes widely applicable to more than one field of inquiry. These section titles unambiguously distinguished the classificatory scheme in *Brush Talks* from conventional ones in encyclopedias, and they signaled Shen's readiness to build epistemic norms. "Precedents" and "Identification and Verification" were the opening sections of *Brush Talks*, an indication of their importance.

In "Precedents," Shen gathered accounts describing old objects and practices associated with the imperial bureaucracy; more importantly, on the basis of these descriptions he demonstrated a methodological consistency, that drawing on reliable knowledge from the past was an intellectually virtuous process to ground new knowledge. "Precedents" contained a range of "things" concerning all aspects of bureaucratic life in the Tang and early Song, from erasing methods used by book compilers at the Imperial Libraries to grandiose state rituals. The selection of items was the product of two methodological concerns. For one thing, Shen included reliable precedents (in his own judgment) meant for contemporary reference; for another thing, he showcased instances in which valid precedents grounded new knowledge in his own times.

For example, Shen described how the historical practice of civil examinations in the Tang validated the current Song system. In his day, when candidates for the *jinshi* degree took their tests, they sat in a chamber adorned with fine furniture and were served tea; candidates for the "classical erudition" (*mingjing* 明經) degree, on the other hand, sat in a barren room sealed with plain drapery. Many people, including Ouyang Xiu, assumed that this was due to the greater cultural-political privilege *jinshi* degree holders enjoyed in Song society. Shen discovered from Tang precedents that such was not the case. The differentiation of treatment had started in the Tang times, when the classical erudition degree was much more favorably considered than the *jinshi* degree in the cultural world. *Jinshi* candidates were served tea because the skill tested in this examination did not preclude human contact (tea servers, for instance); classical erudition candidates were never provided beverages— many of them resorted to drinking from the inkpots when thirsty— because they were supposed to recite classical passages verbatim on the test and thus have no interactions with outsiders. By uncovering the Tang practice, Shen was able to confirm that the current practice continued well-reasoned test protocols and did not owe to social biases, and hence were a right course of action.

As with "Precedents," "Identification and Verification" is a section where all items converged to demonstrate a unitary epistemic virtue that entailed reliability: to correctly match a "thing" with its name. Shen once criticized his peers for "loving to cite words of the ancients and yet did not fathom their meanings," and he strove to bring attention to the issue of lexical accuracy.[26] In Shen's reckoning, an accurate match between an object/process and a linguistic convention premised the production of reliable knowledge regarding "things." An effort to coordinate the two was precisely his central pursuit in this section.

For example, in one item Shen proposed to redefine a term, *yema* 野馬 by associating it with a different phenomenon.[27] The word came from the famous old text *Zhuangzi* 莊子, and its conventional interpretation as "wild horses" had long remained a *faute de mieux*. The original line containing the term is as follows:

The "wild horses," the dust clouds, and the living things blow their breaths among themselves.

野馬也，塵埃也，生物之以息相吹也。[28]

Many exegetes found the presence of real wild horses in this context to be awkward and bizarre. They thus interpreted "wild horses" as a figurative reference to dust clouds—because clouds could sometimes assume the shape of horses—and rendered the first two phrases of the quotation synonymous parallelism.[29]

Shen decided that this term denoted something else, namely, moving mists in the field, hence the first step: "identification" (*bian* 辨). He then "verified" (*zheng* 證) the new explanation with clues from observations: "If viewed from afar, [moving mists] resemble a herd of horses or water waves."[30] In addition to his visual judgment, he drew circumstantial evidence from a Buddhist sutra, which included the line "it is just like wild horses and solar flames (fluctuating sun beams) arising when [the weather] is hot."[31] Correlated with "solar flames" in the context of scorching heat, "wild horses" most likely gave a figurative expression to wavering hot air. Shen thus confirmed the new match by relying on both direct observation and textual testimony.

In another item, Shen exposed the inaccuracy of a concept by showing that the object it designated no longer existed. "Short-back robe" (*duan hou yi* 短後衣) was a popular reference when describing the uniform of a military man in the Song times.[32] Often cited, yet never defined, the word confused Shen because all nonmilitary robes in his times featured a short back. He traced the origin of term to *Zhuangzi* and retrieved its historical meaning. All Warring States aristocrats dressed in robes with a long back, and occasionally someone who wore a short-back piece— often a military man with a humble background—would stand out. Therefore, the designation was highly specific to the ancient context and found no experiential correspondence in the Song. Any statement employing this concept would thus fail as unreliable knowledge. A name without proper experiential match was not valid.

Shen's promotion of reliability-inducing methods was by no means limited to the section titles of *Brush Talks*. In fact, a few other methods were so pervasive in his practice that they proliferated through multiple sections if not the entire text. The most prominent one regarded the employment of one's sensory perception, or as he often straightforwardly claimed: "I have seen it with my own eyes" (*yu qin jian zhi* 予親見之).[33] As previously discussed, Shen paraded his constant reliance on visual witness even into the fantastic realm. In more quotidian cases, he developed complex observational procedures to fully implement the power of the visual modality.

Shen's discussion of rainbows is a good example to demonstrate his employment of systematic observation. He opened this item with a statement asserting a reliable instance of knowledge: "People say that a rainbow dives into and drinks from a river. [I find it] trustworthy (*xin*)."[34] "Rainbow drinking water" was an anthropomorphic metaphor alluding to a possible connection between the formation of a rainbow and nearby waterways. Through observations made on his diplomatic mission to the Liao, Shen was able to ascertain this connection. One day after a rain shower, a rainbow appeared, dipping into the creek in front of his camping tent. Shen and his fellow emissaries stepped out to examine it closely. They found out that both ends of the rainbow indeed plunged into the creek, showing a convincing physical proximity. He sent a colleague to repeat the observation on the other side of the creek and confirmed the closeness.

Afterward, Shen applied systematic observations to testing a hypothesis concerning sunlight as another potential causal factor for rainbows. He discovered that "if one were to observe in the west–east direction, [the rainbow was visible and thus] it was a sunset rainbow"; if observed in the east–west direction, the rainbow became invisible because it was "masked by the sunlight."[35] The formation of a rainbow was dependent on the projection of rays of sunlight. Shen repeated the procedure the next day and again saw the same set of phenomena, confirming the hypothesis with the evidence of repeatability.

Although vision was usually the primary sense at work in Shen's inquiries, he also demonstrated effective employment of other sensory modalities as processes able to produce reliable knowledge. For instance, he once encountered an ancient bronze mirror, which, when pressed at the center, generated a sound resembling the noise of burning turtle shells (a procedure used in prognostication). Others attributed the cause to the possibility that this mirror had double surfaces with hollowed space in between. Shen rebutted this hypothesis with auditory clues: because a double-walled mirror was surely a product of welding, this ancient mirror did not emit a sound as muffled as a welded instrument normally would have.[36]

In items concerning food and medicine, Shen often resorted to taste to form judgments, such that orange peel tasted sweet, and yet citron peel had a bitterness.[37] In recognizing a rosemary plant, he depended on

olfaction, noting that the leaves could be "extremely fragrant"; in identifying licorice, he resorted to tactility, as this plant distinguished itself
with pointed leaves that were rough to the touch.[38]

In addition to sensory perception, Shen demonstrated good reasoning as an extensively applicable epistemic virtue. In his praxis, good
reasoning came in a variety of forms, such as precise and logical quantitative thinking he employed in mathematical (or, more broadly construed, number-based) inquiries. Shen's most acclaimed achievement
in mathematics emerged from the effective exercise of this epistemic
virtue.[39] The two prominent examples are the "method of circle measurement" (*huiyuan shu* 會圓術) and the "method of volume with interstices" (*xiji shu* 隙積術).[40] Essentially the equivalent of the arc-sagitta
method, the first technique aimed for the determination of the lengths
of circle arcs.[41] The method of interstices was for the calculation of the
frustum of a solid rectangular pyramid with stacked articles.[42] Besides
what we conceive as mathematics, Shen applied logical quantitative reasoning to other number-based schemes, such as the so-called incorporation of pitch names, in which he integrated the sixty musical pitches into
the matrix structured by the Five Processes and the sixty pillars.[43] Items
of a purely quantitative nature occupied just a fraction of *Brush Talks*,
but they lend robust testimonials to Shen's devotion to precision in quantitative reasoning, one kind of reliability.

Another type of good reasoning Shen often engaged in involved causation. Causality, broadly construed, comes in a variety of forms, and in
Shen's times, deep orders provided dominant frameworks for identifying
cause and effect. While keeping them in sight, Shen demonstrated a salient interest in investigating local and immediate causes associated with
sensory knowing. Some of his causal analyses provided an alternative perspective for explaining a phenomenon in terms of its relation to a larger
order. Pursuing causality through sensory knowing led Shen to make
some discoveries similar with modern scientific conclusions, contribut-
ing to his reputation in the twentieth century.

For example, Shen argued that the physical dimensions of an instrument were responsible for the qualities of sound it produced. In his
times, an anecdote had it that a Tang scholar, Li Sizhen 李嗣真 (ca. late
seventh century), once managed to recover the lost *zhi* pitch, one in the
pentatonic scale (*gong* 宮, *shang* 商, *jue* 角, *zhi* 徵, and *yu* 羽), from an

ancient stone. According to this story, Li found the stone through yinyang-based calculations and reproduced the *zhi* note with a fraction of the stone after he had cut it into four pieces.[44] The "antiquarian" attempt Li was believed to have made was a common line of action in music making at this time. It followed the stipulation that perfectly harmonious music descended from an all-encompassing cosmos and became accessible through ancient sages' translations (onto a stone that met specific requirements, for instance).[45] This cosmos was a unifying universe built on elaborate correlative patterns consisting of yinyang and the Five Processes. In other words, it was a number-related scheme, and the search for a pitch should be conducted along the structural lines determined by numerical relations. Although not directly countering the system, Shen called the story absurd and reminded his readers that even if this piece of stone carried the lost *zhi* note, one piece of it could by no means reproduce the sound, because "length and thickness are the factors determining the pitch."[46] The sound produced by the whole stone and that by a fraction of it were bound to differ. Here he primarily associated the cause of a sound with the particular, sensory qualities of the musical instrument itself.

Shen's well-celebrated discovery of curved mirror optics is another good illustration of his reliance on local causes perceived by the visual modality. He made an attempt to explain why a "burning mirror" (*yangsui* 陽燧), that is, a concave mirror, formed an inverted image of an object.[47] The "cause" (*gu* 故) he explicitly identified was an "obstruction" (*ai* 礙) standing between the mirror and the object.[48] Shen observed that if he put his finger a small distance from the mirror, the image was normal; as he moved it further from the mirror, he found a range of distances within which no image could be seen; and as he moved his finger beyond this range, the image was inverted. The blank region where no image was formed was the location of the "obstruction." In such a case, the obstruction caused the inversion, serving as the latter's immediate and local antecedent.

To affirm the existence of the obstructive mechanism, Shen compared this case to two other examples. He first likened it to the resistance generated from the pivot of an oar. Then, more elaborately, he analyzed a similar example of a kite and its shadow. For a kite flying in the sky, its shadow shifted in the same direction as the kite. However, if one sat inside a room and observed the shadow of the kite formed through a small aperture in a paper window, the kite and its shadow appeared to move in opposite directions. If the kite went east, its shadow went west, and conversely. In Shen's reckoning, the small aperture was a mechanism of obstruction because of its visible constriction of the passage of light. The change in the direction of the kite's shadow was a phenomenon parallel to that of the inverted image. In all three examples, the consequent "inversion" followed the condition "obstruction." Therefore, Shen was convinced that "obstruction" was the cause of "inversion." The entire line of reasoning was firmly grounded in the "inversion" phenomenon and its sensible local conditions.

In the entirety of *Brush Talks*, Shen navigated the pursuit of causation freely in and out of the sensory realm. While investing great attention to antecedents he could access with perception, he also admitted limitations of sensory knowing in illuminating causality. For instance, a mysterious "thunder fire" once greatly puzzled him and led him to contemplate possibilities beyond human cognition.[49] A thunder ball struck an official's house and caused a fire, which melted a piece of silverware contained in a wooden box. Intriguingly, the box remained intact without a burn mark. A precious knife made from quality steel became totally softened by the fire, yet the safety cover was not affected. The best explanation Shen could find for this instance was a reference from a Buddhist treatise, which stated that beyond mundane human fire, a so-called dragon fire existed and interacted with the world differently. Shen concluded that humans were able to comprehend only "things in the human realm" (*renjing zhong shi* 人境中事), admitting that it might be extremely difficult to calibrate things "outside the human realm" (*renjing zhi wai* 人境之外) with "worldly wisdom and situational perception" (*shizhi qingshi* 世智情識).[50] Shen associated this fire with "ultimate patterns" (*zhili* 至理), the most profound of deep orders, by definition inaccessible by sensory knowing.[51] Note that he resorted to inarticulable profoundness without lessening his intensive interest in observing the "human realm" in detail, which gave rise to his rich way of understanding causality with a sensory bent.

## *The Nonsystem and System Compared*

This chapter has so far revealed the basics of Shen's nonsystem: it was an epistemology driven by the pursuit of reliability, and it consisted of a series of reliability-inducing methods, such as using precedents, naming things accurately, relying on sensory perception, and employing good reasoning. Shen demonstrated these methods in specific sections or across sections through examples. In a number of important ways, his nonsystem differed from systems, which I explicate here.

First and foremost, Shen's nonsystem rejected a total-view type of unity, which, as analyzed in chapter 7, most saliently defined a system. Shen did not mention in words—nor indicate by action—that he was pursuing any encompassing unity in *Brush Talks*. He was silent on the unitary *dao* and chose to embrace a medley of objects and processes that did not coordinate with one another into a meaningful whole. The instruments he provided for controlling this chaotic world were specific means of attaining reliable knowledge, which neither integrate into a homogeneous program nor command a uniform type of normativity Wang Anshi or Shao Yong envisioned. In addition, it was impossible to reduce the range of reliability-seeking skills into one homogeneous cognitive-conative process, such as the heart-mind's "intuitive" discernment suggested by Su Shi and Cheng brothers (see chapters 7 and 10). Although Shen paid heed to certain deep orders, which presumably paved the way to the *dao* in a system, he did not administer them in reductive ways that led to unity. Instead, he assigned high epistemic guiding power to sensory knowing and constantly resisted the normative power of existing rules.

Second, by demonstrating a new epistemological stance, Shen wished to vie for epistemic leadership—an aspiration similar to system builders'—yet in the role of an evaluator instead of an authority. An epistemic authority prompted (occasionally commanded) others' beliefs with testimonies regarding orders of relations. Instead, Shen extrapolated and evaluated epistemic virtues from what he personally did and what he witnessed others do. In the capacity of an evaluator, Shen applied the test of reliability to the broad community he inhabited.

The extensive evaluative power Shen aspired to also found logical corroboration in another feature of the new nonsystem: *Brush Talks* demonstrated mostly commonsense epistemic virtues rather than methods restricted to highly specialized studies.[52] Reliance on sensory perception is a spontaneous cognitive action shared among all humans. Good reasoning applied to mathematics and print making, and thus served to bond the two subjects, as well as a scholar and an artisan, in the coherence of a single treatise. Shen did not invent any of these commonsense processes; he inherited them critically from the community to which he belonged. His contribution lay in efforts to distinguish commonsense epistemic virtues from vices and demonstrate them through textualization.

Third, in *Brush Talks* the nonsystem gave rise to an account of the phenomenal world broader than in many other systems, such as Wang Anshi's; it also contributed a rich amount of original knowledge regarding the natural world. I present this difference with the caveat that what distinguished Shen from his peers was not a nature-culture division. Shen did not intend a nature-culture demarcation. He was equally committed to seeking reliable knowledge in the natural realm and in the human world, and he engaged both natural phenomena and cultural activities with the same epistemic virtues. This makes *Brush Talks* a good reference not only for "science" but also for ritual procedures, court paraphernalia, and painting skills.

To complete this point, I note that no Song literatus ever openly announced a nature-culture divide as a serious philosophical argument. The reason was simple: the *dao* was supposed to be all-embracing and all-inclusive. The concept closest to a nature-culture distinction in historical use regarded the relationship between "the *dao* of Heaven" (*tian dao* 天道) and "the *dao* of humans" (*ren dao* 人道).[53] The former referred to all that was outside human control, thus including the natural world, and the latter designated the sphere of human action. The distinction was an ancient inheritance, and scholars since antiquity held different opinions as to how the two should be connected, specifically how much autonomy the latter had from the former.[54] Despite their diverse conclusions, heavenly and human *dao*s remained unified in one cosmos. Just as Cheng Yi stated, "The *dao* is one. How [can you say that] the *dao* of humans is the *dao* of humans, and the *dao* of Heaven is the *dao* of Heaven?"[55] Sometimes this unity took the form of a direct connection; for example, the Han cosmology posited "resonances" between meteorological occurrences and human politics, assuming an immediate controlling influence of Heaven over humans.

In the Song, literati spoke less of direct resonance, and many acknowledged that the two realms had separate orders. The acceptance of concrete differences, however, did not mean demarcation. Humans did not have to duplicate a natural process or have constant contact with it to remain a harmonious part in the interdependent cosmos. Many Song system builders, including Shao Yong, Zhang Zai, Su Shi, Cheng Yi, and eventually Zhu Xi, found ways to accommodate concrete diversities between the heavenly and human spheres and integrate nature and culture in the purview of their systems. In this regard, the broad account Shen drew of the world was not the result of an unusual interest in nature, but of one among other intellectual designs that successfully united nature and culture.

## The Nonsystem and Its Enlarged Epistemic Community

Another implication of the nonsystem was the enlarged epistemic community it entailed, a significant characteristic that deserves a separate section. Without a doubt, Shen managed to involve a new, augmented epistemic community, which constituted the foundational source on which he extracted epistemic virtues and critical skills.

In addition to himself, Shen included other responsible knowers in *Brush Talks*—some of them Song peers and others historical figures. Within the sphere of the cultural elite, those who were not primarily known for being statesmen, literary magnates, or philosophers would find their voices articulated by Shen and joined the same world as those who were. The so-called "scientists" in modern scholarship are an especially prominent example. In relating methods for reliability, Shen named a number of people known for their expertise in technical studies, for example, Zhang Heng 張衡 (78–139), Lu Ji 陸績 (188–219), Zhang Zixin 張子信 (ca. seventh century), Yixing, and Shu Yijian 舒易簡 (fl. 1050s).[56]

These historical figures surely left traces in the textual tradition, largely in genres devoted to their technical fields (such as sections on calendars and astronomy in official histories). In *Brush Talks*, Shen discussed their work and epistemic virtues/vices in juxtaposition with statesmen and writers, annexing all into one epistemic community.

Even more broadly, Shen's epistemic community extended beyond the elite circle and reached commoners. As a result of his inclination to present anyone who productively exercised epistemic virtues, *Brush Talks* introduced a number of hands-on practitioners who otherwise would have left little trace in elite written history. The famous ones included Bi Sheng 畢昇 (ca. 970–1051), father of movable-type printing technology; Yu Hao 喻皓 (fl. late tenth century), the renowned architect and alleged author of the famous *Classic of Carpentry* (*Mujing* 木經); and Wei Pu, the legendary blind calendrical expert. None were members of the literati class.[57] Shen's inclusion of them was less a conscious challenge to the established social order than a logical extension of his effort to promote his nonsystem. The opening of a new terrain of knowledge called for new heroes, and those who might claim candidacy with designated epistemic virtues could come from elite or nonelite backgrounds. In fact, with less exposure to textual knowledge, hands-on practitioners might possess a greater proclivity for engaging "things" through their own hearing and seeing.

In addition to the roster of figures Shen recorded in *Brush Talks* as practitioners of piecemeal epistemic virtues, the discussion of epistemic community raises the question of who, especially in the literati circle, shared his overall epistemological stance. Two historical categories are helpful in identifying Shen's kindred spirits, "broad learning" (*boxue* 博學) and the genre *biji*.

The concept "broad learning," or "broad" (*bo* 博) for short, became an increasingly popular term in the Northern Song and beyond for highlighting the breadth of one's exposure to knowledge.[58] In some contexts, being broad in learning implied that one held interests beyond classical commentaries, state policies, and literary prose.[59] With a wide command of diverse topics across the phenomenal world, Shen won a reputation for broad learning from contemporaries, such as Zhang Lei, Zhu Xi, and Ma Duanlin 馬端臨 (ca. 1254–1323).[60] Many other scholars with various interests, in Shen's time and after, also enjoyed this reputation in peer opinion. Indeed, to learn broadly was an important intellectual trend that Shen belonged to and exemplified.

Some subtleties of the concept are worth tending to, however, especially in regard to the discussion of epistemology. Broad learning, as a goal and an achievement, could be associated with different epistemological agendas and thus represent varied stances—a nonsystem or a system. Although "broad learning" could refer to a generic extensive coverage of information and stand as an intellectual merit as such, it often appeared in conjunction with another concept, "essentials" (*yue* 約). The breadth/essential dyad was a prevalent scheme in system-building discourse. Su Shi famously exhorted students to "browse in breadth and select the essentials" (*bo guan er yue qu* 博觀而約取) while pursuing learning.[61] To a great extent, the dichotomy resonated with the contrast between hearing and seeing and modeling. In many cases the essentials overlapped with the orders of reality, which resided in the boundless breadth of vicissitudes floating on the surface of the world. Those who saw and heard broadly were supposed to eventually return to the deep orders essential to reality. Thus broad learning became subordinated to the quest for unity. The pursuit of breadth for its own sake was seen as pointless and unproductive; one should avoid indulging in excessive superficial exploration and remain committed to the fundamentals.[62]

It is worth mentioning that among broad learners, some more closely resembled Shen in terms of an overarching epistemological stance. Those less committed to the pursuit of unity and more attentive to sensory knowing, the empiricists, acquired their breadth in a Shen-like spirit. In other words, these people paid fastidious attention to sensory details and resisted the normativity of essentials. Their breadth included not only extensive mentions of varied phenomena in the world but also how they mentioned them.

A prominent example of this category is Su Song, a contemporary analogous with Shen in many ways.[63] Like Shen, he engaged a wide range of technical inquiries—most notably astronomy and medicine—and enjoyed a reputation for broad learning.[64] Su waxed enthusiastic about astronomical observation and built a new and more advanced armillary sphere.[65] He attributed great significance to sensory knowing in medicine, illustrating his pharmacological treatise with pictures to record the "shapes and colors" (*xingse* 形色) of medicinal herbs and collecting only recipes he found experientially efficacious.[66] Besides, his career trajectory was similar to Shen's: holding a high-ranking position in the officialdom and serving during the period of New Policies (though more as a reform antagonist).[67] The two were acquainted, and when Su set out to develop his armillary sphere, Shen's was an immediate precedent. Perhaps Su's anti-reformist ties caused Shen to remain largely quiet about him or his work, but the commonalities were hard to deny: both men achieved broad learning primarily through engaging hearing and seeing in a productive manner.[68]

Another group of erudite learners were in fact system builders committed to unity yet suspicious of uniformity. As I mentioned in chapter 7, a representative figure of this stance was Su Shi, who resisted Wang's uniform prescriptions with an intuitive unity. Su was also known for his wide exploration of the phenomenal world, writing extensively on plants, food, and medicine.[69] He regarded such breadth as a subordinate part of his system, however—an early step leading to the understanding of unity. A disciple of his, Zhang Lei, articulated this stance clearly. Following his teacher's inclination to "learn broadly," Zhang composed a text titled *Mingdao Miscellany* (*Mingdao zazhi* 明道雜誌). This treatise shared noticeable similarities with *Brush Talks*, covering a wide range of miscellaneous knowledge Zhang learned through seeing, hearing, and reading on his own.[70] Like Shen, Zhang was interested in applying critical skills and made repeated efforts to rectify mistakes incurred by the negligence of these skills (including correcting Shen twice).[71] Moreover, Zhang seemed to appreciate Shen and his work, applauding that *Brush Talks* "featured many merits."[72]

However, Zhang viewed his practice of broad learning as a preliminary endeavor rather than a final goal. He demonstrated this secondary significance by presenting the entire scheme of learning and made a statement of purpose saliently absent from Shen's writings:

There is a *dao* to learn, and a progression to [reaching] the *dao*. If one follows the progression and accumulates, he will be able to go far when he proceeds, and climb high when he passes through. On the bottom [he] comprehends all patterns of birds, beasts, insects, and fish, as well as of instruments, machines, clothing, and objects. In the middle [he] cultivates himself, rights the family, and orders the world, establishing all foregoing accomplishments. On the top he reaches his nature and destiny, comprehends life and death, manages Heaven and Earth, administers the ten thousand things, to the extent that he stands above the ten thousand things and no longer stays among their ranks. And then, he has accomplished learning.[73]

Zhang's text strikes a few familiar notes. For one thing, the description of a step-like ascension of learning toward the *dao* was utterly similar to Wang Anshi's account of modeling. For another, his engagement with "things" for the purpose of surpassing them resonated with every practitioner of modeling who valued the place of a "thing" more than its individuated existence. In this scheme, the inquiries Zhang made in *Mingdao Miscellany* were just an initial stage, in which he broadly explored birds, beasts, insects, and fish as well as instruments, machines, clothing, and objects. Zhang expected to eventually be able to "stand above the ten thousand things and no longer stay among their ranks," transcending piecemeal inquiries of "things" with the command of unity.

Zhang Lei engaged in a broad and productive use of hearing and seeing in his epistemic praxis, yet conceded to the philosophical confinement of this concept when it came to contextualizing his doings within the scheme of pursuing unity. This is how a thinker aspiring for a system differed from Shen when articulating the value of "broad learning." Readers may wonder: if the epistemic praxis recorded in *Mingdao Miscellany* was similar with that in *Brush Talks*, did the aforementioned difference truly matter? It did, in the further development of system building. The emphasis on unity led to more elaborate conceptualizations of a unified, spontaneous process of knowing by the heart-mind and further marginalization of sensory knowing as well as critical skills. The divergence became more marked in a more rigorous system builder, a point I pursue in the examples of the Chengs and Zhu Xi (chapters 10 and 11). The designation "broad learning" aptly captured a key intellectual merit Shen shared with many peers, yet it lumped divergent philosophical positions, a fact we should not lose sight of when examining Shen's epistemological stance in a comparative light.

The second historical category helpful in identifying Shen's kindred spirits was the genre *biji*. As Ellen Zhang points out, the content of *biji* was often associated with knowing from hearing and seeing and the ideal of broad learning.[74] According to the bifurcate classification of *biji* in his times, Shen belonged to a group of authors devoted to producing a more credible kind of notebook literature. Taken together, many *biji* authors shared Shen's proclivities to see, hear, and exercise reliability-inducing skills in piecemeal inquiries, and the genre—especially the more credible subgroup—provided fertile soil for identifying other nonsystems.

Nevertheless, a precaution is in order if one is to make sound comparisons between Shen and other notebook composers. Though the use of "hearing and seeing" flourished in the discursive sphere of *biji*, authors did not always conceive its utility in the same way and often framed their methodologies in different schemes. In Shen's nonsystem, he valued epistemological reliability as an end in itself, a position shared by some *biji* writers and not others. To accurately present their commonalities and differences, I divide the comparative narrative into two phases. In Shen's times and earlier (late tenth century through late eleventh century, approximately Northern Song), many *biji* authors framed their pursuit of credibility with conceptual schemes other than epistemological reliability, which they often treated as a means rather than an end. In the second phase, roughly twelfth century and after (approximately Southern Song), some *biji* writers began to explicitly address reliability as the central goal of their work and value critical skills for their own sake. In the following I present the first phase and save the second for the last chapter, where I discuss Shen's historical reception.

From the late tenth century onward, a dominant rhetoric for evaluating the credibility of a *biji* was to compare it to official histories and in many cases the language of assessment reflected authorial intention. A good number of eleventh-century *biji* authors produced personal notes for supplementing official historical records, and they pursued accuracy as a meritorious practice well established in the historiographical tradition.

For example, in the postscript of *Records of Returning to the Fields*, Ouyang Xiu stated that he endeavored to emulate Li Zhao's 李肇 (fl. early ninth century) method of compiling *Supplement to the History of the State* (*Guoshi bu* 國史補), a Tang text famously purported to "repair the omissions" in official histories.[75] In the preface to the *Records from the Dongzhai Study* (*Dongzhai jishi* 東齋記事), Fan Zhen specifically invoked his experience serving on the editorial board of the *History of the Tang*, speaking of his wish to follow the Tang examples and rectify state records with personal observations.[76]

Furthermore, the quest of historical credibility often served as a form of political intervention. Many serious eleventh-century *biji* authors were keen to assert their political/moral judgments as a supplement or a challenge to the official historical discourse. For example, Ouyang Xiu proclaimed in the *Records of Returning to the Fields* that he intended to record virtues only and steer clear of vices, because "to bury the evil and promote the good is the resolution of the superior man."[77] In *Random Chats by the Mian River* (*Mianshui yantan lu* 澠水燕談錄), Wang Pizhi 王闢之 (*jinshi* 1067) adopted such section titles as "Virtues of Emperors" (*dide* 帝德), "Upright Words" (*danglun* 讜論), and "Prominent Ministers" (*mingchen* 名臣)—all evaluative designations parading the political judgments of the author.[78]

A more detailed investigation reveals that a serious *biji* author often intervened to favor a particular political cause. Some early scholars were subjects of the bygone Five Dynasties, and they meant to preserve the regional voices increasingly diminishing in the unifying discourse of the Song empire.[79] *New Writings of the South* (*Nanbu xinshu* 南部新書) by Qian Yi 錢易 (fl. ca. 1000s) was an example emerging from the region where the Southern Tang (937–976) used to be. According to the preface written by his son Qian Mingyi 錢明逸 (1015–1071), the text was committed to disseminating political values as much as preserving historical memories. Qian Yi mainly gathered what he considered "worth moral precautions" (*zu wei jianjie zhe* 足為鑒誡者), for instance, "loyalty and filial piety for educating subordinates, kharma and retribution for warning the simple-minded and vulgar, institutions and rituals for understanding the structure of governance, as well as integrity and modesty for encouraging a [noble] character."[80]

From the midpoint of the eleventh century onward, the call for writing reliable history in notebook form gradually became a reaction to factional struggles, specifically to disruptions imposed by political coalitions on the composition of official histories.[81] Personal *biji* afforded an instrument to combat perceived biases, and "credibility" was the equivalent of perceived impartiality or an outright counterattack on the opposing opinion. For instance, Tian Kuang 田況 (1005–1063) composed *Fair Comments on Literati* (*Rulin gongyi* 儒林公議) to recount major political events from the time of the founding monarch through Renzong. Tian sympathized with the reform initiative of Fan Zhongyan 范仲淹 (989–1052) and Ouyang Xiu. Thus, although he endeavored to present what he considered to be bias-free history, he insinuated defenses of Fan and Ouyang against partisan attacks from time to time.[82] *Biji* of a similar nature became more prominent after the New Policies, as factional strife continued and court chroniclers had no choice but to tailor their work along the ideological lines set down by the faction in power. For example, works such as *Mr. Sun's Garden of Chat* (*Sun gong tan pu* 孫公談圃) sought to protest factional biases by promulgating the author's political view.[83]

When comparing Shen to the aforementioned *biji* authors, some subtle differences are worth noting despite their general commitment to credibility. First, the pursuit of reliability recorded in *Brush Talks* was too multifarious to be contained in the historiographical framework, as Shen included many topics beyond the boundary of conventional historical composition. Second, *Brush Talks* was more of an epistemological argument than a political intervention, especially when viewed against the background of factional politics. The text was a statement of Shen's intellectual stance vis-à-vis the system-building trend—thus political in a generic sense—but it dutifully avoided promoting concrete ideological opinions. Shen's actions were as good as his words in the preface: he avoided making moral/political judgments of peers and spoke of the New Policies only in terms of the technical details of how things were done. He surely wanted to persuade system builders that his nonsystem was a worthy alternative, but this intellectual challenge was not the equivalent of an anti–New Policies agenda (nor, for that matter, a piece of pro–New Policies propaganda).

Thus, Shen's nonsystem found resonances with many contemporary *biji* authors in terms of concrete inquiry. However, it took time before literati methodologically recognized epistemological reliability as a merit valuable for its own sake. The epistemic community Shen's nonsystem could possibly engage solidified and expanded after his times, a point I elaborate in the final chapter.

## *The Empirical Stance*

Let me conclude this chapter by taking stock of the characteristics of Shen's nonsystem under the rubric of empiricism. Shen's nonsystem was not only historically interesting but also philosophically significant. It was a version of empiricism he concocted with intellectual resources available in the eleventh century, and it bore the stamp of its times in a number of ways.

First, due to its close relationship with knowing from hearing and seeing, this empiricism had two aspects in accordance with the two dimensions of the concept. The philosophical denotation of hearing and seeing rendered Shen's empiricism highly comparable to today's conventional definition of this term, which regards sensory perception as the main source of knowledge. In *Brush Talks*, Shen demonstrated an emphasis on sensory perception in various forms throughout the text. Among the multiple epistemic virtues he promoted, reliance on sensory perception stood out as the most prominent. As shown earlier in this chapter, Shen reposed on all five sensory modalities while contemplating objects and processes. In a few cases he even devised experimental procedures based on sensory observation. Furthermore, his emphasis on sensory experience infiltrated other processes. For instance, in suggesting using precedents as a reliable process of knowledge production, his most favored "precedent" was often a sense-based one. When setting out to identify a cause, he often searched in local conditions using his own eyes and ears. His impulse to seek accurate matches between "things" and names bespoke constant attention paid to the sensory content of linguistic conventions and a philosophical insistence that if a concept contained sensory content, this content must be properly grasped to render the concept valid.

The more expansive capacity of knowing from seeing and hearing as a textual practice moved Shen's empiricism beyond a simple philosophical consideration of sensory perception and permitted it to become a broad assertion of one's epistemic autonomy. *Brush Talks* recorded varied information he extracted from observation, social interactions, and textual learning through the use of critical skills, witnessing him switching deftly between observing the sensible and coordinating it with the deep structures of the cosmos. By identifying a range of skills aiming to generate reliability in different contexts, Shen acquired a new epistemological foothold and to a great extent rendered epistemic authorities superfluous.

On a related note, Shen's empiricism was a scholarly informed variant of empiricism, which in many ways reflected his cultural preferences as a learned man of his times.[84] In the world of *Brush Talks*, book learning was not antithetical to experience, and Shen did not renounce texts—he embraced them. He trusted sensory observation and reliable textual testimonies when approaching a "thing," and he frequently resorted to textual records of the same subjects to inform his experiential judgment. He also applied his sensitivity to sensory experience to the reading of classical texts, using observation to aid philological reconstruction. These propensities were manifest in "Identification and Verification," as in the examples of "wild horses" and "short-back robes." The text/observation dichotomy bore no decisive significance in Shen's empiricism, because it related only contingently to the core problem in his designation. Although Shen was reluctant to concede to the normativity of existing rules regarding the deep orders—a popular subject in textual knowledge—he did not reject textuality itself. He privileged observation as an effective means of capturing attributes of "things," yet he did not take it as the only, exclusive means. In fact, reliable textual testimonies often aided observation.

Another characteristic of Shen's empiricism concerned its relationship with deep orders, a feature modern readers must recognize to grasp the subtleties of this empirical stance.[85] Shen's nonsystem was situated in a cosmos structured by certain orders of relations. As I discussed in chapter 4, Shen lived immersed in a world structured by yinyang and $qi$, hence the section "Number and Figure" in *Brush Talks*. His empirical stance allowed him to detach from a rigorous epistemological commitment to number, which nevertheless was not a renunciation. In his case, the liberation was more total in some areas (for instance, astronomical observation) where he largely relied on sensory phenomena for guidance; in other contexts, Shen paid heed to both the normative order and sensory data and worked to coordinate them.

My previous discussion of the Course-$Qi$ system already demonstrated one example of the latter kind. Shen effectively navigated the Course-$Qi$ scheme by juggling between meteorological phenomena and numerical patterns. Specifically, he assigned high epistemic guiding power to meteorological phenomena, and on this basis rendered the existing numerical principles into better accord with sensory experience. Shen pivoted on the sensory while considering a deep order, parading a robust empirical spirit throughout a structured cosmos.

The relationship between Shen's empiricism and deep orders deserves elaboration with one further example: medicine, a field of inquiry dependent on the senses and associated with number-based cosmology.[86] A devoted student of medicine, Shen greatly valued sensory knowing in medical praxis. In identifying medicinal herbs, he observed the herb's appearance, noted its odor, sampled its taste, and touched it to obtain the haptic feeling. Shen thought it vital for physicians to understand through experience the varied possible efficacies of an herb in different combinations. For example, two similar ingredients, *zhongru* 鐘乳 (stalactite powder) and *zhu* 朮 (*baizhu* 白朮, *Atractylodes macrocephala*), would interact in opposite fashion in two different combinations.[87] Thus, in the preface to *Efficacious Prescriptions*, Shen stated:

> What I call "efficacious prescriptions" have to be [those] which I have witnessed to be effective and then committed to writing. Hearsay has no role [in my identification of them].
>
> 予所謂良方者，必目睹其驗，始著于篇，聞不預也。[88]

Shen's stress on experience nevertheless remained in tandem with his attentive consideration of systematic doctrines in medicine. In the same preface he emphasized with equal strength that current medical practitioners should better educate themselves in the cosmological orders recorded in medical classics, such as "the cycles of yinyang and of time," "the exhalations of [the *qi*] from mountain, forest, river, and marsh," as well as the Five Circulative Courses and Six *Qi* (now in medical use).[89]

In his discourse on diagnosis, Shen's penchant to heed sensory qualities and number-based principles was also salient. He exhorted a contemporary healer to "pay attention to every sound the patient made, his complexion, his movements, the condition of his skin, his temperament, and his preferences," to "ask what he did and investigate his behaviors," as well as "thoroughly read the man-welcome and inch-mouth pulses and all twelve pulsating vessels."[90] This was an elaborate program to collect relevant perceptible signs.

He reiterated fundamental rules for diagnosis: "While disease breaks out in one of the five yin viscera, the five colors respond to it; the five sounds change accordingly; the five tastes become correspondingly unbalanced; the twelve pulsating vessels move accordingly."[91] This statement delineated the general correlations between pathological alterations and functional changes in the viscera, systematic correspondences organized according to principles of yinyang and the Five Processes.[92] The five yin viscera were "basic systems of visceral functions" loosely connected to biomedical organs but not equivalents of them.[93] Together they mapped the physiological structure in which correlations took place, and they constituted, as in my metaphor, the patterns of the medical body as a net. The "Five Colors," "Five Sounds," and "Five Tastes" were classificatory modules loosely associated with sensory signs instead of literal descriptions of them. For instance, the color "black" (*hei* 黑) did not mean that one's facial complexion had to be as dark as ink; rather, it indicated that the patient had a blockage of *qi* in the kidney.[94] A kidney-black correlation was a cord in the net.

Similar to his employment of the Course-*Qi* system in weather forecasting, what Shen proposed to do, first and foremost, was to seek the most "elaborate" (*xiang* 詳) observations he could.[95] The rich body of perceptual clues provided him with many "knots," which, once woven into the net, would contribute to further developing the patterns of cords and eventually enable the net to better map onto experience. In so doing, Shen demonstrated his commitment to sensory perception as a critical epistemic guide and the foundation of his elaboration/revision of normative principles. This is why he was attentive to numerical doctrines and at the same time fathomed sensory phenomena more deeply than many.

Let me conclude this point by responding to two questions readers may raise. First, in the examples of Course-*Qi* and medicine, if Shen was indeed concerned with improving the systems, how was he different from a system builder, such as Shao Yong? Two distinctions are important. For one thing, Shen attributed more guiding power to sensory perception and on that basis held a more critical stance when operating number. He ameliorated the system by expanding and complicating the rules in light of hearing and seeing, rather than staying attached to the existing rules and their mechanical self-duplications. Shen's consideration of deep orders was not driven by a desire to return to the few "essentials" widely circulated in textual testimonies. Instead, his use of number was open-ended and expatiating, infinitely expanding according the flow of sensory clues. For another, beyond the specific contexts of weather forecasting and medicine, Shen made no commitment to a unitary *dao* on the foundation of a fully articulated numerological scheme. Thus, in *Brush Talks* he could comfortably juxtapose "Figure and Number" with a chaotic range of "things" floating beyond numerical patterns (or any patterns).

Second, if Shen employed number and sensory knowing as epistemic guides, how was he different from a practitioner of modeling, who followed the guidance of an order of reality with (sanctioned) flexibility germane to his own experience? The difference was also dual. Shen's epistemological commitment, again, might gravitate toward sensory perception and grant it power to structurally revise (his understanding of) the order. A more important distinction lay in the cognitive-conative process. The psychology of modeling relied on the heart-mind and (in an ideal situation) involved no deliberative activity. Shen, in contrast, placed intensive emphasis on the exercise of critical skills and attributed less significance to the spontaneous heart-mind.

The final characteristic of Shen's empiricism was that it was a stance rather than a normatively articulated philosophical position. In *Brush Talks* he gave descriptive accounts of epistemic praxis rather than dictating a formal theory. His empirical stance can be conceptualized as a sum of practical commitments and approaches. The true difference Shen could have presented, as he might have imagined, lay in his actual praxis and the results that ensued. This is how *Brush Talks* became the textual expression of Shen's empiricism.

Let's take stock of how Shen articulated his last intellectual statement through writing *Brush Talks*. Drawing on *biji*, a genre particularly congenial toward presenting knowledge of diverse and nonofficial kinds, *Brush Talks* in its entirety presented a nonsystem. The core of the nonsystem, and the central message it evinced vis-à-vis systems, was the pursuit of reliability, which consisted of a range of epistemic virtues Shen identified in concrete inquiries. This nonsystem constituted his version of empiricism, which privileged sensory perception as an epistemic guide and accentuated an individual knower's epistemic autonomy. Although Shen's nonsystem operated in a world where deep orders provided the basic structures, this empirical stance visibly deviated from modeling. The difference grows more salient in a comparison between Shen and further developments in system building, an issue I examine in the next chapter.

# CHAPTER 10

# *Farewell to System*

This chapter concludes the comparative narrative between Shen Gua and systems by incorporating some further developments in the system-building mainstream and raising a few more philosophical questions. These comparisons merit their own chapter for a couple of reasons. First, the second stage of system building—represented primarily by the Cheng brothers—featured several innovative features that went beyond the general characteristics of systems discussed in chapter 7. Second, the new comparisons are slightly different in nature. They are more philosophical than historical in the sense that unlike Wang Anshi, the Cheng brothers did not seem to have much personal interaction with Shen.[1] Making these philosophical comparisons in a historical work is imperative for a historiographical reason. Given that the Cheng brothers were responsible for heralding Zhu Xi's neo-Confucianism, the dominant thinking system ruling the Chinese intellectual world until the twentieth century, an analytical narrative that contrasts them with Shen helps better place the latter in the long-term development of Chinese thinking and illuminate further implications of his empiricism.

Following the comparative initiative, this chapter introduces how the Chengs and Shen engaged two key concepts, *li* (pattern/Coherence) and *xin* (heart-mind). The first four sections cover how the Cheng brothers built their system on the foundation of these concepts, and the following two analyze Shen's different approach.

# Li *as Coherence*

As I briefly introduced in chapter 7, the system proposed by the Cheng brothers surpassed stage-one systems in at least two ways: it expanded to include all relevant orders of relations, and it substantiated the psychology of modeling with a concrete plan. Here I lay out in detail the philosophical designs that made these goals possible.

Intellectual historians generally agree that the Cheng brothers' formulation of the new concept *li*—translated as "Coherence"—marked a breakthrough in Chinese intellectual history.[2] In the current discussion, I address the emergence of *li* with an emphasis on its significance specifically for system building: it provided a new metaphysical structure to acknowledge the relevance of all orders of relations in the cultural repertoire. By teasing out the continuity between *li* and other concepts of order, I seek to make historical sense of *li*'s new metaphysical claim and complete my narrative of system building.

The word *li* 理 was a common and versatile semantic unit in the vocabulary of Chinese thought. It has made multiple appearances throughout the course of this book. First, in *Mengzi*, the ancient thinker called for attention to the "*li* of the ten thousand things" (see chapter 2). In the eleventh century, *li* maintained a pervasive presence in the literati's writings. All key Song figures I have covered so far—Shen Gua, Wang Anshi, Su Shi, Shao Yong, the Cheng brothers, and more—referenced the word in one way or another. In different contexts *li* registered different meanings and represented distinctive concepts.

In most cases prior to the Chengs, *li* can be translated as "pattern." The earliest mention of *li* referred to natural patterns engrained in jade or divisions of farming land.[3] From antiquity up to the eleventh century, the term liberally designated patterns in the phenomenal world. Su Shi used to contemplate the patterns of water: "As vast as rivers and as deep as the ocean, [ . . . ] there are necessarily patterns (*li*) to the myriad changes [of water]."[4]

The Cheng brothers turned the unassuming "pattern" into a philosophically fecund concept, "Coherence." In a word, *li* as Coherence meant "the valuable and intelligible way(s) that things fit together."[5] Unlike an inconsistent pattern, Coherence was supposed to be omnipresent. It existed at every possible level in the world: a chair had its Coherence, a human activity had its Coherence, and the vast Heaven and Earth also possessed its Coherence. As Cheng Yi claimed, "each thing has to have a Coherence" (*yi wu xu you yi li* 一物須有一理).[6]

The most salient characteristic of Coherence was its unique promise of unity. In Cheng Yi's well-known stipulation, "Coherence is one and its allotments are different" (*li yi er fen shu* 理一而分殊).[7] That is, there existed many distinct Coherences in different forms, but eventually they came to unify in one grand Coherence. The philosophical reasoning behind this claim was twofold. For one thing, the unity of Coherence did not depend on selfsameness of any kind. Since *li* was "fitting together," one had to admit that there were numerous distinctive ways of fitting together in the phenomenal world. The legs of a chair fitted together with the seat; a father and a son also fitted together, albeit in a vastly different sense. The designation of Coherence meant to embrace this infinite diversity.

For another thing, each specific Coherence was ultimately identical with another "because of the ways in which the individual Coherence of anything systematically interrelates with the Coherence of other things."[8] Put differently, one Coherence connected and harmonized with another infinitely across the world, which gave rise to one grand Coherence. A knower was supposed to keep "extending" (*tui* 推) Coherences from one case to another until he attained the ultimate unitary Coherence.[9] This was the process the Chengs labeled "investigating things and extending knowledge" (*gewu zhizhi* 格物致知), the epistemological tenet famously associated with neo-Confucianism.[10]

## Li *as Any and All Orders of Relations*

Now I link the new concept *li* back to the big picture of system building. The Chengs' coinage of Coherence was no isolated intellectual invention; it was a new response to the collective concern of ordering the world and another attempt to scour the cultural repertoire for organizational norms. What differentiated the Chengs from other system builders was their metaphysical upgrade, accomplished by envisaging a unity independent from uniformity. Coherence, the embodiment of such unity, was able to accommodate any relation in the cultural repertoire and acknowledge its normative force in any particular context. Metaphysically speaking, Coherence claimed conceptual priority to any other existing orders, and its content, if articulated, was as extensive as the assemblage of all orders of relations. The *li*-centered system was able to overwrite all stage-one systems by acknowledging the multitudinous *dao*-seeking process with a valid philosophical design.[11]

Below I list a set of concrete examples to demonstrate the variety of *li*, which grew out of its all-encompassing connection with existing orders of relations. For the sake of rigor, each example features the Chengs' explicit usage of the term *li* (thus excluding many examples in which they spoke of a *li* without explicit identification).

Starting from the most concrete level, the Chengs sometimes defined a *li* as the "function" of a "thing." As discussed in chapter 7, such a function often related the "thing" to a human purpose. For instance, Cheng Yi once stated: "Divination can invoke responses [from deities], and sacrifices can invite appreciation [from ancestors]—such is a *li*."[12] The *li* at issue was the capacity of divination to communicate with deities and the function of ancestral sacrifices to reach forefathers. In each case, the *li* concerned the relation between a "thing" (divination/sacrifice) and the human utility it fulfilled.

The Cheng brothers extensively identified *li* in the orders of moral and ethical relations. Cheng Hao famously called "humaneness" (*ren* 仁) a *li*. He advised that "once you have recognized this *li*, preserve it with sincerity and respect; there is no need to defensively guard it or to extensively search it."[13] Cheng Yi directly made *li* an equivalent of "ritual propriety," incorporating all ethical relations into the purview of the new concept of order.[14]

In Cheng Yi's conceptualization, a *li* could also be a numerical relation. In the Song, a widely shared interest in the *Change* annexed a large group of literati into the world of number, and Cheng Yi was one of them. Because of his deep commitment to the classical text, Cheng frequently employed the basic language of number—most notably the yinyang alternation—in characterizing the structure of the universe. In his opinion, "between Heaven and Earth there is not a single thing without yinyang."[15]

Not surprisingly, Cheng Yi extensively invoked yinyang while artic-ulating *li*. He once contemplated a *li* concerning meteorological phe-nomena as follows:

> In Chang'an a westerly wind often brings down rain, a *li* I have not yet grasped. It has to be the case that northerly and easterly cause rain, and southerly and westerly do not. Why? The east and north are yang. When yang sings, yin would chime in, hence a rainfall. The west and the south are yin. When yin sings, yang would not chime in.[16]

Cheng Yi was making an effort to identify a *li* although he had not yet reached a satisfactory conclusion. Despite the temporary bafflement, he seemed certain that this *li* resided in yinyang. In fact, his association of *li* and number far surpassed sporadic cases like this; he made an encom-passing claim for the *Change* that "the meanings (embedded in the sys-tem of *Change*) extensively cover the *li* of the ten thousand things."[17]

Readers may question the foregoing argument by citing Cheng Yi's criticism of Shao Yong, which seemed to imply Cheng's objection to num-ber and the incompatibility between *li* and number as two conflicting concepts. I argue that this is a misunderstanding. Cheng and Shao both engaged number and endorsed its ordering effect. The discord between them was dual: first, they held a more specific methodological disagree-ment within the realm of the *Change*; second, they differed over the hier-archical order between *li* and number in terms of conceptual priority.

Two exegetical methods were well known among Song literati in the study of the *Change*: "meaning-principle" (*yili* 義理) and "figure-number" (*xiangshu* 象數). The former focused on excavating ideas and meanings from the text of the *Change*, a line of inquiry generating a discourse con-cerning morality and political order. The latter concentrated on the analy-sis of hexagram structures, which entailed complicated mathematical correlations of the trigrams as well as elaborate diagrams.[18]

Cheng Yi, a representative of the meaning-principle school, disagreed with Shao Yong, a prominent adherent of the figure-number tradition, for concrete methodological reasons.[19] Cheng surely preferred manifest moral messages over cryptic numerical diagrams. He worried that the figure-number method denigrated the study of the *Change* by maintain-ing implicit connections with the occult.[20] Both concerns challenged

Shao's concrete praxis, his certain renditions of number, rather than the validity of the whole concept.

Against the background of the study of the *Change*, Cheng Yi provided a metaphysical reason that *li* was capable of taking over number. He assigned priority—in ontological and conceptual senses—to *li*:

> First there is *li* and then, figure. First there is figure, and then, number. The *Change* illuminates *li* with figure, and renders figure with number. Once [one] obtains the meaning [of the *Change*], [he would find] figure and number in there as well.[21]

Cheng further described *li* as the "utterly intricate" (*zhi wei* 至微) and figure as the "utterly manifest" (*zhi zhu* 至著). He asserted that figure (such as a hexagram) and number (such as a numerical relation Cheng endorsed) were both articulations of a correspondent *li*. Speaking of the numerical relation was the same as referring to *li*, except that *li* was conceptually prior to its articulated content. *Li* thus successfully incorporated number.

Similarly, within the conceptual framework of the *Change*, Cheng Yi also incorporated *qi* into *li*. *Qi* was the focal order Zhang Zai chose in structuring the universe (see chapter 7). *Qi* assumed an intimate relation with number by providing the latter with a unified material base (see chapter 4). With this connection, Cheng Yi naturally subsumed *qi* into the hierarchical sequence where *li* held conceptual priority: "There is *li* and then there is *qi*; there is *qi* and then there is number."[22] In so claiming, Cheng subverted Zhang Zai's stipulation that *qi* assumed priority and *li* (in the sense of "pattern") was just a property of *qi*.[23] Cheng's revision later evolved into Zhu Xi's famous formulation of the *li-qi* dyad, for which Zhu elaborately argued for the primacy of *li*.[24]

Because of the connection between *li* and other concepts of order, Coherence inherited a number of philosophical traits I discussed previously: the conflation of fact and value, a causal influence in structuring reality, and resultant normative force, which served as a form of epistemic guidance. Most importantly, when one spoke of a *li*, he always established his foothold in relationality. The Coherence of a "thing" was not simply what the "thing" was (e.g., its substance as an individuated entity) but a relational "fitting together" by which it was as it was, or in Cheng Yi's words, "by which it was so" (*suoyiran* 所以然).[25]

To clarify this point, let me cite a potentially misleading example, a famous reference from the later Zhu Xi. He spoke of a *li* of a bamboo chair as follows:

[A bamboo chair] has to have four legs to be stable and suitable for sitting. If it lacks a leg, one definitely shouldn't sit on it. If one is not yet aware [of the *li*] and only makes fuzzy guesses, two legs may do, and three legs may do. But at the time of sitting, it simply is not possible for sitting.[26]

At first sight, this *li* seemed to regard the chair as an individuated object: it addressed the number of legs—the internal structure of the chair. But as Zhu strongly implied by focusing on the suitability for sitting, he was in fact concerned with the relation between the chair and its human utility (or "function"). The *li* was the way the legs and seat of the chair fitted together to serve an intelligible and valuable function (to-be-sat-upon). This *li* might as well provide grounds for a geometrical understanding of the seat–leg structure as an individuated phenomenon, but the latter inquiry differed by a paradigmatic distinction.[27] "Function" as a deeper order was distinguishable from the properties of the chair in its individuated existence, although in this case the difference was indeed slight in terms of concrete conclusions.[28] The paradigmatic distinction grew in salience as the Coherence seeker (Zhu Xi) kept "extending" to larger orders, and the empiricist (Shen Gua) demanded more details in the world of individuated entities.

In sum, the Cheng brothers' system emerged with a new conceptualization of unity that overcame the restraints of systems envisioned by Wang and Shao. It absorbed the multiform deep orders into a unitary framework that yet left the diversity unmolested. The Chengs ushered in a new metaphysical horizon, on which the ultimate total-view system, neo-Confucianism, readily emerged.

The new prospect may invite readers to imagine that the Chengs prescribed a large body of "Coherences" unprecedented in width and variety; obviously, they were able to do so. In reality, they did not. The aforementioned examples of *li*, though indeed diverse, constituted only a small portion of the Chengs' writings. Compared to other system builders, the Cheng brothers were curiously restrained in prescribing norms. They did not seem to view articulating specific Coherences as the core task of their system.

How so? An immediate reason has to do with a practical concern: how was this system supposed to guide its followers? Incorporating all eligible relations meant that the new system relinquished the kind of leading authority derived from uniform orderliness. Making specific prescriptions on an all-embracing scale was a logical solution but hardly a viable one. The Cheng brothers took an "inward turn" by reorienting the guiding function of their system from the external to the internal, a point I explicate next.

## *Heart-Mind*

In my brief introduction in chapter 7, I present the heart-mind with two key features. First, it was the epistemic organ that supported the mainstream way of knowing, modeling. Second, it enabled one to come to good knowledge spontaneously. This type of "spontaneous discernment" was supposed to surpass the less desirable alternative that one knew by following an explicit prescription.[29] As I show with evidence shortly, these philosophical stipulations were an ancient heritage. Any eleventh-century thinker who intended to systematically address the issue of knowing was obligated to accommodate these arguments in his conceptual design.

As readers may have noticed, stage-one systems had scant capacity to meet these requirements. Featuring fully specified norms as their main content, these systems exhorted adherents to follow their guidance in the form of concrete prescriptions. In the extreme case of Wang Anshi, followers simply had to obey the rules he prescribed; in the more lenient example of Shao Yong, adherents still needed to rely on the complex numerical formula he offered. Despite the fact that Wang and Shao made mention of the heart-mind and its special features (see later quotations), they made little evident effort to reconcile the conflict between prescriptions and the heart-mind's putative independence from them.[30] The Cheng brothers proposed to expound the heart-mind psychology with a substantial philosophical design; in so doing, they were able to rectify the issue of excessive normativity.

Before delving into the details of the Chengs' design, let me first introduce the history of the heart-mind in broad strokes. One focal question that ran through this history concerned the source of the heart-mind's special knowledge capacity. Most thinkers believed that the capacity derived from a connection between the heart-mind and larger cosmic orders. The first prominent thinker who mentioned the connection was Mengzi. According to him, the heart-mind—potentially anyone's—was supposed to connect to Heaven via one's "nature."[31] In chapter 1 I cited the famous Mengzian argument: "To exhaust one's heart-mind so that one knows his nature. To know one's nature, one knows Heaven." A key implication of this stipulation is that if the heart-mind remained in unobstructed contact with Heaven, any spontaneous response from it should be appropriate. Therefore, the key to good knowledge reposed on a well-functioning heart-mind.

One should not assume that the heart-mind would always stay connected with cosmic perfection. As part of an individual human, the heart-mind could deviate from the cosmic perspective with which it was originally endowed.[32] A responsible knower should vigilantly discipline his heart-mind, and for this purpose eleventh-century literati suggested several actions. Wang Anshi aptly exhorted people to "discipline the heart-mind" (*zhi xin* 治心).[33] He described that when one "emptied his heart-mind" (*xu qi xin* 虛其心) and "pacified his intentions" (*ping qi yi* 平其意), he would see "principles" (*yi* 義).[34] Shao Yong compared the ideal state of the heart-mind to "still water" (*zhishui* 止水) and analogized that "the still water is stable; being stable, it will then be tranquil; being tranquil, it will then be illuminating."[35] Zhang Zai engaged the discussion of the heart-mind with notable enthusiasm.[36] In various contexts he called for people to "enlarge the heart-mind" (*da qi xin* 大其心), "preserve the heart-mind" (*cun qi xin* 存其心), "maintain this heart-mind" (*weichi ci xin* 維持此心), and "empty the heart-mind" (*xu xin* 虛心).[37]

Following Zhang Zai's steps, the Cheng brothers presented the connection between the heart-mind and cosmic orders with the strongest possible emphasis. Besides a link with Heaven, Cheng Yi further argued that the heart-mind was in unity with the *dao*, mandate/destiny, *li*, and nature. Regarding the *dao*, he stated that "the heart-mind is indivisibly

one with the *dao*."[38] To Cheng Yi, this unity was not simply a close connection. From an absolutely unhindered omni-perspective, the heart-mind
and these other entities were different existences of the same thing. In
his characterization: "In Heaven it is mandate/destiny, in meaning it is
Coherence, in human it is nature, when commanding the human body
it is the heart-mind. In fact, they are all one."[39] From the human point of
view, that the heart-mind and *li* were one meant that all Coherences were
in the heart-mind. When one encountered a "thing" and wondered about
its *li*, he should assume that the *li* was already there in his heart-mind,
albeit in a dormant condition. The next move was to "illuminate" (*ming*
明, *zhu* 燭) it.[40]

## *Spontaneous Discernment*

Cheng Yi's illumination metaphor effectively characterized how one
should discern a Coherence: an immediate process involving little delib-
eration. More specifically, this immediacy meant that the normative force
of a Coherence directly prompted the knower to a correspondent belief/
action without having to invoke his awareness of the articulated content
of the Coherence. In other words, the knower reached good knowledge
without following an explicit guideline.

Cheng Yi further elaborated on the difference between the two types
of knowing:

> There is a difference between acting after discerning [a *li*] and acting after
> thought (*lü* 慮). If you have discerned it in yourself, the action will be as
> simple as using your hand to lift a thing, and then everything else will fol-
> low. If you have to think, however, it is not yet within yourself, and action
> is like holding one thing in your hand to take another. Such is not good
> for knowing.[41]

In the process of discernment, as soon as a specific *li* dormant in one's
heart-mind had been illuminated, an appropriate belief/action would
spring from inside the knower. In contrast, if he had to guide himself
with thinking (on some explicit rule), the belief/action that ensued would

not count as real knowledge because all causal factors of this knowledge lay external to himself.

The distinction Cheng drew most easily made sense in a moral context (not surprisingly, the realm that interested him the most). For example, a son took a three-year leave from work to properly mourn for his deceased father. In an ideal scenario, he did so because he immediately knew that it was the right thing to do; his commitment to mourning sprang automatically from his heart-mind. But if he took the action simply to comply with a ritual prescription, the nature of his decision would be very different from and inferior to the former case.

Beyond the realm of morality, one was supposed to spontaneously discern Coherences existing extensively in the world. An intriguing conversation between Cheng Yi and Shao Yong (recorded from Cheng's point of view) illustrates this epistemic praxis in a case concerning the natural world:

> Shao Yaofu (Shao Yong) once said to Master Cheng: "Although you are a bright man, things in the world are numerous indeed. Can you know them all?" The Master [Cheng] said: "there are certainly many that I don't know. But what do you mean by 'not know?'" At that moment a thunder struck, and Yaofu asked: "Do you know where the thunder comes from?" The Master answered: "I do, but you don't." Yaofu asked in astonishment: "How so?" The Master said: "If you already know, then why bother to calculate it with number? Precisely because you don't know yet, you have to first calculate and then come to knowledge." Yaofu said: "So where do you think it comes from?" The Master said: "It comes from where it comes from." Dumbfounded, Yaofu extolled the brilliance of the answer.[42]

Why was Cheng Yi's tautological answer brilliant? The reason was precisely his emphasis on the distinction between discernment and knowing by means of a rule. In this case, he stated explicitly that to calculate through number, that is, to follow a guideline articulated in numerical terms, provided a lesser alternative when one had not yet reached the state of true knowing. True knowledge appeared spontaneously when the certain *li* concerning the origin of the thunder became illuminated in one's heart-mind, a process involving no deliberative activity concerning the content of this *li*.

Readers may wonder: if to discern a Coherence was a matter of maintaining a marvelous heart-mind and experiencing an internal illumination, why should one bother to explore orders in the external world, the arduous task that had thus far engaged system builders? In Cheng Yi's system, the internal psychology of knowing was not supposed to contradict or retreat from the exploration of the external world, a stance he made conceptually plausible in at least two ways.

In one way, although the psychological process of discerning a *li* did not require an awareness of its content, a *li* nevertheless had articulable content. In the example regarding thunder, Cheng Yi rejected the clumsy action of following a rule, but he did not deny the possibility that the *li* of thunder could have any content beyond tautology. In fact, he once made his own effort to articulate the *li* of thunder; in his understanding, "thunders arise from mutual attacks between yin and yang."[43] Thus, he agreed with Shao that the *li* of thunder had content and that this content could be articulated with a numerical formula (except that Cheng's formula was not as elaborate as Shao's due to their methodological distinction).[44]

In another way, Cheng Yi incorporated the external world into the procedures of cultivating internality, in two different senses. First, he believed that the potential of one's heart-mind to discern Coherence depended on an active exercise in exploring Coherences among "things."[45] Second, as a logical extension, Cheng endorsed the written culture—where the repertoire of articulated accounts on orders resided—as a pedagogical tool in nurturing the heart-mind. In his opinion, "to nurture the heart-mind with principles" (*yili yi yang qi xin* 義理以養其心), though not the only choice (much less the best) in ancient praxis, remained an effective educational model.[46] In sum, due to Cheng Yi's explicit efforts to conjoin the internal with the external, the psychology of spontaneous discernment did not pertain to mysticism in any sense.[47]

Cheng Yi's two innovations—loosening the fixation on one focal order and substantiating the heart-mind psychology—joined to form a new solution to the problem of excessive normativity. Moreover, a simultaneous adjustment he managed was to redefine the guiding function of his system: instead of building his authority on prescribing norms in the external world, he asserted power in guiding followers' cultivation of the heart-mind, among other aspects of the self. At this concluding point, my narrative of system building is reaching the prominent, panoramic observation of the Song intellectual world that sees the rise of neo-Confucianism as signaling an "inward turn" of Chinese thinking.[48] In addition to social and political factors, the Song literati held convincing epistemological reasons for moving the focus from the state toward the self.

## *Shen's Empirical* Li

In contrast to the Cheng brothers, Shen Gua expressed far less interest in *li* and *xin*, which unsurprisingly claimed no dominant significance in his nonsystem. Although the comparison commences in asymmetric terms, it yields critical results: Shen's disinterest in these unity-related terms provides new angles for exploring the depth of his empirical spirit.

In general, Shen's use of *li* was unsystematic and inconsistent. In his more formal writings, such as classical commentaries and official essays, he invoked the term in vague resonance with the Chengs a couple of times. He once referred to "the *li* of the ten thousand things" in his commentary on *Mengzi* (see chapter 2). In a memorial, Shen made a humbling remark to the emperor, calling himself "terribly incapable of illuminating *li*" (*zhu li shen bei* 燭理甚卑), alluding to the heart-mind's capacity of discerning Coherences.[49] In *Brush Talks*, he used the term most often as a loosely construed "pattern," and the kinds of *li* in his accounts varied markedly. It could be a calculation principle in the system of the *Change*, a proper sequence of pitch standards, the medicinal quality of cinnabar, or a cause for the formation of earth.[50]

Shen also once explicitly denied that *li* stood for any omnipresent order in the world, a curious claim that would immediately cancel out the laconic homage he paid to Mengzi's "*li* of the ten thousand things."[51] In a letter to a friend, he asked with striking honesty, "where does *li* lie," and made a statement as follows:

And then here comes my doubt on *li*. Where does *li* lie? I do not really know. There are sayings of this and that kind. One should cautiously evaluate the right and wrong and choose accordingly. [As for you,] just hold onto your heart-mind, believing only in the right [*li*] instead of obsessing over the great many [possibilities] out there. Even though [you] may miss [some *li*], I would say that [you] still for sure have made some gains.

況而又吾之所疑理。理之所在，某不知也。彼說焉，此說焉，審別其是非而取之。以吾子之心，信其是，無信其多，雖失之，某猶必謂得之。[52]

The query of the whereabouts of *li* was unusual even among thinkers who did not use the term as systematically as Cheng Yi did. Nevertheless, the bizarre question exposed Shen's stance with clarity: he did not believe that a claim for the omnipresence of *li* was viable or, at the very least, meaningfully productive. On this basis, he announced that his approach to grasping *li* would be judicious rather than inclusive; he would believe only in the ones he was convinced of and ignore the rest.

Later, in Shen's multiform uses of *li* in *Brush Talks*, at least one kind had a strong resonance with the selective stance he stated in the early years: what I call the empirical kind. Such *li* was either descriptive or explanatory. In the former case, it stood for an observable (and often repeatable) phenomenon of a particular "thing," and in the latter, an empirical cause. Both derived from observation, and both passed Shen's reliability tests, thus remaining limited in number and yet trustworthy in content.

Examples of descriptive empirical *li* abounded in *Brush Talks*. Shen saw a *li* in that a magnified needle always pointed south and a cypress always faced west, a pattern salient and yet hard to fathom.[53] Similarly, he identified a *li* with the herb cinnabar, which, depending on dose and usage, could turn into a lethal toxin or a nourishing medicine.[54]

Shen connected *li* with his thinking on causation and presented a series of local, experiential causes under its rubric. His discussion of an apparent malfunction of the clepsydra serves as a good example. Calendrical experts at the Bureau of Astronomy had been long concerned with a glitch, that on the ruler of the clepsydra, the length of a day in winter months exceeded one hundred marks (the entire range of calibration covering the span of a day), whereas its counterpart in summer months fell under one hundred marks. Practitioners conventionally attributed the cause to water, presuming that liquid dripped through the tanks more smoothly in the summer than in the winter.

Shen disagreed and proposed to "explore it with *li*" (*yi li qiu zhi* 以理 求之).[55] He argued that the cause of the discrepancy—the *li* at issue— concerned a different phenomenon: the sun's apparent movement. Instead of brooding over the instrument, Shen suggested that the length of an apparent solar day was indeed different.[56] From decade-long observations, he noticed that the sun moved more rapidly and finished its daily travel among the stars faster on the winter solstice, and it slowed down and took longer to conclude the trip on the summer solstice. In Shen's consideration, these fluctuations would cause changes in the length of a day, which was the interval between two successive appearances of the sun at its highest point in the sky (noon). When subject to measurement by the same graduated ruler, the days on the winter solstice and summer solstice would certainly entail different readings.[57] Thus, the *li* Shen identified in this case was the causal relation between one sensory phenomenon and another.

Shen also referred to a "constant *li*" (*changli*), which highlighted certain patterns for being more established in repeatability. For instance, that a concave mirror formed an inverted image of an object was a constant *li*, and so was the inversion of a kite's shadow as it passed through a small aperture.[58] Another *li* that merited the constant status was the resonance between two string instruments when they simultaneously produced sounds of the same pitch.[59] Shen noticed a "constant *li*" in the connection between blooming season and temperature: while spring flowers in general were on the wane in the fourth month of a year, peach blossoms on a high mountain just started to bloom around that time.[60]

"Constancy" was not a foreign concept to Shen. In tackling number, he addressed a numerical relation as a "constancy," and a meteorological phenomenon that deviated from a constancy as an "aberration." Despite the hierarchical implications embedded in the dichotomy, he emphasized that for a person to know the world an aberration was as indispensable as a constancy. In chapter 4, I discussed this line of argument as Shen's effort to individuate "things" (meteorological phenomena) from larger orders (number). His invocation of "constancy" in "constant *li*" signaled an even more aggressive promotion of sensory knowing. By assigning the "constant" status to a sensory phenomenon he repeatedly observed, he practically created a deep order by virtue of extrapolation, which demonstrated a confidence in sensory observation and a robustly empiricist spirit.

## *Toward the Rise of Subjectivity*

In this section I discuss Shen's use of the heart-mind—or rather, his choice not to use it much—when discussing his own way of knowing. I argue that Shen's departure from the discourse on the heart-mind signaled a nascent division between subject and object, which posed an even deeper challenge to the premises of the mainstream way of knowing. Like "representation," the subject-object schism was an epistemological assumption Shen held rather than a concept he engaged with normative articulation (see chapter 4). Nevertheless, the assumption made such a decisive difference in guiding his epistemic decisions that it deserves a detailed investigation no less than any articulated concepts.

To a modern audience, the difference between a knower in modeling and a subject is salient.[61] To start with, while a subject holds his individual point of view at the center of a knowing activity, a practitioner of modeling does not. The propositions regarding the heart-mind specifically amplify this distinction: due to the connection between the heart-mind and cosmic orders, any creditable knowledge a knower obtained from the process of modeling had to reflect a cosmic perspective and not a personal point of view. To place a "thing" in a relation was essentially to actualize (part of) the cosmic order. It is fair to say that in modeling, there was a knower but not a subject.

Second, in modeling there was no real object either. Because knowing was to place a "thing" in a relation to generate a belief/action, neither the "thing" nor the relation was strictly an object. Furthermore, neither "thing" nor relation could assert independence from the knower. Cheng Yi viewed the heart-mind and Coherence as the two different existences of the same thing. Similarly, "things" also had a conflated relationship with "the self" (*wo* 我, literally, "I"). The Cheng brothers held that "things" were in unity with the self. Cheng Hao famously asserted that to possess the true quality of a moral human being, one should "form an undivided body with things" (*hunran yu wu tongti* 渾然與物同體).[62] Cheng Yi took a less radical stance, but he also objected to alienating "things" from the self in any substantial sense. When a student approached Cheng Yi and inquired whether he should distinguish "things" external to his nature from "things" internal to him, Cheng Yi advised him to drop the distinction and investigate all "things," because "every thing had a Co-herence" (*wu wu jie youli* 物物皆有理).[63] For the Chengs or any serious practitioners of modeling, there was no philosophical reason to demarcate humans from "things," as both were equal parts of the interdependent cosmic dynamism.

In Shen's epistemic praxis, however, a thinking self with a resem-blance to a subject was emerging. First and foremost, without being held responsible for carrying a cosmic perspective, the knower in each par-ticular inquiry projected an individuated, local point of view. The im-mediate cause of the change undoubtedly lay in the fact that "hearing and seeing" remained the main epistemic apparatus. A knower held full ownership of his sensory modalities, which, unlike the heart-mind, claimed nothing more than an egocentric viewpoint. All other critical skills in Shen's demonstration were oriented toward an individual knower as well. The knower perceived, believed, and doubted, exercising all the epistemic virtues for his own reliable epistemic praxis. Therefore, Shen's epistemology represented a mathematically singular point of view, and any follower of this nonsystem was reasonably comparable to a subject.

Shen's knower had a new known: "things" in individuated existences. Instead of trying to vanish into larger cosmic orders, the knower claimed an epistemically privileged position and demanded individuated "things" to be present to his perception and thinking. In so doing, he turned "things" into objects and became the "relational center" of them all.[64] Taken together, a subject–object relationship was thus arising from Shen's epistemological horizon.

In addition to the general characteristics, Shen demonstrated a few concrete predilections that spoke to an arising self-awareness as a thinking subject. One prominent example was his intensive interest in matching "things" with their names. Although it was common for a responsible scholar to pay attention to the choice of words, Shen's me-thodical, chapter-length commitment to lexical accuracy was unusual among his peers, and the epistemic connotations of such devotion were even more unconventional.

Shen's sensitivity to the issue of "names" was closely related to his na-scent awareness of a subject–object divide. Generally speaking, lexical accuracy becomes a particularly critical issue when one views language as a medium. In Shen's case, the name of a "thing" served as a semantic medium that connected the "thing" (object) to him (the subject). The word "short-back robe" articulated the idea of an article of clothing and made available what was essential about it (a short back, instead of a regular long back). The accuracy of such communication immediately affected the subject's ability to access the object and formulate reliable knowledge of it.

It is interesting to note that the semantic relationship between a "thing" and its name, though a natural assumption of a modern audience, was at odds with the mainstream conceptualization of language. For instance, Cheng Yi argued that the name of a "thing" was a natural constituent of it: the name drew on the *qi* of the "thing" and derived from its Coherence. A "thing" and its name thus had no ontological distinction. In Cheng's words:

> The character(s) in the name of a "thing" is in continuum with (*xiangtong* 相通) the pronunciation and meaning (of the character), as well as the *qi* and Coherence (of the thing). Besides those "things" which have easily identifiable qualities and thus can be named accordingly, [some cases are less obvious]. For instance, Heaven is called Heaven, but before it was named, Heaven was just some infinite vastness that had no name. How did it obtain this name? [The name] derived from its self-so Coherence, and the pronunciation came from the *qi*, hence the name and the character.[65]

In addition to Cheng Yi, a number of his peers, such as Wang Anshi, made similar statements.[66]

Such a naturalistic view of language traced back to ancient thought and, most importantly for the current discussion, served as a premise for the famous classical argument "rectifying names" (*zhengming* 正名). Though bearing a deceptive resemblance to Shen's pursuit of lexical accuracy, "rectifying names" in the mainstream discourse pursued a vastly different purpose. A "name" (*ming* 名) formed a dichotomy with "actuality" (*shi* 實), and they formed an ontologically undivided whole (just like a "thing" and its name in Cheng Yi's discussion).[67] A name was a social, political catalyst with a "performative force," and applying a correct name was to "form its relation to reality and support its emergent actuality."[68] For instance, calling one "father" was to invoke the norm of being a father and exhort the person to fulfill his fatherly responsibilities. Assuming salient moral and political functions, "rectifying names" was thus no simple textual exercise of searching for the right words.

Shen's intense pursuit of lexical accuracy was essentially a disparate commitment in a different intellectual context; the semantic relation between language and reality, although just an implication in this case, offered the most plausible explanation for his motive. Whereas peers such as Cheng Yi might share a concern for diction in the praxis of writing, they had little philosophical motivation to do what Shen did. It was not until several hundred years later that Shen's proclivity found enthusiastic resonance among scholars interested in philology, a development I elaborate in chapter 11.

Another prominent example attesting to Shen's recognition of a subject–object division was his stance on foreknowledge. Many system builders in Shen's age—including Cheng Yi and Shao Yong—believed in foreknowledge because they viewed it as a marvelous function of the heart-mind. Given that the heart-mind maintained a connection to cosmic orders, it had the ability to know the place of a "thing" in an order now and at any particular temporal point in the past or future. In fact, the omni-perspective held by the heart-mind rendered time a relative concept. As Shao once argued, "the present is called 'present' when observed from the standpoint of the present; when it is observed in hindsight, the present is also called 'past'."[69] The future, just like the present and the past, was another temporal point completely accessible to the perspicacity of the heart-mind. Cheng Yi once applauded Shao Yong as someone "superiorly intelligent" who "had foreknowledge of things."[70] The reason, as Cheng reckoned, was precisely because Shao was able to keep an "empty and illuminated" (*xuming* 虚明) heart-mind, through which all cosmic orders became discernible.[71]

Shao held a subtly different stance on foreknowledge. On one hand, he practiced divination by operating number-related schemes (such as Course-*Qi* and the *Change*) and thus certainly believed that that one could access information about the future by deciphering number. On the other hand, he made a critique of the concept of "foreknowledge" (*qianzhi* 前知), implying that the so-called fore-knowledge should be categorically separated from "knowledge" (*zhi* 知). He said:

> People who [claim to] have foreknowledge can speak of things which are supposed to happen hundreds and thousands of years from now; sometimes they attain it in dreams. Based on this it is believed that everything is predetermined. I do not agree with this: things are not predetermined. The moment when one knows [something] is a "today," with which any time in between ("today" and the moment the thing happens) is the same. There is no earlier and later to start with. This pattern is evident, and if you frequently make observations, you shall understand it. Some people may argue: "With foreknowledge one can avoid things of harm." This is also not the case. If you are able to avoid something, the moment you fore-know about it, you have already seen the thing you are avoiding. Since you have not yet seen the thing you are avoiding, you cannot claim that you have fore-known it.
>
> 人有前知者，數十百千年事皆能言之，夢寐亦或有之，以此知萬事無不前定。予以謂不然，事非前定。方其知時，即是今日，中間年歲，亦與此同時，元非先後。此理宛然，熟觀之可諭。或曰："苟能前知，事有不利者，可遷避之。" 亦不然也。苟可遷避，則前知之時，已見所避之事；若不見所避之事，即非前知。[72]

Shen presented two arguments to make the case that "to know" and "to foreknow" were conceptually incompatible. First, the moment one knew that something was happening and that the event was happening had to be the same temporal point. Second, one had to actually "see" (*jian* 見) the thing he was avoiding in the future to claim that he foreknew it. Normally, if someone successfully avoided a misfortune by consulting the oracle, he would never "see" (or more generally, "experience") the incident.

Shen emphasized that "to know" was an experience that required the knower to be present at the occurrence of the event. The knower had to effect the presence with his cognitive capacities, such as to see. In other words, Shen preferred to understand *zhi* as a cognitive act distinguishable from the marvelous function of the heart-mind. He was not claiming that prognosis was not possible; instead, he was challenging the conceptualization of prognostication as a kind of "to know," which deviated from the essential meaning of *zhi* in his definition. Once again, Shen spoke from the position of a perceiving self who subjected everything, including time, to his own scrutiny.

Shen's less systematic mention of the heart-mind was a natural consequence of the development of his epistemology rather than deliberate sedition. In certain contexts, he embraced the utility of the heart-mind with an acknowledgment of its connections with deep orders. For instance, when debating on moral options, he pointed to the heart-mind and its connection with Heaven as ultimate justification (see chapter 1). While practicing divination, Shen fully endorsed the heart-mind's capacity of spontaneous discernment because the senses could not access the numinous realm. For example, he once commented that in divination humans had to rely on certain inanimate objects such as turtle shells, because "humans are unable to reach [the state of] having no heart-mind, so that they depend on things (turtle shells), which have no heart-mind, to speak [about the future]."[73] Here, "having no heart-mind" (*wu xin* 無心) was equivalent with Wang Anshi's "emptying the heart-mind" and Shao Yong's metaphor of a "still-water" heart-mind, all referring to an ideal state in which the heart-mind maintained an undisrupted connection with a deep order. In discussing the insufficiency of calendrical systems, he referred to "true number" (*zhenshu* 真數), the most "intricate" aspect of number, claiming that "the technique can be obtained by the heart-mind and yet cannot be articulated in words."[74] Here he alluded to the heart-mind's connection with number in its most profound state. Shen paid due respect to the hierarchical order between the heart-mind and sensory knowing in the philosophical discourse, yet in his epistemic praxis, the former worked in segregated, special niches, and the latter claimed pervasive utility and validity.

Shen occasionally invoked the heart-mind to refer to the exercise of his nonsystem, thus alienating the term from the conceptual framework of modeling. He once claimed that medical doctors should "obtain [their techniques] from the heart-mind" (*de zhi yu xin* 得之於心) rather than "relying on books [only]" (*shi shu yi wei yong* 恃書以為用).[75] In the rest of the passage, Shen discussed how the efficacy of a medicinal ingredient could vary in different recipes, and he exhorted doctors to rely on their observations and critical thinking—both residing in *xin*—to attend to these experiential vicissitudes. In this case, the heart-mind embodied the nonsystem, and the mechanism via which it worked also switched from spontaneous discernment to exercising critical skills.

In sum, a comparison between the Cheng brothers and Shen entails a gamut of interesting revelations about the intellectual world at the dawn of neo-Confucianism. The Chengs ushered in a new phase of system building by making fresh metaphysical claims centered on Coherence and the heart-mind. At the same time, by engaging or disengaging these concepts in his own way, Shen developed the depth of his empirical nonsystem. The diversity among thinkers grew deeper, wider, and greater in salience before the next dominant trend peaked and eventually flattened the differences.

# CHAPTER II

# Reverberating in the World
## (1100–1800)

While Shen Gua's thinking had an interesting life journey, it has had an even more intriguing afterlife. Although the nonsystem did not inspire a revolution in its own time, it left a long, reverberating sound that has grown increasingly louder with time. In charting this history, I highlight three key trends, which featured system builders in the twelfth century, reliability seekers in the twelfth and thirteenth centuries, and "evidential" (*kaozheng* 考證) scholars in the seventeenth and eighteenth centuries. We can view the appreciation of Shen's thinking as a long process of appropriation and reconstruction motivated by different historical agendas.

In this chapter, I first introduce the general characteristics of the reception of Shen and *Brush Talks*. I follow by presenting the three major episodes in which different historical actors appreciated Shen for distinctive reasons. To conclude, I offer a summary of the new findings Shen has illuminated for us in the broad world of Chinese thinking.

## General Reception: The Book and the Man

The afterhistory of Shen bifurcates into two interlocked yet distinguishable narratives: the reception of *Brush Talks* and the appreciation of the man. The popularity of the text was remarkable, as evident in the abundant citations it attracted. From Shen's times through the eighteenth century, the book was cited over a thousand times across the whole collection of formal writings in China. Authors who referenced *Brush Talks* came from all four bibliographical divisions, namely, "classics" (*jing* 經), "history" (*shi* 史), "masters" (*zi* 子), and "literature" (*ji* 集). The majority of writings featuring quotations from *Brush Talks* belonged to "masters"; many of them were reference collectanea and notebooks in the subsection "Miscellany" (*zajia* 雜家). Next were classical commentaries in the category of Classics, where exegetes cited Shen's opinions with the second-highest frequency. Some authors quoted *Brush Talks* in essays, which were further included in their collected works under the heading of "literature."

Among other genres, large-scale, multisubject collectanea most frequently cited *Brush Talks*, ironically blurring the distinction Shen drew between derivative compilations and his auctorial work. Encyclopedias prominently featuring citations of Shen include the *Collection of Classified Things in the Song Dynasty* (*Song chao shishi leiyuan* 宋朝事實類苑, twelfth century), *Classified Assembly of Things and Words in the Past and Present* (*Gujin shi wen leiju* 古今事文類聚, thirteenth century), *Barbican of Talks* (*Shuo fu* 說郛, fourteenth century), *Compendium on the Arts* (*Yilin huikao* 藝林彙考, seventeenth century), and *Mirror of Origins Based on the Investigation of Things and Extending Knowledge* (*Gezhi jingyuan* 格致鏡原, eighteenth century). The high citation frequency Shen enjoyed among these collectanea greatly bolstered the visibility of *Brush Talks*. Many items thus became semi-official lexical definitions of things, widely circulated as standards and guidelines in the intellectual world. The partiality of encyclopedia compilers toward *Brush Talks* demonstrated a popular view of the text: readers treated it as a reference book and in a way valued its reliable quality as Shen would have wanted.

The appreciation of Shen the person is a more nuanced narrative. On one hand, Shen's influence was quiet. From his own age through the nineteenth century, he did not enjoy special prominence; he was not seen as the founder of a line of thinking or a key member of a specific intellectual lineage (such as neo-Confucianism). For scholars who focused on the neo-Confucian mainstream, Shen was a good source of piecemeal knowledge who did not invite much evaluative attention. The full-throated appreciation of Shen is not a pre–twentieth-century phenomenon. On the other hand, this quiet influence was steady and became increasingly systematic over time in certain areas. Some scholars, especially *biji* writers in the Southern Song and the so-called evidential scholars in the seventeenth and eighteenth centuries, valued Shen for his reliability-centered epistemology as a whole. For these people, Shen served as a historical reference point to ground new methodological claims. In the following sections, I break down Shen's historical reception into three distinctive voices: those of the neo-Confucians, the reliability seekers, and the evidential scholars.

## *The System Builder's View*

The most immediate appreciation of Shen came from Song peers who lived in the latter half of the dynasty. During this time, the trend of system building continued and was about to culminate in the hands of Zhu Xi. Drawing on the work of the Cheng brothers, Shao Yong, Zhang Zai, and other Northern Song system builders, Zhu was in the process of formulating a new system—famously known as neo-Confucianism—on the foundation of *li* (Coherence). All key features of the Cheng system I introduced in chapter 10 persisted and underwent further development in Zhu's new arrangement.

Unsurprisingly, scholars keen on system building applied the two-tier epistemological discourse to frame the evaluation of Shen and associated him with knowing from hearing and seeing in the narrow meaning dwarfed by higher learning goals. Such a view seemed to enjoy some currency in the Southern Song. For instance, an editor of a twelfth-century edition of *Brush Talks*, Tang Xiunian 湯修年 (ca. 1100–1200), prefaced this version with the following message:

> *Brush Talks* records the activities of the emperors in the golden era of our dynasty, the endeavors of councilors and ministers in the age of peace, as well as the marvelous material forms (*zhizuo* 制作) produced in previous times. Although the treatise mostly consists of [things] seen by the eye and heard by the ear (*mu jian er wen* 目見耳聞), it surpasses most miscellaneous jottings by providing something useful to our own age.[1]

On one hand, Tang was clear and a little apologetic that *Brush Talks* was just a work of hearing and seeing. On the other hand, he spoke highly of the "useful" quality of the information Shen provided and appraised the book as exceptional in its kind. Obviously, Tang made his positive and yet partial endorsement of Shen from the vantage point of a pursuer of higher learning, a stance consistent with the Northern Song system builders.

Zhu Xi, the ultimate system builder, held a similar view of Shen. Zhu regarded Shen as a man of "broad learning," and he appreciated the variety of knowledge Shen recorded, sometimes as the basis for his discussion of "Coherence." Nevertheless, Zhu did not identify Shen as a member of the lineage of neo-Confucianism he constructed, nor did he make any systematic evaluation of Shen, such as mentioning his methodology. In a way, he treated Shen just as a system would respond to a nonsystem: the former would incorporate piecemeal insights offered by the latter but regard it as no more than a preparatory stage.

Zhu's appreciation of Shen was notable. In the twelfth century, Shen's dubious political reputation was still vividly remembered, and for this reason some literati openly disparaged *Brush Talks*. Zhu, however, defended Shen's intellectual contribution:

> Lü Bogong (Zhu Xi's friend) disliked *Brush Talks* and thought it to be reckless talk (*luanshuo* 亂說). I told him, "You can't really say that, and I'm afraid you can't beat him on this. It's just that he wasn't really a good person."[2]

Zhu's frequent and positive references to Shen attested to the genuineness of his appreciation. He cited Shen on twenty-two different topics, with few repetitions, mostly from *Brush Talks*.[3] He showed a distinctive preference for particular bits of knowledge from hearing and seeing, for example, movements of heavenly bodies, oyster shells discovered in mountains, and materials of ritual attire.[4] Among these topics, astronomy seemed to be the one that most drew Zhu's interest. Not only did he contemplate the astronomical knowledge Shen provided, he put some into practice: for instance, he once tried to observe the Pole Star using Shen's method.[5]

A number of *Brush Talks* items regarding astronomical phenomena gained popularity in later ages because of Zhu's promotion. One item he especially appreciated was on the nature of the "Yellow Road" (ecliptic) and the "Red Road" (equator).[6] Song calendrical experts tended to view the Yellow and Red Roads as experiential entities with fully tangible material forms rather than as representational models of the sun's movements. Shen contended that they were merely mathematical constructs devised for the convenience of calculation (see chapter 4). Zhu endorsed Shen's opinion by giving the item a full paraphrase. Following Zhu, a series of authors from the Song through the late imperial period cited this item. To name one example, Ma Mingheng 馬明衡 (*jinshi* 1517), a literatus in the Ming Dynasty (1368–1644), quoted this idea in his commentaries *Questions and Explanations on the Classic of Documents* (*Shangshu yiyi* 尚書疑義).[7] This case suggests that a few hundred years after Shen's contention, the issue had become so prominent that even scholars with no particular interest in technical issues felt the need to cite it.

Another entry concerning eclipses attracted Zhu's attention and gained great visibility in later literature as a result of his influence. Zhu agreed with Shen that an eclipse was the consequence of "overlap" (*xiangdie* 相疊) of the moon and sun, an optical trick rather than a real change of the moon's physical shape.[8] Shen's original proposal of this idea was rather iconoclastic, but Zhu's endorsement enabled it to override the conventional discourse on eclipses and become mainstream thinking.

Zhu praised the "breadth" of Shen's learning when evaluating the man; in his own words, "Shen Cunzhong (Gua) was broadly read (*bo lan* 博覽); the instruments (*qi* 器) and calculations (*shu* 數) he examined in *Brush Talks* are extraordinarily accurate (*jing* 精)."[9] The invocation of *bo* revealed an assessment beyond a generic praise of wide coverage. As I discussed in chapter 9, *bo* often appeared in partnership with "essentials," a dichotomy that resonated with that between "hearing and seeing" and modeling. "Broad learning" served as the preparation for the extraction of "essentials" and thus remained a mere intermediary stage in the transition to the higher learning goal. Zhu was a vocal advocate of this dichotomy.[10] In his diction, broad learning was not an all-positive assessment, which he unfavorably compared to "investigating things" (*gezhi*

格致), the search for deep orders.[11] In his view, essentials constituted the ultimate goal for learners who pursued the "great fundamentals" (*daben* 大本).[12] Here both essentials and fundamentals referred to larger orders in which "things" could be well placed, and they were connected with the multistage *dao*-seeking program repeatedly invoked by eleventh-century scholars (see chapters 2 and 7).

Zhu intended a good balance between breadth and essentials. He considered breadth to be an important initial step and thought it a mistake for beginners to start by brooding over essentials instead of venturing into the broad base of possible knowledge.[13] This was why he favored reading and engaging works like *Brush Talks*. Nevertheless, he criticized the failure to rise to the level of essentials after proper exposure to a breadth of "hearing and seeing," as that would distract the learner from the unitary prospect of the *dao*. As Zhu once commented,

> To examine one institution today and another tomorrow, one pointlessly (*kong* 空) devotes efforts to mere utility. And this is a problem I regard more serious than [fixating with] essentials without breadth.[14]

Thus, by his standard, a "board learning" person like Shen offered no more than a supply of piecemeal information, which would claim significance greater than utility only after it was channeled into the search of deep orders. At its strongest, the system set all rules when compared to a nonsystem and easily absorbed the latter into its own dominion.

## *The Reliability Seeker's View*

Another group of literati in the Southern Song appreciated Shen for a different reason. This voice came from those Shen might have regarded as the truly like-minded: *biji* writers who sought to produce reliable knowledge from hearing and seeing in its expanded capacity.[15] Compared with Shen's times, two changes had occurred: the size of the community had visibly enlarged, and its members had started to apply a rhetoric that addressed epistemological reliability as a distinctive intellectual virtue in assessment and self-assessment.

Of special note, while the Southern Song *biji* authors viewed Shen with respect as a forerunner of their methodology, they did not see him as a unique and unchallengeable authority. It is fair to say that an increasingly systematic appreciation of epistemological reliability constituted a movement on its own, which prompted the participants to recognize Shen. Thus, in the following examples, readers will see that Shen did not occupy a central place in the movement, nor was he even consistently appraised in a positive way, but he did provide a solid precedent in the immediate past that Southern Song literati were eager to invoke.

The Southern Song witnessed the production of a number of prominent *biji*, among which *Random Notes Made in the Rong Study* (*Rong zhai suibi* 容齋隨筆, hereafter *Random Notes*) by Hong Mai 洪邁 (1123–1202) and the *Recorded Observations from Arduous Study* (*Kun xue ji wen* 困學紀聞, hereafter *Recorded Observations*) by Wang Yinglin 王應麟 (1223–1296) were especially famous. Together with *Brush Talks*, these works later became known as the "Three Major Notebooks in the Song." Although this grouping was primarily the product of bibliographers' reflections in the late imperial period, it resonated with the Song understanding of the three works in important ways, one of which concerned reliability.

In the world of *biji*, an evaluative rhetoric centered on the pursuit of reliability by applying critical skills became increasingly distinctive. Although the claim of accuracy grounded in the historiographical tradition still enjoyed wide acceptance among notebook authors, this new assessment language was gaining autonomy. In other words, when authors and readers discussed the credibility of *biji*, they no longer exclusively referred to historical composition as the methodological horizon; instead, they started to address a broader reliability based on a range of critical skills.

For example, in the preface He Yi 何異 (*jinshi* 1154) wrote for *Random Notes*, he praised the merits of the text as follows:

> [The notes written by Hong Mai] enable [one] to examine references (*ji diangu* 稽典故), to broaden hearing and seeing (*guang wenjian* 廣聞見), to identify errors (*zheng e'miu* 證訛謬), to lubricate the tip of one's brush (*gao biduan* 膏筆端, to improve one's literary style). They are indeed where a literatus can advance his learning (*jin xue* 進學).[16]

He highlighted the value of the notes as a key resource for a literatus to pursue learning: not only did they provide a broad base of knowledge from hearing and seeing (in the narrow sense), they also supplied critical tools able to rectify errors. In the postscript to *Random Notes*, Qiu Su 丘橚 (*jinshi* 1202) appraised in more specific terms Hong's use of critical skills to forestall inaccuracies and promote reliability:

> In this book, [the author] seeks (*gou* 鉤) and investigates (*suo* 索) [on topics such as] yinyang and figures, as well as rectifies (*jiu* 糾) and remedies (*bian* 砭) [problems] in the Classics and commentaries.[17]

Qiu applauded Hong's efforts "to rectify errors and examine anomalies, to inspect falsehoods and anatomize details" (*zhengwei kaoyi, hewei pouwei* 正譌攷異，核偽剖微).[18] Here he invoked a series of verbs—to "investigate," "rectify," "remedy," "examine," "inspect," and "anatomize" (analyze)—as ways to specify Hong's efforts to ensure reliability among a wide range of topics, including subjects as important as number and the Classics.

Beyond Hong and his *Random Notes*, assessment language centered on reliability appeared in a large group of notebooks whose authors ranged from the famous to the little known. For instance, Dai Biaoyuan 戴表元 (1244–1310) once commented on *Wild Words from the East of Qi* (*Qi dong yeyu* 齊東野語, hereafter *Wild Words*) by Zhou Mi 周密 (1232–1298), a prolific participant of the notebook genre:

> As for Master Zhou's book under discussion, the language is rigorous (*he* 覈), the [records of] affairs are certain (*que* 確). The official titles [he] has researched are as accurate (*jing* 精) as if [he is] emulating Tanzi.[19] The maps [he] has modified are as prudently (*shen* 審) made as if [he is] emulating Bozong of the Jin [Dukedom].[20] The [details] regarding literary writing as well as ritual and music are as insightful (*zhan* 瞻) as if [he is] emulating Zha, the young master of Wu.[21]

Dai enumerated a range of topics in which Zhou successfully produced reliable knowledge, such as details in the bureaucratic system and cartographical compilation, in addition to prose, ritual, and music.

As readers may have noticed, although these *biji* were still designated as knowledge from hearing and seeing, the authors often engaged such topics as ritual, music, and the Classics—subjects frequently associated with the search for deep orders. In other words, the pursuit of reliability boldly advanced into subjects mainly associated with modeling as a distinctively new methodology. Shen's broader application of hearing and seeing found much resonance in these further developments.

Some Southern Song scholars made explicit their intentions to offset negative connotations associated with hearing and seeing and to promote reliability as a fresh intellectual merit. For instance, in a preface he prepared for *Written Notes from the Jieyin Studio* (*Jieyin biji* 芥隱筆記) by Gong Yizheng 龔頤正 (fl. late 1100s), Liu Dong 劉董 (fl. 1210s) wrote: "For literati, it is not broad learning that is difficult; rather, it is being able to reflect prudently (*shensi* 審思) and to make clear identifications (*mingbian* 明辨) that is difficult."[22] Here Liu was alluding to a binary scheme, but he did not pair "board learning" with "essentials," a concept often invoked by system builders. Instead, he pointed to "prudent reflections" and "clear identifications" as the higher goal for a broad learner to emulate. In his formula, critical skills aiming for reliability replaced the search for deep orders.

Another similar example concerns Zhang Shinan 張世南 (fl. late 1100s and early 1200s) and his *Records of Official Travel* (*Youhuan jiyou* 遊宦紀遊). In the postscript, Zhang stated:

> Gaining broad knowledge of things and comprehensive hearing [and seeing] (*bo wu qia wen* 博物洽聞) is the occupation of literati (*ru zhe shi ye* 儒者事也). If one has not set his foot on (*zuji suo jingli* 足跡所經歷) [something], or to witness it with ears or eyes, [yet one goes ahead and distributes it], that would be spreading doubtful information when one is still in doubt. If one does not even trust his own work, how will he be able to win others' trust?[23]

The statement started with the robustly confident assertion that "broad learning" and "comprehensive hearing [and seeing]" should unapologetically be the "occupation of literati." Thus the evaluative framework hinged on essentials or modeling was out of the picture. Zhang then presented "setting foot on" and "hearing and seeing" as the means to achieve reliability, the intellectual merit that enabled him to cast breadth and hearing and seeing in plain positive light. Again, the author placed reliability under the spotlight as the central merit of the project.

Pursuit of reliability not only prompted a corresponding language of assessment in prefaces and postscripts, it entered the classificatory schemes of notebook literature. In the Southern Song, more and more *biji* authors adopted sectional titles standing for critical skills rather than subjects of knowledge, similar to what Shen did in *Brush Talks*.

For example, Wu Zeng 吳曾 (fl. 1140s), the author of *Scrambled Notes from the Nenggai Study* (*Nenggai zhai manlu* 能改齋漫錄, hereafter *Scrambled Notes*), employed a number of division titles with salient methodological messages. The first category was "Origins of Things" (*shishi* 事始), in which he enumerated early meanings of words and designations, origins of certain bureaucratic practices and policies, and early references to certain objects and phenomena.[24] This section was comparable to Shen's "Precedents," which required effort to seek and examine antecedents in experience and textual records. Another category Wu designated was "Identifying Errors" (*bianwu* 辨誤), in which he focused on correcting mistakes in earlier texts, a section similar to Shen's "Errors [and Corrections]."[25]

The Southern Song reliability seekers were aware of *Brush Talks* and paid homage to Shen. They viewed him as a forerunner in the reliability-seeking cause and cited his claims as good precedents. Nevertheless, they did not grant Shen a fundamental significance in establishing this methodology, and their endorsement sometimes took the form of critiquing Shen for not always exercising his critical skills effectively. Shen might not have earned greater esteem among this community of scholars for a variety of historical reasons, but his tarnished political reputation, which had considerable currency in the Southern Song, seemed an ineluctable factor at play in the background.

For example, in *Random Notes* Hong Mai quoted Shen a number of times and, in his unique, critical manner, revealed a high regard for Shen. In one passage, Hong criticized three specific mistakes previous scholars had made in their *biji*, two of which came from *Brush Talks*. The first concerned a story of Xiang Minzhong 向敏中 (949–1020), where Shen asserted that Xiang was the first official Emperor Zhenzong promoted to the position of Vice Director of the Right (*you puye* 右僕射).[26] Hong refuted this belief by presenting new evidence that six other officials had received such a privilege prior to Xiang.[27]

The second mistake occurred in Shen's account of Ding Wei 丁謂 (966–1037), another prominent official who served during Zhenzong's reign. Ding once accompanied the monarch on an inspection tour, and at the completion of the mission, Zhenzong ordered that jade belts be bestowed on eight ministers in the entourage. The administrators had only seven jade belts on hand, and Ding, who longed to receive one, realized to his dismay that seven colleagues ranked above him were the more plausible recipients. So as not to lose the opportunity, he spoke with the emperor in private, confessing that he happened to have brought along a small jade belt of his own and would like to wear it to the bestowal ceremony as a placeholder. The emperor could, Ding suggested, grant him the real jade belt after they were back in the capital city.[28] Hong carefully examined the records of three inspection tours Zhenzong conducted, pointing out that none could possibly be the occasion for Shen's account.[29]

Although in this passage Hong's point was to rectify Shen's errors, his closing statement indicated that his criticism in fact derived from an appreciation of Shen as a like-minded reliability pursuer. Hong said: "Wei Tai 魏泰 (ca. 1040s–1100s) is not worth mention, but Shen Cunzhong should be better than this."[30] Wei Tai was the author who had committed the third error (in addition to Shen's two). While unsurprised at Wei's mishap, Hong expressed regret for Shen, whom he obviously held in higher respect.

A number of other Southern Song *biji* writers treated Shen in ways similar to Hong. They engaged *Brush Talks* deeply and critically, which revealed a common underlying consensus that Shen was a kindred spirit who heralded the new methodological trend and deserved to be subject to peer scrutiny. Wu Zeng, the scholar who adopted methodological section titles similar to those in *Brush Talks*, is one example. In *Scrambled Notes*, he challenged Shen in a number of cases for his failure to achieve reliability. For instance, he questioned the origin and meaning of the term "Wu Hook" (*Wu gou* 吳鉤), which Shen took as a curved knife, a weapon often used in the Tang.[31] Wu argued that it was in fact a hook, made by artisans in the Wu Dukedom in the Spring and Autumn period.[32] Wu was also suspicious of Shen's explanation of a so-called dark ghost (*wugui* 烏鬼) as a cormorant (*luci* 鸕鷀) and asked rhetorically, "[I] wondered what evidence [Shen] had" (*buzhi you he suo ju ye* 不知又何所據也)?[33] In this case, Wu was questioning not only the result but also Shen's proper use of a critical skill (identification of evidence), a challenge he raised with obvious methodological consciousness.

Evaluation of Shen within the framework of pursuing reliability continued from the Southern Song well into subsequent times. A number of Ming bibliographers, for instance, commented on Shen and *Brush Talks* in similar terms. Ma Yuandiao 馬元調 (1576–1645), the compiler of a major Ming edition of *Brush Talks*, made the following remarks:

> Books on examining evidence and making arguments (*kaoju yilun* 考據 議論) are particularly abundant in the two Song periods. The Northern Song featured the three Liu and Shen Gua, and the Southern Song, the Wenmin brothers.[34] People like Ou[yang Xiu] and Zeng [Gong] apparently could not measure up to them.[35]

Ma made a number of interesting points. First, any Song audience would be surprised to see Shen ranked above Ouyang Xiu and Zeng Gong, as the latter two, known as cultural giants and members of Eight Masters of the Tang and Song, enjoyed considerably more cultural prestige in their own times. Also, the juxtaposition of Shen with the "three Liu" was bizarre. Liu Chang 劉敞 (1019–1068) was famous for his classical scholarship, authoring the *Modest Commentaries on the Seven Classics* (*Qijing xiaozhuan* 七經小傳). Liu Ban 劉攽 (1023–1089, Chang's younger brother) and Liu Fengshi 劉奉世 (1041–1113, Chang's son) were known for their work in historical compilation.[36] In Ma's view, these classical and historical scholars should be grouped with Shen and Hong Mai, who were known for participating in the less serious *biji* genre. Ma's rationale in making these two judgments undoubtedly had to do with his endorsement of reliability as an intellectual merit, which, in his words, consisted of practices such as "examining evidence" and "making (critical) arguments." Shen superseded Ouyang and Zeng because of the attention he paid to evidence seeking and other critical skills. Shen and Hong were entitled to the same acclaim as esteemed classicists and chroniclers in their times because they also produced reliable knowledge through evidence-based, critical inquiries.

Ma's invocation of the term *kaoju*, "examining evidence," heralded the next phase of the reception of Shen, when a large group of scholars explicitly embraced evidence seeking as a distinctive methodology of textual studies, a topic I address in the following section.

## *The Evidential Scholar's View*

The seventeenth century witnessed the rise of a new mode of learning, "evidential scholarship," and the evidential scholars endorsed Shen by coopting his reliability-seeking nonsystem into their methodological paradigm. Compared to the Southern Song reliability seekers, evidential scholars embraced Shen with more enthusiasm, put more emphasis on his methodology than on his piecemeal insights, and identified him as a precursor of their own paradigm in more explicit terms. In their view, Shen was one among a group of viable prophets of the evidential methodology. Thus, although still not occupying a central place in this evaluative discourse, for the first time since his own era, Shen enjoyed an outright positive appraisal.

Evidential scholarship was essentially a critical mode of textual studies. Scholars in this tradition endeavored to rectify previous misinterpretations of old texts with a range of philological techniques centered on the use of "evidence" (*zheng* 證). They sought to "construct an argument or interpretation about the content of texts to which readers could respond on the basis of the evidence offered and to which further evidentiary material could be adduced to support or refute any particular claim."[37]

In the seventeenth-century vocabulary, the evidential model was meant to counter the so-called Song learning (*Song xue* 宋學), a loose concept not limited to the Song Dynasty. At its broadest, Song learning included mainstream learning in Northern and Southern Song and post–Zhu Xi development of neo-Confucianism up to the early Qing (often excluding Zhu Xi as an exception for a variety of reasons). In the view of evidential scholars, the central thread that strung all these varieties together into a coherent category was the common subject "meanings and principles" (*yili* 義理), which, somewhat continuous with Song systems, addressed deep orders leading to a metaphysical unity.[38] The evidential scholars regarded themselves as heirs of "Han learning" (*Han xue* 漢學), the earlier scholarly practice mainly represented by classical exegesis in the Han and Tang Dynasties. Contrary to meanings and principles, they mainly focused on "evidential study" (*kaozheng* 考證, literarily, "examining evidence"). In their perception, followers of Song learning often pursued "meanings and principles" in an ungrounded fashion, offering nothing but speculation and "hollowed" (*zaokong* 鑿空) arguments.[39]

A more concrete reason evidential scholars appreciated Shen resided in a historical narrative they constructed for the new methodology. Many of them believed that in the Song times, while the principles-and-meanings approach was prevalent, the evidence-based Han learning still persisted, albeit lurking in the less prominent notebook literature. As one scholar stated: "Therefore, the evidential learning of the Han literati became dispersed in the miscellaneous and notebooks."[40] By implication, these subterranean streams later reemerged and led to the revival of evidential study in the seventeenth century. In this narrative, Shen provided one of the earliest Song examples of Han learning, often the only Northern Song example.

Therefore, evidential scholars frequently grouped Shen with a number of Southern Song *biji* authors as representatives of the evidential type of *biji*, and they compared this subgenre favorably to other Song writings. Qian Daxin 錢大昕 (1728–1804), for instance, credited Shen along with Wu Zeng, Hong Mai, Cheng Dachang 程大昌 (1123–1195), Sun Yi 孫奕 (fl. 1220s), and Wang Yinglin with the most famous methodological motto of evidential scholarship, "seeking reliability in facts" (*shishi qiu shi* 實事求是).[41] Qian accorded them the highest acclaim one could receive as an original thinker: "standing out with established words of one's own school" (*zhuoran cheng yijia yan* 卓然成一家言).[42] Acknowledging their commitment to grounded knowledge, he suggested that these Song scholars should be differentiated from "rambling and ungrounded scholars" (*youtan wugen zhi shi* 遊談無根之士) and works like *Brush Talks* lifted out of the category of "unofficial history and lesser sayings" (*baiguan xiaoshuo* 稗官小說).[43]

In an affiliated phenomenon, during this time *Brush Talks* became more frequently associated with two specific Southern Song *biji*: Hong Mai's *Random Notes* and Wang Yinglin's *Recorded Observations*, which together constituted the "three major Song notebooks." For instance, Hong Jing 洪璟 (fl. 1700s), a descendant of Hong Mai, grouped the three together when celebrating the occasion of reprinting *Random Notes*:

> The book *Random Notes* by my forefather Master Wenmin, Mr. Rongzhai, became famous in the world along with *Brush Talks from Dream Brook* by Shen Cunzhong and *Recorded Observations from Arduous Study* by Wang Bohou in a consecutive manner. This book records everything including references in the Classics and history, the words of the Various Masters, poetry and literary writings, as well as medicine, divination, astronomy, and calendar. And, much [of the content] has been [clearly] identified and verified. In the past people praised [this book] for being accurate in evidence seeking (*kaoju jingque* 考據精確) and succinct in argumentation (*yilun gaojian* 議論高簡). [The use of evidence in the book] is like to hold measuring tools to calibrate the ten thousand things—[so accurate that] there is not even a millet's discrepancy. People like Ou[yang Xiu] and Zeng [Gong] cannot measure up to this.[44]

In this quotation, Hong Jing presented *Brush Talks* along with two other notebooks as landmark works representing critical skills such as "accuracy in evidence seeking" and "succinctness in argumentation." Once again, he suggested that these forebears of evidential scholarship surpassed famous cultural magnates such as Ouyang Xiu and Zeng Gong, a point Ming scholar Ma Yuandiao made in almost identical terms.

More examples abounded. The compilers of the *Four Treasures*, mostly proponents of evidential scholarship, repeatedly invoked Shen along with Hong and Wang. In the editorial preface to the *Old History of the Five Dynasties* (*Jiu Wudai shi* 舊五代史), they addressed Shen, Hong, and Wang together as "broad-learning scholars representative of a generation (the Song)" (*yidai boqia zhi shi* 一代博洽之士).[45] In introducing the *Trivial Comments on the Classics* (*Jingbai* 經稗), which they considered a work of evidential inquiries, the editors again identified Hong, Wang, and Shen as the Song representatives of this line of learning. Their introduction of Shen was particularly interesting. Toward the end of the passage, they restated that although evidential work might appear trivial, it eventually served greater intellectual purposes as an indispensable aid. To make this point, the editors invoked a comparison between Zhu

Xi and Shen Gua, stating that "this is why Master Zhu did not abandon (*bu fei* 不廢) Shen Gua's *Brush Talks from Dream Brook* while annotating the *Mean* (*Zhongyong* 中庸)."[46] Clearly, the Qing scholars recognized the difference between Zhu and Shen, and they capitalized on this to promote evidential study as an outgrowth of Shen's legacy.

The evidential scholars' appreciation of Shen must be understood with the caveat that their assessment was not—nor was it intended to be—an accurate historical understanding of Shen.[47] Shen's pursuit of reliability in the scope of *Brush Talks* did not map precisely onto the evidential methodology. First, the wide range of epistemic virtues in *Brush Talks* did not reduce to a consistent "evidence-seeking" theme, either in Shen's definition or by extrapolation. Specifically, textual evidence was just one analytic tool Shen used, perhaps best expressed in the section "Identification and Verification."

Second, evidential scholars in the Qing posited specific epistemological claims that differed from or even contradicted Shen's propensities. For instance, some highlighted the significance of bookish learning and criticized Song scholars for not reading. The *Four Treasures* compilers commented that "scholars in the Song [ . . . ] most of them made judgments based on Coherence without reading much (*bu shen guan shu* 不甚觀書)."[48] When Qian Daxin praised Shen and other "evidential" Song literati, he pointed out that their arguments "were mostly derived from book reading" (*yao jie cong dushu zhong chu* 要皆從讀書中出).[49] None of these claims accurately captured what Shen and his peers did. The role of textual learning in Song intellectual praxis was complex. First, it remained an important component up to the time of Zhu Xi. For system builders, the textural tradition provided a repertoire of discourse on orders of reality, and reading served as an important means for cultivating one's heart-mind as preparation to perceive "Coherences" (see chapters 7 and 10). It is fair to say that Song system builders did not use textual learning in the same way evidential scholars did, but it is inaccurate to assert that they paid no attention to books.[50] For Shen, textual learning was important, too, but it did not claim philosophical significance in his epistemology, as it pertained only marginally to his empiricism (see chapter 9).

During the age of evidential scholarship, a cross-cultural encounter contributed to Shen's positive image: the arrival of Western science on Chinese soil in the eighteenth century. In the wake of the first encoun-

ters between China and Europe, eighteenth-century literati with technical expertise—some of them simultaneously evidential scholars—keenly
felt the challenge Western science posed; in reaction, many endeavored
to prove that Western science in fact had origins in China.[51] In their efforts to scour the past for "Western precedents," these literati found Shen.
Ruan Yuan 阮元 (1764–1849), for example, in his famous *Biographies of
Calendrical Experts* (*Chouren zhuan* 疇人傳), highlighted the observation-
based methodology in calendar making as a long-standing tradition
rooted in Chinese antiquity and described the elaborate observational
technology in early modern Europe as derivative of this origin. In charting the historical genealogy, Ruan selected Shen as one of the upholders
of this tradition, aligning Shen with Western science as its predecessor.[52]
Again, Shen was regarded by a Qing scholar as one among a group of
forefathers of a methodology, this time specifically regarding science, an
endeavor some evidential scholars chose to engage.

The connection between Shen and science returns us to the opening
of this book: in the twentieth century, comparisons between *Brush Talks*
and modern scientific conclusions fueled a feverish discourse that swept
through scholarly and popular imaginations. Indeed, the millennium between the Song and today has witnessed various transmutations of Shen
in the historical imagination, from a despondent exile to an anachronistic cultural hero.

## Chinese Thought under the Searchlight

In the introduction to this book I proposed to use Shen Gua as a searchlight and follow his lead to explore the conditions he illuminated in the
Chinese intellectual world. Now let me take stock of what this searchlight has revealed.

First, Shen has illuminated—for us as much as for himself—a vibrant, nonunitary world primarily perceived through one's hearing and
seeing. In this world, one touched objects, navigated processes, and engaged myriad "things" with particular shapes, colors, and other changing qualities; one might see, hear, and read and, with the assistance of
critical skills, reach for a reliable understanding of this lively yet chaotic

universe. This world was shared by artisans, farmers, and hands-on workers of various kinds, who brought to bear their aptitudes and skills; it was similarly appealing to some cultural elites, who held an interest in critically inquiring about "things" floating in the world and endeavored to find proper ways to organize this type of inquiry beyond the discourse centered on unity and relationality. *Brush Talks* shed light on this world to reveal the empirical impulses hidden in daily epistemic praxis, and it gave rise to a consistent stance couched in textual terms.

Shen also draws into critical light the accuracy of some of the conventional analytical categories in Chinese thought. For instance, was his empiricism the result of a cleft between theoretical and practical knowledge?[53] Reasons for a negative answer are multiple. For one thing, the premodern Chinese ontology provided no grounding for such a division. Speaking from his own point of view, a Song scholar could delve into abstract construction as deeply as possible without thinking that he was theorizing; instead, he was projecting deep insights into the intricate mechanisms of reality.

For another thing, such a division was not historically true even if we endorse the modern terminology as a reasonable analytical tool. None of the conventional ways of distinguishing theoretical and practical knowledge applies to the case at hand. First, this was not a difference between "know that" and "know how."[54] In fact, the mainstream and Shen were both concerned with "know how" in their separate epistemic regimes. The practitioners of modeling wanted to know how to properly order "things" and how to live in accordance with the *dao*. Shen sought to know how to pursue reliable knowledge beyond the dominion of systems. Both sides might harbor some "know that" intuitions, but neither of them depended on "know that" theoretical propositions for building their systems/nonsystem.

Second, most serious practitioners of learning were prone to action; they were as eager as Shen to engage praxis (especially of statecraft), except that they subscribed to a methodology that depended more on the written culture. Therefore, none intended to demarcate praxis from theory—or any similar concepts in the same family, such as textual learning, conceptual learning, or classical learning. Similarly, Shen did not mean to distinguish himself with an exclusive commitment to praxis.

A study of Shen and his middle-period peers also reveals the absence of a rigorous boundary between nature and culture. The coherence between what a modern audience sees as two separate fields of inquiry was as true for Shen as it was for his mainstream peers. Living in a world characterized with an all-embracing, cosmic *dao* precluded any philosophical grounding for an arbitrary separation of humans from the cosmos. Shen exercised his nonsystem in all inquiries that drew his attention and applied his set of epistemic virtues to all relevant domains. His peers also proposed a variety of systems to account for activities in the human world and the natural realm—without drawing a boundary. The famous proposal "investigating things," from its original concoction by Cheng Yi to its various dispersions up to the early twentieth century, had always included objects and processes in the natural world in its purview.[55] It is true that the majority of literati held a preference for discussing human activities in their formal intellectual discourse, but it is a fundamentally faulty assumption that premodern Chinese elites held no commitment to understanding the natural world. Shen was different in a variety of ways because of his epistemological stance, but he was not that exceptional in terms of having an interest in nature.

Last, the reflections on culture-nature unity cast more light on the issues of science in Chinese history. The classical question posed by Joseph Needham, in one way or another, still prompts modern readers' curiosity: why did the scientific revolution not take place in China? This is the type of inquiry that has led many modern minds to associate Shen with Western science: maybe at some point there was a possibility of this happening! For historians, the Needham question is "not a question that historical research can answer."[56] Any phenomenon as concrete as Western science has either happened or it has not. Instead of wondering why something familiar in the modern, Western experience did not occur in a different context, a historian should ask how reliance on familiarity might have blinded us, in this case, to how a modern audience swiftly captured Shen's interest in nature while largely ignoring his mainstream peers' similar commitment under the rubric of "investigating things." Shen's findings surely shared some commonalities with modern science, but incidental similarities between a historical subject and modern experience demand even more rigorous efforts in reconstructing the historical context and rendering the idea in its own terms. Such is the central mission I set out to achieve on this book-length journey.

# Major Editions of *Brush Talks* *from Dream Brook*

The main text of *Brush Talks* was circulated immediately after Shen Gua finished the manuscript, circa 1093. The known major editions are shown in table 2.

In addition to the main text, there existed two sequels to *Brush Talks*, *Bu Bitan* 補筆談 (*Supplement to Brush Talks*) and *Xu Bitan* 續筆談 (*Sequel to Brush Talks*). Although they were published as printed texts as late as in the Ming, internal evidence shows that at least the first sequel, *Bu Bitan*, was likely Shen's own work, rather than a posthumous compilation. In this text, annotations in small fonts were attached to a number of jottings, stating which chapter and what part of the chapter of *Brush Talks* these notes supplemented for. Presumably only the author with a clear plan of the textual organization would make such detailed instructions. *Xu Bitan* was a slim volume containing merely eleven items. The note attached to the beginning of the text stated that it might have come from Shen Gua's oldest son, Shen Boyi.[1] The known major editions of these additions are shown in tables 3 and 4.

---

1. Preface to *Sequel* of the Shang edition, *Bai hai*, 288.1118.

## Table 2
### Major Editions of *Brush Talks*

| Year | Edition | Details |
| --- | --- | --- |
| ca. 1087 | Zou | Edited by Zou Hao 鄒浩 (1060–1111) and lost |
| ca. 1100 | Gongku | Compiler unknown |
| 1166 | Tang | Edited by Tang Xiunian 湯修年 (ca. 1100–1200) on the foundation of the Gongku edition |
| 1305 | Chen | Edited by Chen Renzi 陳仁子 (ca. 1250–1350) on the foundation of the Tang edition |
| 1495 | Xu | Edited by Xu Bao 徐瑤 (ca. 1490–1590) on the foundation of the Tang edition |
| 1602 | Shen | Edited by Shen Jingkai 沈敬炌 (1554–1631) |
| 1602 | Shang | Edited by Shang Jun 商濬 (fl. 1591–1602) |
| 1630 | Mao | Edited by Mao Jin 毛晉 (1598–1659) |
| 1631 | Ma | Edited by Ma Yuandiao 馬元調 (1576–1645) on the foundation of the Tang edition |
| 1778 | SKQS | Edited by Ji Yun 紀昀 (1724–1805) et al. on the foundation of the Ma edition |
| 1805 | Zhang | Edited by Zhang Haipeng 張海鵬 (1755–1816) on the foundation of the Mao edition |
| 1906 | Tao | Edited by Tao Fuxiang 陶福祥 (1834–1896) on the foundation of the Ma edition |
| 1916 | Liu | Edited by Liu Shihang 劉世珩 (1875–1926) on the foundation of the Tang edition |
| 1934 | SBCK | Reprint of the Mao edition |
| 1957 | Hu | Modern standard edition; edited by Hu Daojing 胡道靜 (1913–2003) on the foundation of the Tao edition |
| 2006 | Hu Jingyi | Edited by Hu Jingyi 胡靜宜 on the foundation of the SBCK edition |
| 2011 | Hu | Updated version supplemented with Hu Daojing's further notes (edited by Jin Liangnian 金良年) |
| 2011 | Yang | Edited by Yang Weisheng 楊渭生 on the foundation of the Chen edition |

SOURCES: All texts mentioned above, Shen, *Mengxi bitan jiaozheng*, edited and annotated by Hu Daojing, 778–91, and Chiang, "Shen Gua zhushu kao," 32–39.

Table 3
Major Editions of *Supplement to Brush Talks*

| Year | Edition | Details |
|---|---|---|
| ca. 1600 | Chen Jiru | Edited by Chen Jiru 陳繼儒 (1558–1639) and included in *Baoyan Tang miji* 寶顏堂秘集 |
| 1602 | Shang | Edited by Shang Jun |
| 1631 | Ma | Edited by Ma Yuandiao on the foundation of the Shang edition |
| 1778 | SKQS | Same as the main text |
| 1805 | Zhang | Edited by Zhang Haipeng on the foundation of the Shang edition |
| 1906 | Tao | Same as the main text |
| 1957 | Hu | Same as the main text |
| 2006 | Hu Jingyi | Edited by Hu Jingyi on the foundation of the Ma edition |
| 2011 | Hu | Same as the main text |
| 2011 | Yang | Edited by Yang Weisheng on the foundation of the Ma edition |

SOURCES: See Table 2.

Table 4
Major Editions of *Sequel to Brush Talks*

| Year | Edition | Details |
|---|---|---|
| 1602 | Shang | Edited by Shang Jun |
| 1631 | Ma | Same as *Supplement to Brush Talks* |
| 1778 | SKQS | Same as the main text |
| 1805 | Zhang | Same as *Supplement to Brush Talks* |
| 1906 | Tao | Same as the main text |
| 1957 | Hu | Same as the main text |
| 2006 | Hu Jingyi | Same as *Supplement to Brush Talks* |
| 2011 | Hu | Same as the main text |
| 2011 | Yang | Same as *Supplement to Brush Talks* |

SOURCES: See Table 2.

# APPENDIX 2

# Shen Gua's Other Writings

1. **Yi jie** 易解 (*Explications on the* Book of Change), two chapters, lost

Dates unavailable. This was a brief text including commentaries on a few hexagrams, such as "Great Domestication" (*daxu* 大畜, Hexagram 26), "Lesser Domestication" (*xiaoxu* 小畜, Hexagram 9), "Major Superiority" (*daguo* 大過, Hexagram 28), and "Minor Superiority" (*xiaoguo* 小過, Hexagram 62).[1]

2. **Sangfu houzhuan** 喪服後傳 (*Extended Discussion of Mourning Apparel*), number of chapters unknown, lost

Composed during the Xining reign (1068–1077). A treatise on the details of apparel in mourning rituals.[2]

---

The key references I rely on in compiling this appendix are three accounts: Hu Daojing's summary in Shen, *Mengxi bitan jiaozheng*, annotated by Hu Daojing, 816–20; Yang Weisheng's summary in Shen, *Shen Gua quanji*, vol. 3, edited by Yang Weisheng, 963–74; and Chiang, "Shen Gua zhushu kao." The current list also benefits from other scholars' piecemeal explorations of Shen's writings, as I specify in references under each item. Also, the following titles are sometimes mentioned as Shen's work by historical bibliographers or modern scholars. I do not include them because for each title there is solid historical evidence that the author is not Shen.

*Hemen yizhi* 閤門儀制

*Huimin yaoju ji* 惠民藥局記

*Chonghe xiansheng kouchi lun* 沖和先生口齒論

1. Chen Zhensun, *Zhi zhai shulu jieti*, 1.15a.

2. MXBT, item 46, 3.28.

**3. Yue lun** 樂論 (*On Music*), four chapters, lost

Composed in the early 1060s. Including an introduction, a chapter on musical instruments, a chapter on recovered Zhou melodies, and a chapter on harmonics.[3]

**4. Chunqiu jikuo** 春秋機括 (*Chronicle of the Spring and Autumn Period*), two/three chapters, lost

Composed before the 1070s. A chronological record of significant political events in the Spring and Autumn Period (770–476 BCE). It had either three chapters (according to *Jade Sea*) or two chapters (according to the *History of the Song*).[4] The first chapter was a chronicle of the Eastern Zhou Dynasty and twelve states based on the calendar of the Lu state; the second chapter contained genealogies of the Eastern Zhou and twelve states, and the third chapter recorded the official and variant names and titles of dukes and subordinates.[5]

**5. Zuo shi jizhuan** 左氏記傳 (*Zuo's Commentary on the Spring and Autumn* in Annal-Biography Form), fifty chapters, lost

Dates unknown. A historical treatise and the product of Shen's attempt to transcribe *Zuo's Commentary* from a chronological form into the so-called annal-biography style.[6]

**6. Xining xiangding zhuse ren chu liao shi** 熙寧詳定諸色人廚料式 (*Code of the Food Budget for All Officials in the Xining Reign*), one chapter, lost

An official document composed in 1068–1077.

**7. Xining xinxiu nüdaoshi jici shi** 熙寧新修女道士給賜式 (*New Code of the Ordination Ceremony for Female Daoist Priests*), one chapter, lost

Another official document composed in 1068–1077. The Northern Song state kept an institutionalized grip on the Daoist religion by instituting a Bureau of Daoist Registration (*Daolu yuan* 道祿院). Incumbents

---

3. For more details, see chapter 2.
4. *Yu hai*, 40.36; SS 155.5059.
5. Zhu Yizun, *Dianjiao buzheng Jingyi kao*, 183.856.
6. Ma Duanlin, *Wenxian tongkao*, 183.1576.

in the bureau supervised issues ranging from organizing rituals, issuing ordination certificates, and selecting Daoist officials, among other duties.[7] As far as I know from existing materials, Shen did not have an official affiliation with this bureau, and he might have received the commission in the capacity of an Imperial Library editor.

**8. *Zhu chi shi* 諸敕式** (*Style of Edicts*), twenty-four chapters, lost, dates unknown

**9. *Zhu chiling geshi* 諸敕令格式** (*Style of Edicts and Orders*), twelve chapters, lost, dates unknown

**10. *Zhu chi geshi* 諸敕格式** (*Style of Edicts*), thirty chapters, lost, dates unknown

Three manuals of style Shen prepared for official documents of various kinds.

**11. *Nanjiao shi* 南郊式** (*Code of the Southern Suburb [Ritual]*), 110 chapters, lost

Completed in 1072 and released under Wang Anshi's name.[8] This text was the first effort to codify the Southern Suburb Ritual in the form of regulatory codes, and it was part of Wang's agenda to set up an extensive system of institutions/regulations that reached into the realm of rituals.[9] Although the text no longer exists, the sheer number of chapters indicates how much more elaborate it was than previous texts of a similar nature. For instance, *Continuity and Discontinuity in Rites* (*Taichang yinge li* 太常因革禮), the ritual encyclopedia assembled by Ouyang Xiu, had one hundred chapters and yet covered a much wider range of materials. It included all five categories of state rituals, auspicious rituals (*jili* 吉禮), felicitous rituals (*jiali* 嘉禮), guest rituals (*binli* 賓禮), military rituals (*junli* 軍禮), and inauspicious rituals (*xiongli* 凶禮). The Southern Sub-

---

7. For a brief introduction to this bureau, see Tang Daijian, *Songdai daojiao guanli zhidu*, 150–53.

8. SS 204.5133.

9. SHY, *liyue* 禮樂 25.1. For a discussion of the convergence of ritual and regulatory codes during the Tang and Song periods as a historical trend, see Wang Meihua, "Li fa heliu."

urb Ritual was just one liturgy in the category of auspicious rituals, which occupied a mere five chapters. Nor were later documents on the Southern Suburb Ritual bulky. In 1078, Song Minqiu 宋敏求 (1019–1079) was commissioned to reconstitute a set of ritual codes, which included one section titled "the Code of Southern Suburb Ritual." The entire collection had a total of 190 chapters distributed among nine sections.[10] As a result, the section on the Southern Suburb Ritual was again much slimmer than Shen's treatise. Shen's codification of this ritual was probably the largest in scale.

**12. *Shi lu tuchao* 使虜圖鈔** (*Maps Made on the Mission to the Foreign Land*), two chapters, lost

Atlas made no earlier than 1075, the year Shen went to the Liao for border negotiations (see chapter 3). Also known as *Shi Liao tuchao* 使遼圖抄 (*Maps Made on the Mission to the Liao*), *Shi bei tu* 使北圖 (*Maps Made on the Mission to the North*), or *Shi Qidan tuchao* 使契丹圖抄 (*Maps Made on the Mission to Khitan*).[11] It included an elaborate survey of local particulars in the Liao, including "the altitudes and distances of mountains and rivers as well as their differences in height and width, the slopes and turns of road as well as their changes in direction, local customs, ritual paraphernalia, bureaucratic hierarchies, administrative orders, the military system, economy, major cities, languages, intelligence agencies, and details of [political] transformations."[12] Yang Weisheng recovers part of the text from the *Encyclopedia in the Yongle Reign* (*Yongle dadian* 永樂大典).[13]

**13. *Huai shan lu* 懷山錄** (*Recollections of My Mountain Life*), number of chapters unknown, lost

Shen mentioned this title in his preface to the *Wanghuai lu* 忘懷錄 (see entry 17). He composed it early in his life and wrote about things that "facilitated a joyful mountain life" (*yi zi shanju zhi le* 以資居山之樂). More details in entry 17.

10. SS 98.2422–3.

11. Shen, *Mengxi bitan jiaozheng*, edited and annotated by Hu Daojing, 1153.

12. SS 331.10653.

13. Yang Weisheng, "Shen Gua Xining shi liao tu." Also see Shen, *Shen Gua quanji*, edited by Yang Weisheng, 233–42.

**14.** ***Yimao ruguo zouqing [bing bie lu]*** 乙卯入國奏請[並別錄] (*Memorial on Entering the [Liao] State (with an Appendix) in the Yimao Year (1075)*), one chapter, partly preserved in XCB

A long memorial in which Shen reported on his diplomatic mission to the Liao (see chapter 3). The text was lost, but Li Tao copiously cited it in XCB so that the sections on border negotiations have been preserved.[14]

**15.** ***Tianxia zhouxian tu*** 天下州縣圖 (*Atlas of the Prefectures and Counties All under Heaven*), lost

Completed in 1087. Also known as *Shouling tu* 守令圖 (*Atlas for Prefects and Magistrates*). The atlas consisted of twenty maps: one big map of the state (height: 1.2 *zhang* [approximately 3.80 meters/12.47 feet], width: 1 *zhang* [approximately 3.17 meters/10.39 feet]), one small map of the state, and eighteen maps of the circuits, all drawn to a uniform scale of 1:900,000. Shen submitted two copies: one master copy backed with yellow silk, and a spare copy backed with purple silk.[15]

**16.** ***Mengzi jie*** 孟子解 (*Explications on* Mengzi), one chapter, preserved in CXJ

Composed between the 1050s and 1060s. A small set of commentaries Shen made on the classic *Mengzi*, specifically on 1B:2, 1B:8, 2B:10, 3B:8, 4B:7, and 6A:15.

**17.** ***Wanghuai lu*** 忘懷錄 (*Records of Forgetting and Recollecting*), three chapters, lost

Composed between 1086 and 1095, similar in content to and possibly an augmented version of the *Recollections of My Mountain Life* (item 12).[16] Chao Gongwu 晁公武 (ca. twelfth century) briefly introduced the content of the text as a guide for living in the mountains, covering topics such as food, medicine, agricultural tools, utensils, daily implements employed in a mountain life, and gardening tips, but he did not identify the author as Shen.[17] Chen Zhensun 陳振孫 named Shen as the author and

---

14. XCB 261.6498–513.

15. BBT, item 575, 3.322. For contexts of the atlas, see chapter 8.

16. For the connection between the two texts, see Hu Daojing, "Shen Gua de nongxue zhuzuo," 18.

17. Chao, *Jun zhai dushu zhi*, 12.15a.

regarded this text as a sequel to the previous *Recollections of My Mountain Life*.[18] Modern scholars such as Hu Daojing, Wu Zuoxin, and Yang Weisheng have retrieved a considerable part of this text from extant materials.[19]

**18.** *Qingye lu* 清夜錄 (*Records on Clear Nights*), one chapter, lost

Composed between 1069 and 1076.[20] It was a collection of strange tales on ghosts, spirits, and paranormal occurrences. The text was extant until the early Ming.[21]

**19.** *Xining Fengyuan li* 熙寧奉元曆 (*The Oblatory Epoch System in the Xining Reign*), seven chapters, lost

**20.** *Xining Fengyuan li jing* 熙寧奉元曆經 (*Calculation Methods of the Oblatory Epoch System in the Xining Reign*), three chapters, lost

**21.** *Xining Fengyuan li licheng* 熙寧奉元曆立成 (*Data and Diagrams of the Oblatory Epoch System in the Xining Reign*), fourteen chapters, lost

**22.** *Xining Fengyuan li beicao* 熙寧奉元曆備草 (*Drafts of the Oblatory Epoch System in the Xining Reign*), six chapters, lost

**23.** *Bijiao jiaoshi* 比較交食 (*A Comparative Look at [the Methods of Predicting] Eclipses*), six chapters, lost

**24.** *Xining guilou* 熙寧晷漏 (*Gnomons and Clepsydras in the Xining Reign*), four chapters, lost

The foregoing six titles, presumably all written in 1068–1077, were treatises Shen composed on the Oblatory Epoch calendrical system. All

---

18. Chen Zhensun, *Zhi zhai shulu jieti*, 10.6b.

19. See Hu Daojing, "Shen Gua de nongxue zhuzuo," Hu and Wu, "Mengxi wanghuai lu go chen," and Shen, *Shen Gua quanji*, edited by Yang Weisheng, vol. 3, 875–916.

20. Shen, *Shen Gua quanji*, edited by Yang Weisheng, vol. 1, 28, and vol. 3, 917.

21. Li Yumin, "Mengxi bitan," 286–87. For a few recovered items, see Shen, *Shen Gua quanji*, edited by Yang Weisheng, vol. 3, 917–19.

of them are lost. Qing scholar Li Rui 李銳 (1768–1817) partially recovered the system through inferences from the two key constants preserved in the *History of the Yuan* (*Yuan shi* 元史): "Accumulated Years" (*jinian* 積年) and "Day Divisor" (*rifa* 日法).[22] The Accumulated Years was the interval between the so-called Superior Epoch (*shangyuan* 上元) and the year the Oblatory Epoch system was launched. This was a great number, amounting to 83,185,070 in Shen's designation. The Superior Epoch was supposed to be a moment when the sun, moon, and Five Planets conjoined at the same degree in the far past. In calendrical calculations, it was a single initial point from which computists counted off all astronomical cycles.[23] The Day Divisor, also known as the Common Divisor (*yuanfa* 元法) in Shen's times, was the denominator used for all fractions in the system. Li was able to reconstruct a series of basic divisions of the calendar on the foundation of these two constants along with other data he derived from the predecessors of this system. For instance, he deduced from the Resplendent Heaven calendar that the so-called Year Cycle (*suizhou* 歲周) Shen used was 8,656,273, from which he derived the length of a mean tropical year:[24]

$$\frac{8656273}{23700} = 365\frac{5773}{23700}\,\text{days}$$

Li then calculated other divisions, such as the length of a solar division:

$$\frac{1}{24} \times 365\frac{5773}{23700} = 15\frac{5178\frac{3}{24}}{23700}\,\text{days}$$

22. Song Lian et al., *Yuan shi*, 53.1183. The use of Accumulated Years and Day Divisor as two central constants was a long-term practice started in the Han and ended by the launch of the Seasons Granting (*shoushi* 授時) system (1280). See Sivin's translation and annotation of the passage where the Seasons Granting calendrical experts traced the history of this practice and explained reasons for its abolishment, Sivin, *Granting the Seasons*, 370–88.

23. For the definition of the Superior Epoch and its significance in calendrical calculations, see Sivin, *Granting the Seasons*, 73–74.

24. For the data and procedures to reconstruct constants regarding the lunation, see Li, "Bu qishuo shu," in Shen, *Shen Gua quanji*, edited by Yang Weisheng, vol. 3, 923–24. Also see Yan Dunjie's summary, "Fengyuan li (fusuan)," 165.

as well as the length of a mean synodic month:

$$\frac{699875}{23700} = 29\frac{12575}{23700} \text{ days}$$

Next Li reconstructed a list of constants related to the tropical year and lunation, in addition to providing procedures for calculating the date of the winter solstice (*dongzhi* 冬至) and the date of regular conjunction (*jingshuo* 經朔), among other significant astronomical data. He computed the lengths of some subdivisions of the day, such as the "double-hour" and "mark," and stated procedures for determining the placement of intercalary months. All results were gathered in two essays: "Method for Pacing the Solar Division and Lunation" ("Bu qishuo shu" 步氣朔術) and "Method for Pacing the Putting Forth and Gathering In" ("Bu falian shu" 步發斂術).

A Ming text, *New Treatise on the Methods of Calendar* (*Lifa xinshu* 曆法新書), by Yuan Huang 袁黃 (1533–1606), preserved a few more constants than the *History of the Yuan*. Li Rui did not see this source, yet the results of his calculations were consistent with it.[25] Modern scholar Li Yong 李勇 also arrives at similar conclusions.[26]

**25. *Bianzhou zhenfa* 邊州陣法** (*The Border-Prefect Formation*), number of chapters unknown, lost

This text resulted from Shen's study of the so-called Nine-Army Formation, a lost Tang infantry tactic in 1075. It was commissioned by Emperor Shenzong and later preserved in the Bureau of Military Affairs (*shumi yuan* 樞密院).[27]

**26. *Xiucheng fashi tiaoyue* 修城法式條約** (*Principles of Fort Building*), two chapters, lost

Shen coauthored this text with Lü Heqing while he was supervising the Directorate for Armaments from 1074 through 1075. Shen submitted the text to Emperor Shenzong in 1075. It introduced building principles

---

25. See Yan Dunjie, "Fengyuan li (fusuan)," 165.
26. Li Yong, "Zhongguo guli jingshuo shuju de huifu ji yingyong." For another summary of the reconstructive efforts, see Chen Jiujin, "Fengyuan li shuping."
27. XCB 260.6342.

for various components of a city wall, such as a "horse face" (*mamian* 馬面, additional fortified perimeter space curving around the outer city wall at intervals) and a "defensive turret" (*dilou* 敵樓), among others.[28] Emperor Shenzong endorsed the treatise and immediately ordered the implementation of the guidelines.[29]

**27. *Tuhua ge* 圖畫歌** (*Song of Paintings*), one chapter, preserved in CXJ
Dates unknown. A rhymed song in which Shen summarized the styles of multiple renowned painters.

**28. *Cha lun* 茶論** (*On Tea*), number of chapters unknown, lost
Composed between 1086 and 1095. A treatise on the knowledge of tea.[30]

**29. *Lingyuan fang* 靈苑方** (*Prescriptions of the Numinous Garden*), twenty chapters, lost
A collection of prescriptions Shen gathered between 1056 and 1067.[31] There was an overlap of content between this text and his later collection of medical recipes, *Liangfang* 良方 (*Efficacious Prescriptions*), though perhaps not a significant one. A number of entries were preserved in the imperially commissioned collection of medical prescriptions, *Classified Pharmocopoeia* (*Zhenglei bencao* 證類本草, 1116). It was quoted in the famous *Compendium of Materia Medica* (*Bencao gangmu* 本草綱目, 1593–1594) and a Qing Dynasty medical text *Constant Words of an Elder* (*Laolao hengyan* 老老恆言, 1773).[32] Some scholars infer from the evidence above that this text was still extant in the late Qianlong reign (1736–1795).[33]

---

28. For the definitions of these components and some illustrations, see Zeng and Ding, *Wujing zongyao*, 12.2a–5b.

29. Chen Zhensun, *Zhi zhai jieti*, 7.35a.

30. MXBT, item 441, 24.170.

31. Wu Zuoxin, "Shen Gua de *Lingyuan Fang*," 78; Li Yumin, "Mengxi bitan," 281–82. Yang Weisheng has recovered 66 items from other existing materials. See Shen, *Shen Gua quanji*, edited by Yang Weisheng, 840–74.

32. The title was later changed to *Yangsheng biji* 養生筆記. See Cao Tingdong, *Yangsheng biji*.

33. Wu, "Lingyuan Fang," and Li, "Mengxi bitan," 282. Li Yumin also identifies a number of *Liangfang* prescriptions preserved in other texts. See Li, "Mengxi bitan," 283–85.

**30. *Liangfang* 良方** (*Efficacious Prescriptions*), ten or fifteen chapters, preserved and merged into the *Su Shen Liangfang* 蘇沈良方 (*Efficacious Prescriptions by Su [Shi] and Shen [Gua]*), see item 31)

Composed between 1088 and 1095.[34] A collection of medical prescriptions. According to Chao Gongwu, the text consisted of fifteen chapters, whereas in the *Catalogue of the Imperial Libraries in the Age of Revival* (*Zhongxing guange shumu* 中興館閣書目) and the *History of the Song*, it had ten chapters.[35] The text was later combined with the medical writings by Su Shi and became part of the *Su Shen Liangfang*. For recovered content of the book, see Yang Weisheng's work.[36]

**31. *Su Shen liangfang* 蘇沈良方** (*Efficacious Prescriptions by Su [Shi] and Shen [Gua]*), ten or eight chapters

Also known as *Su Shen Neihan Liangfang* 苏沈內翰良方 (*Efficacious Prescriptions by Hanlin Academicians Su [Shi] and Shen [Gua]*). This treatise, as the title suggested, was coauthored by Shen Gua and Su Shi. The coauthorship, however, was the work of an anonymous editor who conflated Shen Gua's *Liangfang* with some of Su Shi's medical writings. After the merge, *Liangfang* no longer circulated as an independent text. Modern scholars, represented by Hu Daojing, have made strenuous efforts to differentiate Shen's writings from Su's.[37] Two major editions of this text exist. Besides a received ten-chapter edition, the SKQS compilers culled materials from the *Encyclopedia in the Yongle Reign* and collated another edition of eight chapters.

**32. *Bie ci shanghan* 別次傷寒** (*Differentiating and Ordering Cold Damage Disorders*), number of chapters unknown

Dates unknown. Mentioned in the preface Zhang Chan 張葳 (fl. 1100s) wrote for *Treatise to Save Lives* (*Huo ren shu* 活人書).[38] Presum-

---

34. Hu Daojing, "*Su Shen neihan liangfang* chu shu pan," 195.

35. Chao, *Jun zhai dushu zhi*, 15.18b, and SS 207.5321.

36. Yang recovered 170 items from *Su Shen Liangfang* and 19 from other extant texts. For the latter, see Shen, *Shen Gua quanji*, edited by Yang Weisheng, vol. 3, 827–38.

37. See Hu, "*Su Shen neihan liangfang* chu shu pan" and Yang's work in Shen, *Shen Gua quanji*, edited by Yang Weisheng, vol. 3, 633–838. Yang identified 170 items as Shen's and 50 as Su's, in addition to 28 unidentified.

38. Zhu Gong, *Huo ren shu*, 23.

ably this book was Shen's attempt to participate in reviving the ancient medical approach "Cold Damage" (*shanghan* 傷寒), an endeavor popular among eleventh-century literati interested in medicine.[39]

**33. *Changxing ji* 長興集** (*Collection of Changxing*), nineteen or thirty chapters

*Changxing ji* was one of the three treatises included in *Collected Works of the Three Masters of the Shen Clan* (*Shen shi san xiansheng wenji* 沈氏三先生文集). The other two, *Collection of Xixi* (*Xixi ji* 西溪集) and *Collection of Yunchao* (*Yunchao ji* 雲巢集), featured writings by Shen Gou and Shen Liao. *Collected Works of the Three Masters of the Shen Clan* was first compiled by Gao Bu 高布 (ca. early twelfth c.) with the original title "Collected Works of the Three Shen in Wuxing" (*Wuxing san Shen ji* 吳興三沈集). The Song edition of the *Changxing ji* had forty-one chapters.

Two editions of *Changxing ji* are available to readers today. One is a reprint circa fifteenth century (included in SBCK). This version, however, has nineteen chapters only, with chapters 1–12, 31, 33–41 absent. The other was edited by an early Qing scholar Wu Yunjia 吳允嘉 (fl. eighteenth century) in 1718. Wu supplemented a number of Shen's writings he culled from other texts and expanded the extant nineteen chapters into thirty chapters. Wu's edition was reproduced in 1896 by the Zhejiang Press (*Zhejiang shuju* 浙江書局).

**34. *Jixian yuan shi* 集賢院詩** (*Poems Written in the Jixian Library*), two chapters, lost

Composed in the 1070s. A collection of poems Shen composed during his tenure in the Jixian Library (one of the Imperial Libraries).

**35. *Shi hua* 詩話** (*On Poetry*), number of chapters unknown

Dates unknown. Mentioned in *General Gazetteer of Zheyjiang* (*Zhejiang tongzhi* 浙江通志) and *Gazetteer of Hangzhou Prefecture* (*Hangzhou fu zhi* 杭州府志). Guo Shaoyu 郭紹虞 suspects that the text was com-

---

39. See Goldschmidt, *The Evolution of Chinese Medicine*, 146–50.

piled by later editors, who turned Shen's discussions on poems from *Brush Talks* into a new text.[40]

**36. Shen Zhongyun ji** 沈中允集 (*Collection of Shen Zhongyun*), one chapter

Dates unknown. The chapter contained twenty-five poems and was preserved in the treatise *The Modest Collection of Worthies in the Two Songs* (*Liang Song mingxian xiaoji* 兩宋名賢小集).[41]

**37. Shen Gua shici ji cun** 沈括詩詞輯存 (*Collection of Shen Gua's Poems and Lyrics*)

Modern work collated and edited by Hu Daojing in 1985. It contains fifty-five poems Hu culled from a variety of sources ranging from the Song to Qing.

---

40. Guo Shaoyu, *Bei Song shihua kao*, 189.
41. Chen and Chen, *Liang Song mingxian xiao ji*.

# Notes

## Introduction

1. Wang Wei, "Kexue juren, zhengzhi aizi: hua shuo Shen Gua qiren" (Science Giant and Political Dwarf: The Story of Shen Gua), accessed January 8, 2018, http://www.wenshitiandi.com/html/82/3/3796/1.htm; Li Cuicui, *Kexue juxing: Shen Gua* (A Superstar in Science: Shen Gua); Li Huaxin, *Shen Gua*.

2. Chen et al., *Shen Gua*, 4.

3. Chen et al., *Shen Gua*, 9.

4. Chen et al., *Shen Gua*, 10.

5. Chen et al., *Shen Gua*, 13.

6. Chen et al., *Shen Gua*, 16.

7. Chen et al., *Shen Gua*, 19, 21, and 23.

8. Chen et al., *Shen Gua*, 25. In his original words, the cited "foreign scientist" Joseph Needham referred to Shen as "one of the greatest scientific minds in Chinese history." See Needham, *The Grand Titration*, 27.

9. For instance, famous meteorologist and geologist Zhu Kezhen introduced Shen's achievements under the rubric of geology. See Zhu Kezhen, "Shen Gua duiyu dixue zhi gongxian." For a comprehensive survey of scholarship on Shen during this time, see Sivin, "Shen Kua," 49–52.

10. For the significance of science in twentieth-century China, see Tsu and Elman, "Introduction," 1–14, especially 6–7.

11. On the first generation of native scientists in China, see Reardon-Anderson, *The Study of Change*, 79–131; on the composition of the history of science and its implications for the making of Chinese modernity, see Amelung, "Historiography of Science and Technology," 39–65.

12. Amelung, "Historiography of Science and Technology," 53–54.

13. For example, the University of Science and Technology in China (Zhongguo kexue jishu daxue 中國科學技術大學) and the Iron and Steel Company in Hefei (Hefei gangtie gongsi 合肥鋼鐵公司) coauthored a translation of some parts of *Brush Talks*

(with no attribution of credits to individual authors). See Zongghuo kexu jishu daxue and Hefei gangtie gongsi, *Mengxi bitan yizhu*. For a complete list of these projects (mostly in the form of journal articles), see Bao, "Shen Gua yanjiu," 325–33.

14. Ebrey et al., *East Asia*, 128; Xiaobing Li, *China at War*, 416.

15. For the rise of literati and the transformations they initiated in the eleventh century, see Bol, *Neo-Confucianism in History*, 1–77.

16. For civil service examinations in the Song, see Chaffee, *The Thorny Gates of Learning*.

17. For new learning in the eleventh century, see Bol, *Neo-Confucianism in History*, 43–77, and *This Culture of Ours*.

18. For the antiquarian model, see Bol, *Neo-Confucianism in History*, 61–65.

19. For the development of the examination curriculum from the start to the mid-Northern Song, see Chaffee, *The Thorny Gates of Learning*, 69–70.

20. Bol addresses Wang Anshi's sociopolitical scheme as a "system." See Bol, *Neo-Confucianism in History*, 72–77. In the current book, I define "system" more broadly as a total view that unifies infinite particulars in the phenomenal world, a concept that includes not only Wang's sociopolitical system but also the alternatives that emerged later to replace it, for example, Zhu Xi's neo-Confucianism. My discussion of system builds on Bol's narrative, yet takes this project in a new direction. I am also inspired by Donald Kelly's discussion of system in historical study. Kelly, "Between System and History," 224.

21. Sivin, "Shen Kua," 31.

22. For a survey of the development of mathematics, see Needham and Wang, *Science and Civilisation in China*, vol. 3, 38–48; for that of astronomical achievements, see Chen Zungui, *Zhongguo tianwen*, 233–36; for developments in medical knowledge, see Hinrichs, "The Song and Jin Periods."

23. For the use of magnetic orientation, see Needham et al., *Science and Civilisation in China*, vol. 4, pt. 1, 281–82. For gunpowder, see Wang Ling, "On the Invention and Use of Gunpowder," 141–42, and Needham et al., *Science and Civilisation in China*, vol. 5, pt. 7, 117–26. These three innovations are particularly famous in historiography because Francis Bacon (1561–1626) mentioned them as inventions that defined the modern era, a claim further cited by Needham when presenting them as capstone achievements of premodern Chinese science and technology. See Needham, *The Grand Titration*, 11.

24. Sivin, "Science and Medicine," 54–55.

25. For a study of the Song state's participation in producing technical knowledge, see Xiaochun Sun, "State and Science."

26. For a survey of these publication projects, see Zhang Xiumin, *Zhongguo yinshua shi*, 145–52.

27. A good example is Northern Song literati's management of state-sponsored medical publications. For a detailed study, see Fan Jiawei, *Beisong jiaozheng yishu ju*.

28. For a critical survey of scholarship on Shen since the 1970s, see Sivin, "Recent Publications on Shen Kuo."

29. To avoid the anachronism of the term "scientist," Sivin proposes the term "polymath," which describes a literatus with many intellectual interests and specifically some expertise in technical knowledge. See Sivin, "Shen Kua," 31–32, and "A Multi-

Dimensional Approach," 18–19. I discuss this concept under the rubric of "broad learning" (*boxue* 博學) in chapter 9.

30. For the definition of "framework of taken-for-granted" (or "background"), see Taylor, *A Secular Age*, 13, and *Sources of the Self*, 8–11, 19–20, and 26–32.

31. For example, see Rošker, *Searching for the Way*, 4.

32. For instance, Needham, *Science and Civilisation in China*, vol. 2, 281.

33. Hall and Ames, *Thinking from the Han*, 150.

34. Hall and Ames, *Thinking from the Han*, 168, and Ames, "Meaning as Imaging," 239.

35. Ames, "Meaning as Imaging," 228; Lenk, "Introduction," 1–11; Fraser, "Knowledge and Error"; Allen, *Vanishing into Things*, 103–4 and 197.

36. Harbsmeier, "Conceptions of Knowledge," 14.

37. For an elaborate discussion of correlative thinking, see chapter 4.

38. For a critical assessment of such comparisons, see Peterson, "What Causes This?"

39. Graham, *Disputers of the Tao*, 319–20.

40. Needham, *Science and Civilisation in China*, vol. 2, 281.

41. Thomas Metzger addresses this issue as a kind of "epistemological optimism." See Metzger, *A Cloud across the Pacific*, 21–31. For a critical assessment of this argument, see Allen, *Vanishing into Things*, 7–9.

42. Some historians already use "empiricism" as a handy label to characterize Shen's work, for example, Sakade, "Shin Katsu no shizenkan," 78; Sivin, "Shen Kua," 33; and Schäfer, *The Crafting of the 10,000 Things*, 133. The current study renders this judgment into a philosophically sound thesis, primarily by grounding the term in intellectual construals specific to Chinese thought.

43. Allen, *Vanishing into Things*, 201.

44. On a related note, the Western notion of "experience" had no counterpart in premodern Chinese discourse, either. For how *jingyan* 經驗 became an approximation of the concept of "experience" in early twentieth-century China, see Lei, "How Did Chinese Medicine Become Experiential?" and *Neither Donkey Nor Horse*, 91–93.

45. Ames, "Meaning as Imaging," 234.

46. Asmis, "Epicurean Epistemology," 260–94.

47. Bacon, *Francis Bacon: The New Organon*, 21. For a comprehensive introduction to Bacon's thinking, see Rossi, *Francis Bacon*.

48. Locke, *An Essay Concerning Human Understanding*, 104. For a comprehensive discussion of Locke's empiricism, see Ayers, *Locke: Epistemology and Ontology*, 154–68.

49. For an introduction to the problem of perception and related theories, see Fish, *Philosophy of Perception*, specifically 1–48.

50. Jane Geaney, for instance, has conducted a book-length study of sensory knowing in early China, even though she distinguishes "sense discrimination" from "sense perception" in the ancient context. See Geaney, *On the Epistemology of the Senses*.

51. Cheng and Cheng, *Er Cheng ji*, Yishu, 25.317. Zhang Zai first phrased it as *dexing suozhi* 德性所知. Zhang, *Zhang Zai ji*, Zhengmeng, 7.24.

52. This differentiation carries a distant echo of Aristotle's demarcation between the senses and intellect, except that the Chinese thinker did not intend to alienate the senses from the world or from the heart-mind. See Geaney, *On the Epistemology of the Senses*, 13–14. This argument is also applicable in the Song.

53. In fact, my conceptualization of empiricism as a stance is inspired by a philosopher, Bas Van Fraassen. He chooses the term "stance" due to the difficulty of defining empiricism a priori in the philosophical discourse, an issue distinctive from and yet resonant with the historical problem here. See Van Fraassen, *The Empirical Stance*, 31–63, especially 47.

54. See Crisciani, "Histories, Stories, Exempla, and Anecdotes," 298.

55. For exemplary works that adopt "historical epistemology" as a methodology, see Poovey, *A History of the Modern Fact*; Daston and Galison, *Objectivity*; and Rheinberger, *On Historicizing Epistemology*.

56. Due to the complexity of the New Policies, I present Shen's experience during that period in two consecutive chapters, 5 and 6. Thus, chapters 5 and 6 as a whole correspond with chapter 7 as a "life and thought" pair. Chapter 11, the concluding part where I discuss the reception of Shen in subsequent times, is structurally independent from the dual narratives.

## 1. Peripateting the World (1051–1063)

1. Shen, "Chu Hanlin xueshi xie xuanzhao biao" 除翰林學士謝宣召表, CXJ 23.5a–6a.

2. Wang Anshi, *Linchuan*, 98.1013.

3. For an introduction to the aristocratic families, see Ebrey, *The Aristocratic Families of Early Imperial China*, 1–14. For a study of them in the Tang era, see Tackett, *The Destruction of the Medieval Chinese Aristocracy*, 27–69.

4. Wang Anshi, *Linchuan*, 98.1013.

5. Wang Anshi, *Linchuan*, 98.1013.

6. Shen, "Gu Tianzhang ge daizhi Shen Xingzong muzhiming" 故天章閣待制沈興宗墓誌銘, CXJ 30.6b.

7. See Bossler, *Powerful Relations*, especially her comparative analysis of eulogies in the Tang and Song, 13–14.

8. Zhou Shengchun, "Shen Gua qinshu kao," 50.

9. Zhou Shengchun, "Shen Gua qinshu kao," 50–52.

10. SS 331.10653.

11. SS 331.10651.

12. SS 331.10660.

13. For the political significance of the Imperial Libraries, see Li Geng, *Songdai guange*, 58–60 and chapter 3.

14. For the aforementioned information on Shen Gou's life, see the tomb inscription Wang Anshi composed for him, Wang Anshi, *Linchuan*, 93.961–62.

15. Wang Anshi, *Linchuan*, 98.1013.

16. For a brief summary of the division of the three regions, see Golas, "Rural China in the Song," 292–95.

17. Wang Anshi, *Linchuan*, 98.1013.

18. The political significance of Kaifeng in the Northern Song was partly evidenced by the nature of the most prominent power group in this era, a nationally oriented bureaucratic professional elite who came from different regions and centered their power

and networking in the capital city. See Hartwell, "Demographic, Political, and Social Transformations," especially 406–16, as well as Hymes's further elaboration of this issue in *Statesmen and Gentlemen*, especially 82–123.

19. Wang Anshi, *Linchuan*, 98.1013.

20. Zeng Gong, *Zeng Gong ji*, 45.611.

21. Wang Anshi, *Linchuan*, 98.1013.

22. Ouyang, *Wenzhong ji*, 79.11.

23. Zeng Gong, *Zeng Gong ji*, 45.611.

24. For a brief history of the Xu lineage, see Zhou Chunsheng, "Shen Gua qinshu kao," 53–56.

25. For a comprehensive discussion of Song mothers' significance in educating their young children, see Ebrey, *The Inner Quarters*, 183–87.

26. For the estimate of Xu's dates and a survey of his major life events, see Hu Daojing, "Shen Gua junshi sixiang," 121–22.

27. Zhou Chunsheng, "Shen Gua qinshu kao," 54.

28. For Xu's influence on Shen Gua, see Hu Daojing, "Shen Gua junshi sixiang," 122–23. For Shen Gua's work on the *Spring and Autumn* and the military, see discussions in chapter 2 and appendix 2. According to the tomb inscription Shen Gua wrote for the wife of his other uncle, Shen maintained a close relationship with his mother's family. See Shen, "Gu Xia hou furen muzhiming" 故夏侯夫人墓誌銘, CXJ 26.1a–2b.

29. For an introduction to the Longquan celadons, see Valenstein, *A Handbook of Chinese Ceramics*, 99; for silk production in the south, see Schäfer, "Silken Strands," 52.

30. Jia, *Songdai Sichuan jingji*, 7–8, 86–87.

31. For an introduction to Quanzhou in the Song, see Clark, "Overseas Trade and Social Change."

32. For the underrepresentation of southerners in the civil examinations (up to 1050s), see Chaffee, *Thorny Gates*, 129–34. For similar dominance in the policy-making positions, see Hartwell, "Demographic, Political, and Social Transformations," 414–15. For comprehensive discussions of the situation of southerners in early Northern Song bureaucracy, see Yoshinobu, "Hoku sō shoki ni okeru nanjin," and a series of studies conducted by Aoyama, such as "Sōdai ni okeru kanan kanryō no keifu nitsuite."

33. Bossler, *Powerful Relations*, 43.

34. In the Northern Song, the word *li* 吏 was a polyseme standing for at least three distinguishable meanings. Most generically, it referred to a member of the ranked bureaucracy, an "official." Also a shorthand of *xuli* 胥吏, it denoted a "clerk" who served under the officialdom and was categorically separate from the literati. The third meaning of *li* designated the "specialist" predilection of a literatus. Here I translate *li* as "clerk" because in multiple cases Shen used this term as an antithesis of a literatus, claiming that as a *li* he could no longer partake of the life of the cultural elite. In my understanding, he specifically seized the more derogatory meaning of the term as a way to express his frustration. For the second meaning, see Liu Tzu-Chien, "The Sung Views on the Control of Government Clerks." For the third meaning, see Zuo, "*Ru* versus *Li*."

35. For the definition of the protection privilege, see Chaffee, *The Thorny Gates of Learning*, 190. For a comprehensive book-length study, see You, *Songdai yinbu*.

36. SS 331.10653.

37. Zhou Chunsheng, "Shen Gua qinshu kao," 52.

38. Shen, "Cangwu tai ji" 蒼梧臺記, CXJ 21.10a.

39. For discussions on the distinction between officials and clerks, see Liu, "The Sung Views on the Control of Government Clerks," and Lo, *An Introduction to the Civil Service*, 23–24.

40. Ebrey, "The Dynamics of Elite Domination," 504–5.

41. For details of the salary system prior to 1070s, see Kinugawa, *Songdai wenguan fengji*, 2–13, and Miao, *Songdai guanyuan*, 492–95.

42. SS 171.4109. Song bureaucrats also received grain as part of their salaries. But in the 1050s, after some reforms, grain only occupied a small percentage of an official's earnings, so I do not include it in my calculations. For the changing composition of official salaries in early Northern Song, see Miao, *Songdai guanyuan*, 492–95.

43. Cheng Minsheng, *Songdai wujia*, 136, 170.

44. Cheng, *Songdai wujia*, 402, 407.

45. We do not have records of the exact dates of Shen's first son, Shen Boyi 沈博毅, but his mother died sometime in late 1050s and early 1060s, which places his birth date in that period at the latest. Shen Gua married his second wife around 1068, and we know from his records that he was already a widower years prior to that. See Shen Gua, "Ji Zhang Jianyi wen" 祭張諫議文, in Xie Jin et al., *Yongle dadian*, 14046.15b–16a. Noted and discussed by Xu Gui in "Shen Gua shiji biannian," 88.

46. Shen, "Da tongren shu" 答同人書, CXJ 20.6a.

47. Shen, "Da Cui Zhao shu" 答崔肇書, CXJ 19.3a. For a full translation of this letter, see appendix 3.

48. Shen, "Da Cui Zhao shu," CXJ 19.3a.

49. Shen, "Da Cui Zhao shu," CXJ 19.3a.

50. Shen, "Da Cui Zhao shu," CXJ 19.3a.

51. Shen, "Da Cui Zhao shu," CXJ 19.3a.

52. Shen, "Da Chen Pi xiucai shu" 答陳闢秀才書, CXJ 20.4b.

53. Shen, "Da Chen Pi xiucai shu," CXJ 20.4b.

54. Shen, "Da tongren shu," CXJ 19.4a.

55. Shen, "Wanchun wei tuji" 萬春圩圖記, CXJ 21.1a.

56. An and Wang, *Qi Lu wenhua*, 4–5.

57. An and Wang, *Qi Lu wenhua*, 14–15.

58. SS 331.10653.

59. Shen, "Wanchun wei tuji," CXJ 1a–5b.

60. There was a long misunderstanding that Shen Gua was the supervisor of the project, because Wu Yunjia, the editor of the 1896 version of *Changxin ji*, mistakenly changed the name Shen Pi into Shen Gua in "Wanchun wei tuji." Deng Guangming rectifies the error and argues that Shen Gua had no involvement in the project. See Deng, "Buyao zai wei Shen Gua jinshangtianhua le." Other scholars argue for the possibility that Shen Gua contributed to the project, albeit in a subsidiary role. See Liu Shangheng, "Ye tan Wanchun wei de xingjian," and Zu, *Shen Gua pingzhuan*, 39–40.

61. Shen, "Da Cui Zhao shu," CXJ 19.2b.

62. Shen, "Shanghai zhou tongpan Li langzhong shu" 上海州通判李郎中書, CXJ 19.5b.

63. Shen, "Shanghai zhou tongpan Li langzhong shu," CXJ 19.5b.
64. Shen, "Shanghai zhou tongpan Li langzhong shu," CXJ 19.5b.
65. Shen, "Shanghai zhou tongpan Li langzhong shu," CXJ 19.5b.
66. For the debates on *yi* and *li* in the Northern Song, see Chen Zhi'e, *Beisong wenhua shi*, 260–76.
67. Shen, "Da tongren shu," CXJ 19.6a.
68. Shen, "Da Xu mijiao shu" 答徐秘校書, CXJ 20.5a.
69. Shen, "Da Xu mijiao shu," CXJ 20.5a.
70. This is, of course, a simplified generalization made in accordance with Shen's understanding of the past to explain his frustration. It is fair to argue that the aristocracy (Tang and older) was indeed a landed class and relied on land ownership as a means of maintaining power, although according to Tackett, such a means was not primary and often not enough for building an alternative base of power vis-à-vis the state. See Tackett, *The Destruction of the Medieval Chinese Aristocracy*, 58–61.
71. Shen, "Mengzi jie" 孟子解, CXJ 32.4b.
72. Just to name a few examples, "Da Li Yanfu xiucai shu" 答李彥輔秀才書, CXJ 19.2b–3b, "Da Xu Mijiao shu," CXJ 20.5a–5b, and "Da tongren shu," CXJ 20.5b–6b.
73. Shen, "Da Xu mijiao shu," CXJ 20.5a.
74. Shen, "Da tongren shu," CXJ 20.6a.
75. Shen, "Da tongren shu," CXJ 20.5b.
76. Shen, "Da tongren shu," CXJ 20.6a.
77. Shen, "Da tongren shu," CXJ 20.6a.
78. *Mengzi* 7A:1.
79. Shen, "Mengzi jie," CXJ 32.7a.
80. For an analysis of the role of Heaven in the Mengzian philosophy, see Graham, *Studies in Chinese Philosophy*, 54–57.

## 2. Envisioning Learning

1. I keep "things" in quotation marks throughout this book to distinguish the translation of *wu* from the general use of the English word.
2. *Dao* as the goal of learning in general was different from the more specific "Learning of the Dao" (*daoxue* 道學, also known as neo-Confucianism), a "school of learning and sometimes a political faction with a high degree of ideological self-consciousness" heralded by the Cheng brothers. See Bol, "On the Problem of Contextualizing Ideas," 61–62. The former was the background from which the latter derived. In the following discussion, I purposely choose a diverse selection of thinkers—including Cheng Yi and his antagonists—to prove that the talk of the *dao* and its inarticulability was a general concern among mid-eleventh-century literati.
3. For the Ancient-Style Prose Movement and the transition of learning from the Tang to Song, see Bol, *This Culture of Ours*. For an exemplary case study of the movement leaders, see Hartman, *Han Yü and the T'ang Search for Unity*.
4. Graham, *Studies in Chinese Philosophy*, 426.
5. Cheng and Cheng, *Er Cheng ji*, Yishu, 4.73.
6. Zhang Lei, *Zhang Lei ji*, 47.733.

7. Wang Anshi, *Linchuan*, 72.770, 66.706.

8. Shen, "Dongjing Yong'an chanyuan chi ci Chongsheng Zhiyuan dian ji" 東京永安禪院敕賜崇聖智元殿記, CXJ 23.6b.

9. Su Shi, *Su Shi wenji*, 64.1981.

10. Cheng and Cheng, *Er Cheng ji*, Yishu, 2A.17.

11. For example, see Bol's discussion of Wang Anshi's opinion. Bol, "Reconceptualizing the Order of Things," 688.

12. Many Song literati discussed a sage's spontaneous deployment of virtues. For a few examples, see Bol, "Reconceptualizing the Order of Things," 700, 707, and 715.

13. I define the "cosmos" as the all-encompassing multiplicity including "things," humans, and their activities, not just the natural world. For a detailed discussion of the cosmos vis-à-vis nature, see chapter 9. For further clarification on the different levels involved, see the later discussion in this chapter and chapter 7.

14. For how learning turned into an active ideological enterprise, see Bol, *Neo-Confucianism*, 55–56.

15. See Liu Tzu-Chien, *Ouyang Xiu de zhixue*, 173–74, 229–30.

16. Ouyang Xiu, *Ouyang Xiu quanji*, 104.1590.

17. Some thinkers, such as Shao Yong, Zhang Zai, and Cheng Yi, explicitly identified humans as part of "things." See Smith and Wyatt, "Shao Yung and Number," 101; Zhang Zai, *Zhang Zai ji*, Yulu, 1.313; and Yü, *Zhu Xi de lishi shijie*, vol. 1, 185–86.

18. For a general discussion of "things" in the Song, see Tillman, "The Idea and the Reality of the 'Thing'." For a brief introduction to *wu* and *wanwu* in early Chinese philosophy, see Perkins, "What Is a Thing (*wu*)?," 57–58.

19. See Yü, "Intellectual Breakthroughs in the T'ang-Sung Transition" and Gardner, *Chu Hsi and the Ta-hsuen*, 14–15.

20. A number of scholars have noticed this phenomenon, for instance, Peterson, "Fang I-chih,'" 377; Bol, *This Culture of Ours*, 260; and Elman, *On Their Own Terms*, 29–30, among others.

21. Cheng and Cheng, *Er Cheng ji*, Yishu, 4.372.

22. Perkins, "What Is a Thing?," 58.

23. For a systematic explication of "knowing from virtuous nature"/modeling, see chapter 7.

24. Wang Anshi, *Linchuan*, 68.723.

25. At first glance, the opening sentence can be confusing, which seems to imply that the *dao* was both the origin and the branches. According to the context, Wang undoubtedly viewed the *dao* as the origin and myriad "things" as the branches. His peculiar expression, nevertheless, was philosophically sound, because the *dao* was indeed both one and many.

26. Shao Yong, *Huangji jingshi*, 14.522. Translation after Smith and Wyatt with minor changes, see Smith and Wyatt, "Shao Yung and Number," 106.

27. Shao Yong, *Huangji jingshi*, 14.522.

28. This formula derived from a statement in the "Attached Verbalizations." See Smith and Wyatt, "Shao Yung and Number," 112.

29. Shao Yong, *Huangji jingshi*, 14.522.

30. For a comprehensive study of this formula, including the meanings of all key concepts, see Smith and Wyatt, "Shao Yung and Number," 105–35, and Arrault, *Shao Yong*, 315–22.

31. Liu Mu, *Yi shu gouyin tu*, 1.17; Zhang Zai, *Zhang Zai ji*, Yi shuo, 188.

32. Wang serves as a particularly convincing example to demonstrate the wide acceptance of this thesis. Compared to peers such as Shao Yong, Wang invested less interest in systematically engaging cosmological concepts. This predilection leads some modern scholars to believe that he was apathetic to cosmology. See, for instance, Deng, *Beisong zhengzhi gaigejia*, 118–23. For a summary of scholarly opinions on this issue, see Skonicki, "Cosmos, State, and Society," 446–53. The current example, however, demonstrates that despite a lack of original thinking, Wang was highly aware of—and readily receptive to—the numerical order as an intermediate stage between the *dao* and "things."

33. Wang Anshi, *Linchuan*, 65.686.

34. Ames, "Meaning as Imaging," especially 233.

35. Chung-ying Cheng, "Categories of Creativity," 262.

36. The temporal priority presented in the generative scheme should be understood as a metaphor of weak ontological priority, as number (and other larger orders including the *dao*) did not claim ontological independence from the ten thousand things. For a more detailed analysis, see chapter 4.

37. For example, Bol aptly titles his chapter on Song learning "Reconceptualizing the Order of Things in Northern and Southern Sung." See Bol, "Reconceptualizing the Order of Things," 665.

38. Wang Anshi, *Linchuan*, 39.411.

39. It was a paraphrase of a line in *Mengzi*. See *Mengzi* 7A:45.

40. *Mengzi* 6A:15. Translation after Ivanhoe with minor changes. See Ivanhoe, *Confucian Moral Self Cultivation*, 20.

41. Wang Anshi, *Linchuan*, 66.703.

42. Wang Anshi, *Linchuan*, 66.703.

43. Su, *Su Shi wenji*, 11.356. For analyses of this statement, see Egan, *Word, Image, and Deed*, 159, and Bol, *This Culture of Ours*, 277.

44. For a discussion of *li* as "Coherence," see chapter 10. For discussions of Shao Yong's thinking, see chapter 7.

45. For instance, Zhang Zai discussed the two types of knowing as connected. See Zhang, *Zhang Zai ji*, Yulu, 1.313. Also see Angle and Tiwald's analysis, Angle and Tiwald, *Neo-Confucianism*, 113–14.

46. See appendix 2, entries 1, 4, 5, and 16. Shen's interest in Mengzi seemed to be well known among his peers. See Hu Daojing, "Guanyu Shen Gua pingzhuan," 145, and Le, "Beisong ruxue beijing."

47. For an analysis of his motive, see Bol, *This Culture of Ours*, 228–29.

48. Shen, "Da Cui Zhao shu," CXJ 19.2b.

49. Shen, "Da Cui Zhao shu," CXJ 19.3a.

50. Shen, "Da Cui Zhao shu," CXJ 19.3a.

51. Shen, "Da Cui Zhao shu," CXJ 19.3a.

52. Shen, "Da Cui Zhao shu," CXJ 19.3b.

53. Shen, "Da Cui Zhao shu," CXJ 19.3b. In the original context of the *Analects*, Kongzi sought information on the statecraft of other dukedoms, and the "how" question addressed the demeanor in which he obtained this information. See *Lunyu* 1.10. This original meaning, however, greatly diverges from the content of Shen's letter and cannot serve his intention in any literal sense. It is reasonable to deduce that Shen employed the statement to support the central topic of this letter: contemplation of learning.

54. Shen, "Da Cui Zhao shu," CXJ 19.3b.

55. Shen, "Da Cui Zhao shu," CXJ 19.3b.

56. Shen, "Da Cui Zhao shu," CXJ 19.3b.

57. *Mengzi* 7A:1. Translation after Bloom, *Mencius*, 143. For an analysis of this quotation, especially how *ming* was not fatalism, see Bloom, "Mencian Arguments on Human Nature," 42–44.

58. *Mengzi* 7A:4.

59. Shen, "Mengzi jie" 孟子解, CXJ 32.7b.

60. *Zhouyi zhengyi*, 9.196.

61. Shen, "Mengzi jie," CXJ 32.8a.

62. Hall and Ames, *Thinking from the Han*, 300–301, n.41.

63. *Mengzi* 7A:19.

64. Shen, "Mengzi jie," CXJ 32.8a.

65. Shen, "Mengzi jie," CXJ 32.8a.

66. Shen, "He shumi Xue shiyu qi" 賀樞密薛侍御啟, CXJ 18.3a; Shen, "Dali pingshi Sinong si yuan zhubu Jia jun muzhiming" 大理評事司農寺院主簿賈君墓誌銘, CXJ 28.3a.

67. Shen, "Yangzhou chongxiu Pingshan tang ji" 揚州重修平山堂記, CXJ 21.7a.

68. Shen, "Chizhou xin zuo gujiao men ji" 池州新作鼓角門記, CXJ 23.8b.

69. Shen, "Zhizhigao xie liangfu qi" 知制誥謝兩府啟, CXJ 17.2a.

70. Shen, "Mengzi jie," CXJ 32.6b. For analyses of this statement, see Teraji, "Shin Katsu no shizen kenkyū to sono haikei," 113.

71. Shen, "Mengzi jie," CXJ 32.2b.

72. Ivanhoe, *Confucian Moral Self Cultivation*, 20.

73. Shen, "Mengzi jie," CXJ 32.2b.

74. This is, of course, a limited judgment made on the basis of a corrupted collection of his writings. For details of Shen's extant work, see appendix 2.

75. For a comprehensive introduction to ceremonial music in the Song, see Yang Yinliu, *Zhongguo gudai yinyue*, 380–405.

76. Brindley provides a detailed analysis of the cosmic significance of ritual music in early China. Many of the fundamental notions, such as the belief in harmony, still prevailed in the Song. See Brindley, *Music, Cosmology, and the Politics of Harmony*, 25–85.

77. For comprehensive studies of all Song music reforms, the procedures and personnel, see Hu Jinyin, "Cong Da'an dao Dasheng" and Lam, "Huizong's Dashengyue," 418–27.

78. SS 79.2960–62.

79. Shen, "Shang Ouyang canzheng shu" 上歐陽參政書, CXJ 19.6a–7a. For full translation of this letter, see appendix 3; Shen, "Yu Sun shijiang lun yue shu" 與孫侍講論樂書, CXJ 20.3a–4a; Shen, "Yu Cai neihan lun yue shu" 與蔡內翰論樂書, CXJ 20.1a–2a.

80. SS 202.5054. The compiler of the Southern Song encyclopedia *Jade Sea* (*Yu hai* 玉海) believed that *On Music* was an early (and partial) version of the section "Music and Harmonics" ("Yuelü" 樂律) in *Brush Talks*. See Wang Yinglin, *Yu hai*, 7.30. In "Music and Harmonics," however, Shen repeatedly stated that he did not intend to recycle his old work, which serves as convincing evidence to separate *On Music* from his later writings. For evidence, see MXBT, item 103, 5.48.

81. See Hu Jinyin, "Cong Da'an dao Dasheng."

82. Shen, "Yu Cai neihan lun yue shu," CXJ 20.1a.

83. Shen, "Yu Cai neihan lun yue shu," CXJ 20.1a–b.

84. Shen, "Shang Ouyang canzheng shu," CXJ 19.7a.

## 3. *Measuring the World (1063–1075)*

1. *Qiantang xianzhi*, 401.

2. Shen, "Yanghzou chongxiu Pingshan tang ji," CXJ 21.5b.

3. Shen, "Yanghzou chongxiu Pingshan tang ji," CXJ 21.5b.

4. Shen, "Zhang gong muzhiming" 張公墓誌銘, CXJ 29.8b.

5. Shen, "Zhang gong muzhiming," CXJ 29.10a.

6. Shen provided a detailed account of Zhang's life in the epitaph, from which I extract the commonalities. Shen, "Zhang gong muzhiming," CXJ 29.7a–11a.

7. Shen, "Zhang gong muzhiming," CXJ 29.9a. The Imperial Libraries consisted of four institutions: the Zhaowen Library, the Historical Archive (*shiguan* 史館), the Jixian Library, and the Confidential Archive (*mige* 秘閣). See Li Geng, *Songdai guange*, 48.

8. Shen, "Zhang gong muzhiming," CXJ 29.10a.

9. See Shen, "Zhang Zhongyun muzhiming" 張中允墓誌銘, CXJ 25.2b–4a. This war resulted in the famous Treaty of Chanyuan (Chanyuan zhi meng 澶淵之盟, 1005). The story was recorded by Shen as part of the tomb inscription he penned for Zhang Mu 張牧 (fl. eleventh century)—Zhang Chu's father—at Zhang Chu's request. Zhang Chu already had four sons-in-law at the time, yet he entrusted the task to Shen, again demonstrating their closeness.

10. Shen, "Ji Zhang Jianyi wen," in Xie et al., *Yongle dadian*, 14046.16a, and Shen, "Zhang gong muzhiming," CXJ 29.9a.

11. Three at the time of Shen's marriage to Zhang's daughter, four at the time of Zhang's death. See Shen, "Zhang Zhongyun muzhiming," CXJ 25.4b and "Zhang gong muzhiming," CXJ 29.10b. Shen mentioned the official ranks of his in-laws twice, first in the tomb inscription for Zhang Mu and later in Zhang Chu's. Here I cite the latter reference. Shen, "Zhang gong muzhiming," CXJ 29.10b.

12. To ascertain the location of the libraries, I rely on the map reconstructed by Fu Xinian, "Shanxi sheng Fanzhi xian Yanshan si," 266.

13. Meng Yuanlao, *Dongjing meng hua lu*, 1.40.

14. SS 202.5032.

15. For a detailed account of institutional arrangements in the Imperial Libraries, see Umehara, *Sōdai kanryō seido kenkyū*, 329–422.

16. Bao, "Shen Gua shiji xianyi," 307–8.

17. MXBT, item 129, 7.58–59.

18. For a detailed study of astronomical texts in the Song Imperial Libraries, see Dong and Guan, "Songdai de tianwenxue wenxian."

19. Cai Tao, *Tienwei shan congtan*, 1.15.

20. Li Geng, *Songdai guange*, 49.

21. For a detailed study of the canal in early Song history, see Quan, *Tang Song diguo*, 93–113.

22. MXBT, item 457, 25.175.

23. MXBT, item 457, 25.175–76.

24. According to Shen, this was an old method dated to the Tang. MXBT, item 429, 24.557.

25. XCB 238.5796.

26. MXBT, item 457, 25.176. The numerical values of Song length units varied from case to case. Here I adopt Shen Gua's assertion, which rendered 1 *chi* an equivalent of 31.68 centimeters/12.47 inches (hence 1 *li* = 475.20 meters/1,558.75 feet and 1 *bu* = 158.40 centimeters/62.35 inches). See MXBT, item 68, 3.33, and Wu Hui, "Song Yuan de duliangheng," 18.

27. MXBT, item 457, 25.176.

28. All descriptions, including those of the dimensions of the instrument and ways of operation, come from Zeng and Ding, *Wujing zongyao*, 481–82. Zeng and Ding inherited the accounts from an earlier text, *Taibai yinjing* 太白陰經; these verbal descriptions, however, were incongruent with the illustration they provided. The mistake Zeng and Ding made regarding the shape and use of the sighting board has been corrected by modern scholars. My illustrations follow the corrected version in Wuhan shuili dianli xueyuan et al., *Zhongguo shuili shigao*, vol. 2, 54.

29. For all particular meanings of a Chinese *li*, see Sivin, *Granting the Seasons*, 38–40, and Cullen, *The Foundations of Celestial Reckoning*, 6–29.

30. Sivin, *Granting the Seasons*, 39.

31. Sivin, *Granting the Seasons*, 20.

32. For the factional overtone of Shen's appointment, see Sun Xiaochun, "State and Science," 62–63, and Dong Yuyu, "Cong Fengyuan li gaige kan Beisong tianwen guanli," 205–6.

33. See Dong, "Cong Fengyuan li gaige kan Beisong tianwen guanli," 203–5.

34. MXBT, item 116, 7.53. Yan Dunjie thinks that the "lunation factor" should be "solar-division factor" instead, according to his reconstruction of the Oblatory Epoch System. See Yan Dunjie, "Fengyuan li (fusuan)," 167.

35. For a detailed introduction to the two institutions, see Sun and Han, "The Northern Song State's Financial Support for Astronomy," 19–20.

36. MXBT, item 149, 8.68. Sun Xiaochun suggests that Shen's criticism of the bureau should be taken with a pinch of salt, for it might be affected by factional bias against Sima Guang. See Sun Xiaochun, "State and Science," 62–63.

37. MXBT, item 308, 18.130. Shen's praise might be a little exaggerated. See Li Zhichao, "Shen Gua de tianwen yanjiu er," 52.

38. MXBT, item 308, 18.131, and Zhang Lei, *Mingdao zazhi*, 23.

39. MXBT, item 308, 18.131.

40. Sivin, "Shen Kua," 18–19.

41. For Shen's recapitulation of the history of armillary spheres, see Shen, "Hunyi yi" 渾儀議, CXJ 3.1b–3a. For a critical review of this narrative, see Li Zhichao, "'Hunyi yi' pingzhu," 197–99. Li also argues that the instrument by Yixing was lost long before Shen's times, so his reconstruction mainly depended on textual evidence. See Li Zhichao, "Huangdao hunyi ji Xining hunyi."

42. Shen, "Hunyi zhi qi" 渾儀制器, CXJ 3.8b–9a.

43. In this book I translate *du* as "degree" with an awareness that the stipulation of the Chinese *du* in the early and middle periods followed a different paradigm and was not an equivalent of a degree in Western traditions. For a detailed analysis, see Morgan, *Astral Sciences in Early Imperial China*, 82–85. The complexity of the Chinese *du* has led to modern scholars' different readings of Shen's work; one example, as I show later, is the evaluation of the accuracy of his observation of the Pole Star.

44. Shen, "Hunyi zhi qi," CXJ 3.9a–b.

45. Shen, "Hunyi zhi qi," CXJ 3.9b–10a.

46. Shi, *Zhongguo kexue shi gang*, 169.

47. Shen, "Hunyi yi," CXJ 3.5a–b, and Sivin, "Shen Kua," 18.

48. MXBT, item 127, 7.57. Some modern scholars criticize this result for being excessively erroneous, as the actual distance was 1.52 degrees in modern calculation. See Li Zhichao, "Shen Gua de tianwen yanjiu er," 54. Huang I-long points out that the measure unit *du*, which inadequately translates into English as "degree," referred to a central angle instead of an inscribed angle. Given that the measure of a central angle is twice of that of the inscribed angle that subtends the same arc, Shen's conclusion was actually accurate. See Huang I-long, "Jixing yu gudu kao," 95–100.

49. Shen, "Hunyi yi," CXJ 3.5b. For details on the calculations, see Huang I-long, "Jixing yu gudu kao," 101–4.

50. Needham and Wang, *Science and Civilisation in China*, vol. 3, 352.

51. For comprehensive introductions to the gnomon, see Needham and Wang, *Science and Civilisation in China*, vol. 3, 284–94, and Cullen, *Astronomy and Mathematics in Ancient China*, 101–28.

52. Shen, "Yingbiao yi" 景表議, CXJ 3.13b. My analysis of this method follows Chen Meidong, *Zhongguo kexue jishu shi*, 474, and Guan, "A New Interpretation of Shen Kuo's Ying Biao Yi," 716–17.

53. Shen, "Yingbiao yi," CXJ 3.13a.

54. For detailed analyses of the triple-gnomon system, see Chen Meidong, *Zhongguo kexue jishu shi*, 474, and Guan, "A New Interpretation of Shen Kuo's Ying Biao Yi."

55. For a survey of Chinese clepsydra technique, see Needham and Wang, *Science and Civilisation in China*, vol. 3, 315–29. Scholars such as Zhang Jiaju, Needham, Chen Meidong, and Li Zhichao propose reconstructions of Shen's clepsydra according to his verbal description. For a summary of their differences, see Chen Meidong, "Wo guo gudai louhu de lilun yu jishu." I adopt Li Zhichao's interpretation in his "Fulou yi kaoshi." For the result of a reconstruction experiment based on Li's understanding, see Hua, *Zhongguo louke*, 193–96.

56. A "quarter" approximated 14.4 minutes on a modern clock. See Sivin, *Granting the Seasons*, 82.

57. Shen, "Fulou yi," CXJ 3.10a–b.

58. Shen, "Fulou yi," CXJ 3.10b.

59. Hua, *Zhongguo louke*, 84.

60. Shen, "Fulou yi," CXJ 3.10b.

61. Shen, "Fulou yi," CXJ 3.10b.

62. Shen, "Fulou yi," CXJ 3.10b.

63. Shen, "Fulou yi," CXJ 3.11b.

64. Shen, "Fulou yi," CXJ 3.11a, 12a.

65. For the controversy on his use of jade, see Li Zhichao, "Shen Gua de tianwen yan-jiu yi."

66. For details of the reconstructive work, see appendix 2, entries 19–24.

67. See Dong, "Cong Fengyuan li gaige kan Beisong tianwen guanli," 208.

68. XCB 272.6667, 287.7032.

69. MXBT, item 148, 8.67–68.

70. XCB 287.7032.

71. MXBT, item 116, 7.53.

72. XCB 287.7032.

73. For the Song–Liao relationship in the eleventh century, see Jing-shen Tao, *Two Sons of Heaven*, 2–67, and Wright, *From War to Diplomatic Parity*.

74. XCB 262.6376–86. The Liao missions are discussed in detail in Zhang Jiaju, *Shen Gua*, 77–98 and Lamouroux, "Geography and Politics," 1–28.

75. XCB 261.6362.

76. Although Liao's language was intentionally vague, territorial ambiguities were more likely in the eleventh century than in later times because the so-called border was a wide zone rather than a single line. The conception of a "lineal" border emerged only later in the Song as the product of construction of military defense lines. For this history, see Tackett, "The Great Wall and Conceptualizations of the Border."

77. XCB 261.6368–69.

78. XCB 265.6498.

79. XCB 265.6498–513.

80. XCB 265.6499.

# *4. Individuating Things*

1. To define "individuation" in the premodern Chinese context, I am inspired by Perkins's discussion in "What Is a Thing?"

2. See Wyatt, *The Recluse of Loyang.*

3. Idea of Ziporyn, cited by Skonicki in "Cosmos, State, and Society," 17.

4. Sivin, *Traditional Medicine in Contemporary China*, 61; Hall and Ames, *Thinking from the Han*, 124; Robin Wang, "Yinyang Narrative of Reality," 21.

5. Liu An, *Huainanzi jiaoshi*, 3.246. For clarity I keep the content and simplify the sentence structures. For a full translation and discussion, see Graham, *Disputers of the Tao*, 336–37.

6. Conti, "Realism," 649. I leave out "universals in the thing" (*universalia in re*) for the purpose of keeping my comparisons clear and focused. For the medieval conceptualization of universals, see Conti, "Realism," and Biard, "Nominalism in the Late Middle

Ages." The basic argument I make here is that yinyang (and many similar Chinese universals) conforms with neither realism nor nominalism, a point Ziporyn makes with more elaborate analysis. See Ziporyn, *Ironies of Oneness and Difference*, 1–88.

7. For an introduction to Plato's Forms, see Gill, "Problems for Forms."

8. Cited by Li Shizhen, *Bencao gangmu*, 51.1869.

9. For a recent discussion of the resonance mechanism, see Robin Wang, *Yinyang*, 83–96. For a summary of the exemplary studies of correlative thinking and different definitions of the concept, see Nylan, "Yin-yang, Five Phases, and *qi*," 410–14.

10. Kaptchuk, *The Web That Has No Weaver*, 43.

11. Robin Wang, *Yinyang*, 61.

12. Dong Zhongshu, *Chunqiu fanlu yizheng*, 57.360.

13. Zhang Zai, *Zhang Zai ji*, Zhengmeng, 1.7.

14. Angle, *Sagehood*, 38.

15. Zhang Zai, *Zhang Zai ji*, Zhengmeng, 5.19.

16. I must admit that the idea of an ordinary net does not adequately capture the fluid nature of "things." Imagine the patterns of the net undergoing constant change.

17. For a comprehensive introduction to the system, see Despeux, "The System of the Five Circulatory Phases."

18. Despeux, "The System of the The Five Circulatory Phases," 134–35.

19. My brief introduction in the following pages is based on the following three accounts: Fang and Xu, *Huangdi neijing suwen yunqi*, 17–31; Ren, *Wu yun liu qi*; and Despeux, "The System of the Five Circulatory Phases."

20. MXBT, item 134, 7.60–61.

21. MXBT, item 134, 7.60. Shen drew on some early texts for vocabularies. For instance, *Essential Questions* addressed some terms on the list. See *Huangdi neijing suwen*, specifically 22.1047–122. Shen added more categories and incorporated richer meteorological phenomena.

22. MXBT, item 123, 7.56.

23. MXBT, item 123, 7.56.

24. MXBT, item 123, 7.56. For a detailed analysis of this passage, see Sivin, "On the Limits of Empirical Knowledge," 173–75.

25. MXBT, item 134, 7.60.

26. MXBT, item 134, 7.60.

27. A potential discrepancy is worth clarification here. The "vestiges" might be sensory in nature, and Shen suggested that they did not suffice for illuminating the profundity of number. This was his judgment of the ontological significance of sensory phenomenon vis-à-vis number, which did not entail a rejection of the significance of the sensory in his epistemic praxis. From a practical point of view, sensory experience provided evidence for humans' insufficient understanding of number and the most grounded means to improve it, a point Shen fully embraced.

28. For instance, John Henderson's study shows that the attacks on correlative cosmology from the Han through the Song focused on a couple of recurrent issues: obsession with resonances between natural oddities and political affairs, and the five-phase arrangement of dynastic succession. Both were highly specific issues, and neither challenged the validity of the three basics of number. The most compelling example is Wang

Chong, who, though well known for his skeptical stance on certain aspects of the Han cosmology, remained an "even more consistent correlative thinker than were many of his less iconoclastic contemporaries." See Henderson, *The Development and Decline of Chinese Cosmology*, 86–97, cited 88.

29. For concrete examples of criticisms, see Henderson, *The Development and Decline of Chinese Cosmology*, 86–97. Some scholars thus conclude that "correlative cosmology was marginal to Song Neo-Confucianism." See Henderson, "Cosmology and Concepts of Nature," 191, and Graham, *Yin-Yang and the Nature of Correlative Thinking*, 15.

30. For correlation as a model of causality, see Peterson, "What Causes This?"

31. Bol, *Neo-Confucianism*, 65.

32. See Zuo, "Keeping Your Ear to the Cosmos."

33. For the representational theory of measurement and some important criticisms, see Narens, *Theories of Meaningfulness*, 205–21.

34. In Daniel Morgan's nuanced analysis of astronomical observation in early and middle-period China, he points out that "an act of seeing" was at the same time "an act of inscription," because observation had a lot to do with the model in which perceptual clues were organized and the observational instruments that embodied the model with certain material constraints. The amalgam he describes is precisely where Shen felt the tension between "things" and representation. See Morgan, *Astral Sciences in Early Imperial China*, 50–93.

35. Shen, "Hunyi yi," CXJ 3.4b–5a.

36. Shen, "Hunyi yi," CXJ 3.4b.

37. Shen, "Hunyi yi," CXJ 3.5a.

38. For a comprehensive introduction to this system, see Sivin, *Granting the Seasons*, 90–94.

39. MXBT, item 129, 58–59.

40. MXBT, item 129, 7.59.

41. MXBT, item 129, 7.59.

42. MXBT, item 129, 7.59.

43. MXBT, item 146.67. For a full translation of this item, see appendix 3.

44. Shen, "Hunyi yi," CXJ 3.1b.

45. MXBT, item 123, 7.56.

46. MXBT, item 123, 7.56. Shen was certainly not the first person who critically examined the issue of inaccuracy in calendar making, but he was among a rare few who framed the issue as an ontological problem. For similar discussions prior to him, see Sivin, "On the Limits of Empirical Knowledge," 168–73.

47. MXBT, item 123, 7.56.

48. Ouyang et al., *Xin Tang shu*, 25.533.

## 5. Reforming the World (1071–1075)

1. Deng, *Beisong zhengzhi gaigejia*, 94–98.

2. For exemplary studies of the Green Sprouts, see Sudō, "Ō Anseki no seibyōhō no kigen," "Ō Anseki no seibyōhō," and Li Jinshui, *Wang Anshi jingji bianfa*, 79–141.

3. Paul Smith, "State Power and Economic Activism," 96.

4. For the distinction between "service for local government" (*zhiyi*) and "labor service" (*liyi* 力役), see Sogabe, *Sōdai zaisei shi*, 89–90, and James Liu, *Reform in Sung China*, 99.

5. For details of the Hired Service Policy, see Sogabe, *Sōdai zaisei shi*, 143–61; James Liu, *Reform in Sung China*, 98–113; and Deng, *Beisong zhengzhi gaigejia*, 184–204.

6. For the details of the Mutual Security Policy, see Sogabe, "Ō Anseki no hokōhō"; Lin, "Songdai baojia"; and Wu Tai, "Songdai 'baojia fa' tanwei."

7. For a detailed introduction to the Policy of Farming and Hydraulics, see Li Jinshui, *Wang Anshi jingji bianfa*, 142–214.

8. Previous studies of the relationship between Wang and Shen include Lin Cen's "Lue lun Shen Gua yu Wang Anshi," Li Yumin's "Shen Gua de qinshu," Zu's "Shen Gua yu Wang Anshi," the section "Shen Gua yu Wang Anshi" in Zu, *Shen Gua pingzhuan*, and He Yongqiang's "Shen Gua yu Wang Anshi." Aside from Lin Cen, all other scholars agree that Wang and Shen had developed a friendship since the late 1060s.

9. Cai Shangxiang, *Wang Jinggong nianpu*, 12.403–9.

10. XCB 228.5542.

11. SHY, *zhiguan* 職官 5.8–9. For the significance of this bureau, see also Higashi, *Ō Anseki to Shiba Kō*, 79, and Kumamoto, "Chūsho kenseikan."

12. XCB 283.6934.

13. For Wang's family background and early experience, see Deng, *Beisong zhengzhi gaigejia*, 17–22.

14. For Wang's proclivity to befriend and promote southerners, see James Liu, *Reform in Sung China*, 2.

15. Wang Anshi, *Linchuan*, 41.446.

16. Wang Anshi, *Linchuan*, 39.414.

17. Wang Anshi, *Linchuan*, 39.413.

18. Shen, "Chu Hanlin xueshi xie chuanzhao biao," CXJ 23.6a. Note that Shen made this statement at the end of 1076, years after Wang's launch of the New Policies and a couple of months after Wang's second resignation. At this time, the relationship between Shen and Wang was already publicly contentious (see chapter 6), and Wang's leadership of the reform was drawing to its end. However, Shen did not retract his support of Wang's campaign motto, which demonstrates that he held a genuine belief in activist rule and was not paying lip service to the idea for political convenience.

19. Shen, "Yu xueguan Zhang jietui shu" 與學官張節推書, CXJ 18.5b.

20. Shen, "Hangzhou xin zuo zhouxue ji" 杭州新作州學記, CXJ 24.5a. In this essay, Shen placed concrete affairs in a dichotomy with moral teaching, describing the former as necessarily exigent and the latter as indispensable for long-term interests. The essay was written to celebrate a new state school, so Shen's praise of moral teaching was necessarily the focus. Nevertheless, he chose to frame the contrast in an even-handed dyad, an arrangement that reflected his belief in the value of concrete affairs. The essay was written in 1087, after his service under Wang. Like his reference to "activist rule," Shen continued to emphasize their shared values after the partnership ended, further attesting to the sincerity of his belief. The use of later materials in this chapter is a choice *faute de mieux*, a result of the incompleteness of Shen's extant writings. Nevertheless, these sources prove that Shen invested consistent attention in the

aforementioned ideas and that his earlier collaboration with Wang was built on genuine commonalities.

21. Chen Zhi'e, *Beisong wenhua shi*, 267–73.

22. XCB 219.2023.

23. Chen Shuguo, *Zhongguo lizhi shi*, 38.

24. Chen Shuguo, *Zhongguo lizhi shi*, 49–50.

25. XCB 210.5103.

26. SS 331.10651.

27. Shen recorded some details later in *Brush Talks*. In appendix 3, I include the full translation of the item in which he discusses the order of procedures in this ritual.

28. SS 204.5133. For a more detailed introduction, see appendix 2, entry 11.

29. For information on Jia, see Mihelich, "Polders and the Politics of Land Reclamation," 71–74.

30. Mihelich, "Polders and the Politics of Land Reclamation," 88.

31. SS 96.2381, and SHY, *shihuo* 食貨, 61.100–101.

32. For the details of Jia's financial plan, see Fan Chengda, *Wu Jun zhi*, 19.5b–12b. For the methods of securing funding for hydraulic constructions in the Song period, see Nagase, "Sōdai ni okeru suiri kaihatsu," and Li Jinshui, *Wang Anshi jingji bianfa*, 212–13.

33. For a detailed account of Jia's involvement in this program, see Mihelich, "Polders and the Politics of Land Reclamation," 71–107.

34. XCB 246.5989.

35. SHY, *shihuo*, 61.5924, XCB 246.5990–91.

36. XCB 246.5990–91.

37. XCB 247.6020.

38. For a comparison of Ever-Normal and Green Sprouts systems, see Wang Wendong, "Songchao qingmiao fa."

39. SHY, *shihuo*, 53.19, 53.32, and 58.2. Also see von Glahn, "Community and Welfare," 228.

40. von Glahn, "Community and Welfare," 228–29 and Li Jinshui, *Wang Anshi jingji bianfa*, 65–74.

41. Sima, *Su shui jiwen*, 16.316, and SS 176.4286.

42. SHY, *shihuo* 5.9.

43. Li Jinshui, *Wang Anshi jingji bianfa*, 83–84.

44. Shen was said to have set up *hedi cang* 和糴倉 (granaries of fair grain trade). The term *hedi* in general designated grain trades between the state and farmers since the Tang Dynasty. Thus, *hedi cang* was in essence the same as the Ever-Normal Granaries. For the meanings of *hedi* and other related terms, see Yuan, "Songdai shidi zhidu," and Geng, "Hedi, pingdi guanxi zai tan."

45. XCB 247.6008.

46. For the basic structure of the mutual security units, see Paul Smith, "Shen-tsung's Reign," 408.

47. Paul Smith, "Shen-tsung's Reign," 409.

48. In its original design, the Mutual Security Policy bore a connection with the economic aspects of the New Policies. The *jia*-units (*jia* 甲 as in *baojia*) constituted the

organizational structure for extracting the Green Sprouts money and other taxes. See Lin Ruihan, "Songdai baojia," 16–18. Shen revived the old granary system along the institutional and social lines laid down by the New Policies.

49. XCB 246.5990–91.

50. Paul Smith, "Shen-tsung's Reign," 409.

51. XCB 260.6349–50.

52. XCB 218.5297–98.

53. For the timeline of Wang's actions, see Diao, "Songdai 'baojia fa' si ti," 70–71.

54. XCB 233.5650. For Wang's plan to integrate the mutual security units with existing militia in the northwest, see Sun Yuanlu, "Beisong de yiyong," 8–10, and Chen Xiaoshan, "Beisong baojia fa," 52–53.

55. Sun Yuanlu, "Beisong de yiyong," 8.

56. XCB 257.6272.

57. XCB 257.6272.

58. XCB 257.6272.

59. SHY, *Bing* 兵, 2.11.

60. See Qi, *Wang Anshi bianfa*, 119.

61. Paul Smith, "Shen-tsung's Reign," 413.

62. See Wu Bing, "Beisong Dingzhou junshi," 21–24.

63. XCB 267.6542.

64. XCB 261.6355.

65. For urban control in Tang cities, see Heng, *Cities of Aristocrats and Bureaucrats*, 1–66, and Li Xiaocong, *Lishi chengshi dili*, 152–83.

66. For the transition from the ward system to the open city, see Heng, *Cities of Aristocrats and Bureaucrats*, 205–9; for detailed accounts of the Song city space, see Heng, *Cities of Aristocrats and Bureaucrats*, 117–204, and Li Xiaocong, *Lishi chengshi dili*, 223–44. According to Li Xiaocong, it is not clear whether the physical presence of ward walls completely disappeared in the Song. See Li Xiaocong, *Lishi chengshi dili*, 226. The word "ward" (*fang* 坊) nevertheless remained in the Song vocabulary as an organizational unit and continued to characterize Song urban dwellers, who were known as "households of wards and walls" (*fangguo hu* 坊郭戶). See Wang Zengyu, "Songdai de fangguohu." Shen's revisit of the ward system was likely a revival of the Tang praxis from its conceptual remains in the eleventh century.

67. XCB 267.6543.

68. XCB 267.6543.

69. XCB 260.6350, XCB 267.6543.

70. XCB 267.6543.

71. XCB 260.6349–50. For a comprehensive study of hydraulic defense in the Song, see Lorge, "The Great Ditch of China and the Song-Liao Border." For a survey of the use of arboreal and hydraulic defense in the Dingzhou area in the Song, see Wu Bing, "Beisong Dingzhou junshi," 30–35.

72. XCB 267.6543.

73. For an introduction to this policy and the changes it underwent in the eleventh century, see Chen Zhen, "Lue lun baoma fa."

74. XCB 267.6543.

75. MXBT, items 303, 324, and 331, 18.128, 19.135, and 19.136–37. The first item featured information Shen obtained from his brother, Shen Pi, who was an effective archer. It is reasonable to assume that Pi, like Gua, developed interest in this technique while serving in governmental posts, relatively early in life. In the second item, the recorded incident occurred during the Xining reign—the period under discussion—and involved Li Ding, another participant in the reform. In the third item, Shen stated that he made the observation in this item when he was in Haizhou, ca. 1055.

76. BBT, item 567, 2.219–220.

77. MXBT, item 472, 25.179–80. For translation and analysis of this item, see Holzman, "Shen Kua," 266, and Sivin, "Shen Kua," 22–23.

78. For the claim of primacy, see Wang Yong, "Cong Pei Xiu de ditu zhizuo tan zhongguo ditu," 195, cited in Holzman, "Shen Kua," 266.

79. XCB 260.6341–42. For the content of the text, see Shen's discussion in *Brush Talks*, BBT, items 578 and 579, 3.226, 3.226–27. For full, annotated translations of the two items, see Forage, "Science, Technology, and War," 335–41. According to Hu Daojing, Emperor Shenzong's long comment on the Nine-Army Formation was also based on Shen's understanding of this lost technique. See Shen, *Mengxi bitan jiaozheng*, edited and annotated by Hu Daojing, 731–32 and XCB 260.6339–41. For more bibliographical details, also see appendix 2, entry 25.

80. See appendix 2, entry 26.

## 6. Buffeted by the World (1075–1085)

1. The advent of the New Polices led to a three-decade-long division between reform advocates and adversaries in the imperial bureaucracy. For a comprehensive study of the factional conflicts, see Levine, *Divided by a Common Language*.

2. For exemplary comparative studies of Sima and Wang's political orientations, see Higashi, *Ō Anseki to Shiba Kō*, and Bol, "Government, Society, and State."

3. Paul Smith, "State Power and Economic Activism," 89.

4. Paul Smith, *Taxing Heaven's Storehouse*, 6, and "State Power," 82–88.

5. Bol, "Government, Society, and State," 156.

6. Paul Smith, "Shen-tsung's Reign," 363–65.

7. For divisions in the Wang coalition, see Shen Songqin, *Beisong wenren*, 182–84.

8. For discussions of this dyad, see Chen Zhi'e, *Beisong wenhua shi*, 260–76, and Levine, *Divided by a Common Language*, 2–3.

9. Sima, *Zi zhi tongjian*, 245.7899.

10. For the two dichotomies, see, respectively, Levine, *Divided by a Common Language*, 2–3, and Shen Songqin, *Beisong wenren*, 47.

11. XCB 252.6152–54, 252.6169–70.

12. Shen, *Beisong wenren*, 201–3.

13. XCB 269.6600.

14. XCB 278.6804. For an analysis of Wu and Wang's stances regarding the reform, see Gu, *"Da youwei" zhi zheng*. Note that while Wu and Wang were not reform advocates, nor were they previously adherents of the antireformist coalition.

15. XCB 256.6265–66.

16. XCB 264.6480.

17. XCB 263.6419.
18. XCB 263.6419.
19. XCB 264.6480.
20. The twenty-five included fifteen grades of rural households and ten grades of urban households. See James Liu, *Reform in Sung China*, 103.
21. XCB 283.6935.
22. For detailed information on the division of duty among servicemen, see James Liu, *Reform in Sung China*, 100–102.
23. For all aforementioned content of Shen's polices as well as quotations, see XCB 279.6826.
24. XCB 279.6826, 283.6935.
25. XCB 283.6935.
26. For a comprehensive study of the salt voucher system, see Dai, *Songdai chaoyan zhidu*.
27. For salt monopoly in the Song, see Chien, *Salt and State*.
28. William Guanglin Liu, *The Chinese Market Economy*, 47.
29. For an introduction to all models, see Chien, *Salt and State*, 58–62.
30. Chien, *Salt and State*, 62–65, and Dai, *Songdai chaoyan*, 274–83. Shen also had a short account introducing Fan Xiang and his policy. See MXBT, item 211, 11.91, and full translation in appendix 3.
31. Dai, *Songdai chaoyan*, 289–94; Chien, *Salt and State*, 65–66.
32. XCB 263.6442–43.
33. Li Tao believed that Shen's earlier decision was a reluctant rejection meant only to avoid provoking Wang Anshi. See XCB 263.6443.
34. XCB 274.6717–18, XCB 281.6885, and SHY, *shihuo* 24.13.
35. The Xie salt was produced in Xiezhou (present-day Shaanxi Province). For basic information on the salt industry in Xiezhou, see Dai, *Songdai chaoyan*, 8–9. For the categorization of salt in the Song, see Dai, *Songdai chaoyan*, 1–7.
36. For the triple division, see Guo Zhengzhong, "Beisong qianqi Xie yan," 66–67.
37. For the nature and active time of this bureau, see Cui, "Beisong Shaanxi lu zhizhi Xie yan si kaolun."
38. XCB 280.6872.
39. For the "salt exchange" process, see Dai, *Songdai chaoyan*, 64–67.
40. XCB 280.6872.
41. XCB 280.6871–72.
42. SS 183.4474.
43. XCB 265.6491. In *Brush Talks*, Shen narrated this story with further reinforcement from Emperor Renzong, who promised to lower the salt price in Hebei. Given the late date of *Brush Talks*, it seems that Shen remained unwavering until the end of his life in his centrist stance over the issue of salt. MXBT, item 212, 11.91.
44. For the causes of cash famine in middle-period China, see Gao, *Songdai huobi*, 333–44.
45. von Glahn, "Revisiting the Song Monetary Revolution," 159. For comprehensive data on the Song money stocks, see William Guanglin Liu, *The Chinese Market Economy*, 218–21.
46. Yuan, "Beisong qian huang," 129–31.

47. Miyazawa argues that during the New Policies only about one tenth of the entire output of Song mints was in circulation; meanwhile, the state held the majority of the currency in the treasuries as leverage over the economy. See Miyazawa, *Sōdai Chūgoku no kokka to keizai*, 63. Gao accepts the argument with the caveat that although the Song state absorbed a large amount of currency through taxation, it also released much of it back into circulation due to high state spending during the reform. See Gao, *Songdai huobi*, 339 and von Glahn's comparative summaries, "Revisiting the Song Monetary Revolution," 169–73.

48. Paul Smith, "Shen-tsung's Reign," 441–43.

49. See Araki, "Sōdai no dōkin," especially 12–26.

50. Ye, "Lun Beisong 'qian huang'," 23. The government "misused" copper in similar ways. For instance, Yabuuchi notices that the Northern Song state used eleven tons of copper in the construction of armillary spheres (including the one Shen made). Yabuuchi, "Sō Gen jidai ni okeru kagaku gijutsu no tenkai," 6.

51. XCB 283.6928.

52. XCB 283.6928.

53. XCB 283.6929.

54. XCB 283.6929.

55. XCB 283.6929.

56. For instance, Zhang Fangping described currency as a means to assist "the inexhaustible circulation of ten thousand things." See Zhang, *Zhang Fangping ji*, 15.181.

57. For Cai Que's background and career, see Clark, *Portrait of a Community*, 239–42.

58. XCB 283.6934.

59. Song and Song, *Song da zhaoling ji*, 206.770.

60. XCB 291.7114, 304.7411.

61. For the Song–Xia relation during the time, see Dunnel, "The Hsi Hsia," 191–97.

62. XCB 305.7426.

63. MXBT, item 191, 11.85–86.

64. MXBT, item 200, 11.88.

65. SS 331.10653.

66. MXBT, item 90, 5.43–44.

67. For details of the Song–Tangut war (including this siege), see Forage, "Science, Technology, and War," 59–74, and "The Sino-Tangut War."

68. Shen, "Suizhou xie biao" 隨州謝表, CXJ 16.6b.

69. For details of this case, see Egan, *Word, Image, and Deed*, 39–53, and Hartman, "Poetry and Politics in 1079."

70. See Bao, "Shen Gua shiji," 310.

## 7. In the System (and Then Out)

1. "Order of relations" is an umbrella phrase I use to include all orders eleventh-century literati identified in the process of pursuing the *dao*. It does not have a fixed correspondence in the Song vocabulary; yet as a placeholder, it functions to tease out the central source of knowledge in modeling.

2. Angle and Tiwald, *Neo-Confucianism*, 123.

3. Wang Anshi, *Linchuan*, 66.708.

4. For all quotations in the three foregoing paragraphs, see Wang Anshi, *Linchuan*, 66.707–8.

5. *Zhouyi zhengy*i, 8.184.

6. XCB 241.5886.

7. Wang Anshi, *Linchuan*, 66.706.

8. Wang Anshi, *Linchuan*, 68.720.

9. Wang Anshi, *Linchuan*, 63.673.

10. Munro, "Unequal Human Worth," 132.

11. Wang Anshi, *Linchuan*, 68.720.

12. Wang's further suggestions included limiting the advisory power of remonstrance officials to their own specialties and adjusting the procedure so that remonstrance officials' critiques reach bureaucrats before disseminating as finalized edicts. See Wang Anshi, *Linchuan*, 63.673. These suggestions likely derived from Wang's uneasy relationship with remonstrance officials, who acted as major deterrents to the New Policies. One may argue that this case was more rationalization than genuine knowledge seeking, which does not affect my analysis of epistemic praxis. The main difference between rationalization and knowledge production lies in motives rather than epistemological assumptions. In fact, rationalization that seeks to be successful should faithfully track reliable epistemic praxis.

13. *Zhouyi zhengy*i, 7.161. For the history of the dichotomy, see Cua, "Ti and Yong."

14. For a systematic reexamination of *li* (ritual) as a concept of order in the classical period, see a series of studies by Puett, for example, "The Haunted World of Humanity."

15. For content on institutions in the *Rituals of Zhou* and the foundational role the classic assumed in the discourse on statecraft, see Jaeyoon Song, *Traces of Grand Peace*, 12–15, and Schaberg, "The *Zhouli* as Constitutional Text."

16. For example, see Xiao-bin Ji's analysis of Sima Guang's elaborate thinking on the ruler-minister relation, Ji, *Politics and Conservatism*, 36–49.

17. For Wang's identification of the ancient origins of his policies, see Wang Anshi, *Linchuan*, 440–41.

18. For a methodological discussion of "culture as repertoire," see Swidler, *Talk of Love*, 24–40. My citation of Swidler is inspired by Robert Campany in *Making Transcendents*.

19. Wang Anshi, *Linchuan*, 68.720–21.

20. My definition of "epistemic authority" is inspired by Zagzebski's work. See Zagzebski, *Epistemic Authority*.

21. Allen, *Vanishing into Things*, 199.

22. In theory, a "regulation" (vis-à-vis "ritual") should be different from a generic governmental policy. But in the context of New Policies, Wang did not intend *fa* to be two different concepts. In many cases, a "policy" in Wang's definition carried a salient "regulatory" implication.

23. Wang Anshi, *Sanjing xinyi jikao huiping (3): Zhouli*, 1.1. Translation after Bol with minor changes, see Bol, "Wang Anshi and the *Zhouli*," 235.

24. Wang Anshi, *Linchuan*, 67.711.

25. See Jaeyoon Song's detailed discussion in *Traces of Grand Peace*, 222–44.

26. Wang Anshi, *Linchuan*, 67.711. Another variation was "policies of early kings" (*xianwang zhi fa* 先王之法). See Wang Anshi, *Linchuan*, 39.422.

27. Wang Anshi, *Linchuan*, 39.412.

28. For a comprehensive introduction to Wang's education reform, see Chaffee, "Sung Education," 298–305.

29. Wang Anshi, *Linchuan*, 39.410.

30. Wang Anshi, *Linchuan*, 39.410.

31. For some exemplary uses of the sages by Northern Song literati, see Bol, *Neo-Confucianism*, 67–69.

32. Wyatt, *The Recluse of Loyang*, 121.

33. Historians who examine Song thought in terms of social implications would view Shao as the opposite of Wang, because Wang, the most active statesman, crafted governmental policies and Shao, a recluse, concocted a cryptic system bearing little relevance to statecraft. The current epistemology-centered narrative groups them together for their common epistemic features.

34. In Wyatt's phrasing, number was Shao's "first principle" and "cardinal precept." See Wyatt, "Shao Yong's Numerological-Cosmological System," 22, 33.

35. My very brief summary of Shao's system is based on Birdwhistell's *Transition to Neo-Confucianism*; Wyatt's *The Recluse of Loyang* and "Shao Yong's Numerological-Cosmological System"; Arrault, *Shao Yong*, 230–403; and Bol, "On Shao Yong's Method for Observing Things."

36. Another system builder similar to Wang and Shao was Zhang Zai, who chose *qi* as his focal order. As I introduced in chapter 4, Zhang asserted that *qi* enabled all transformations in the world, which rendered him a stage-one system builder. The reason I do not categorize him as such is that his philosophy focused on the internal rather than the external: Zhang invested more attention in discussing *qi*'s function in transforming human nature than in ordering the universe (although the latter was by all means a key argument in his thinking). Such a different focus makes excessive normativity less of an issue in his system. Nevertheless, Zhang's choice of *qi* as the central order was challenged by Cheng Yi, an issue I will further pursue in chapter 10. For a succinct summary of Zhang's system, see Ong, *Men of Letters Within the Passes*, 51–55.

37. For Wang's reference to *li*, see Bol, *This Culture of Ours*, 231. For a systematic study of Wang's conception of *qi*, see Huang Shiheng, "Wang Anshi Laozi zhu." For Wang's reference to numerical relations, see chapter 2.

38. Su Shi, *Su Shi wenji*, 49.1427. Translation after Egan, *Word, Image, and Deed*, 63.

39. For analyses of Su's political and intellectual protests against Wang, see Bol, *This Culture of Ours*, 269–82, and Egan, *Word, Image, and Deed*, 54–85. For an introduction to the factions led by Su and Cheng, see Shen Songqin, *Beisong wenren*, 145–55.

40. For examples of Su's discussion of ritual, see Su Shi, *Su Shi wenji*, 2.46–47. A good part of Su Shi's discussion of *fa* appeared in the form of criticizing Wang Anshi. See Egan's detailed discussions in *Word, Image, and Deed*, 27–37, 68–73. Besides specific policy interventions, Su devoted much attention to contemplating the general significance of *fa* in forming productive relations in statecraft. A few examples: Su Shi, *Su shi wenji*, 7.218, 7.219, and 9.286. For his reference to *changli*, see Su Shi, *Su*

*Shi wenji*, 11.367. For his engagement with number, see Bol, "Reconceptualizing the Order of Things," 715.

41. Bol, *This Culture of Ours*, 284.

42. For analyses of the swimming metaphor, see Bol, "Chu Hsi's Redefinition of Literati Learning," 177, and Egan, *Word, Image, and Deed*, 54–56.

43. For *li* as Coherence, see chapter 10.

44. Wang Anshi, *Linchuan*, 82.861.

45. Shen, "Hangzhou xin zuo zhouxue ji," CXJ 24.4a–b.

46. Shen, "Fulou yi," CXJ 3.10b.

47. Shen, "Fulou," CXJ 3.10b.

48. See Shen's letter to Wang, "Xie Jiangning fu Wang xianggong qi" 謝江寧府王相公啓, CXJ 17.1a–b.

# 8. Brushtalking the World (1085–1095)

1. Shen, "Xiuzhou xie biao" 秀州謝表, CXJ 16.3b.

2. Shen, "You Xiuzhou Donghu" 遊秀州東湖, *Shen Gua quanji*, vol. 1, 161.

3. Shen, "Xiuzhou Chongde xian jian xue ji" 秀州崇德縣建學記, CXJ, 24.1b.

4. Shen, "Jin Shouling tu biao" 進守令圖表, CXJ 16.4a–5b.

5. BBT, item 575, 3.322. For the relationship between the big map and the small map, see Cao Wanru, "Lun Shen Gua zai ditu xue fangmian de gongxian."

6. For discussions of Shen's techniques, see Needham, *Science and Civilisation in China*, vol. 3, 576–77; Sivin, "Shen Kua," 22; and Fu Daiwie, "On *Mengxi bitan*'s World," 60, n.15.

7. A system allegedly invented by Pei Xiu 裴秀 (224–271); for details see Ding Chao, "Jin tu kaimi." Shen explicitly mentioned that he inherited the old method "Six Principles" (*liuti* 六體) from previous ages in making these maps. See "Jin Shouling tu biao," CXJ 16.4b. But in the *Brush Talks*, he mentioned "Seven Principles" (*qifa* 七法), which Hu Daojing corrects as a typo. See BBT, item 575, 3.224, and Hu, "Gudai ditu cehui jishu shang de 'qifa' wenti," and "*Mengxi bitan* buzheng," 41–43.

8. XCB 413.10033.

9. Shen, "Zi zhi" 自誌, in Lu Xian, *Jiading Zhenjiang zhi*, 11.447.

10. Shen paid thirty strings for the property, a modest price in his times. "Zi zhi," in Lu Xian, *Jiading Zhenjiang zhi*, 11.447. For a survey of home property values in the eleventh century, see Cheng Minsheng, *Songdai wujia*, 44–45 (although Cheng's assessment of Shen's estate is likely a misjudgment).

11. Shen specifically dedicated an essay to this building. See "An lao tang ji" 岸老堂記, CXJ 23.1a–2b.

12. Shen, "Zi zhi," in Lu Xian, *Jiading Zhenjiang zhi*, 11.1b.

13. Shen, "Zi zhi," in Lu Xian, *Jiading Zhenjiang zhi*, 11.447.

14. Shen, "Zi zhi," in Lu Xian, *Jiading Zhenjiang zhi*, 11.447.

15. The word "return" was a long-standing literary trope for the recluse ideal. The best-known work of the famous recluse Tao Yuanming 陶淵明 (365–427), for instance, was titled "Gui qu lai xi ci" 歸去來兮辭, often translated as "The Return."

16. Shen, "Gui ji" 歸計, CXJ 1.7a–b.

17. Shen, "Si gui" 思歸, *Shen Gua quanji*, vol. 1, 169.

18. Shen, preface to "You shanmen" 遊山門, *Shen Gua quanji*, vol. 1, 154.

19. Shen, "Yunzhou Xingguo si Chanyue tang ji" 筠州興國寺禪悅堂記, CXJ 22.2a–3b.

20. For records on visits, see "Runzhou Ganlu si" 潤州甘露寺, CXJ 1.4b, and "You er chanshi daochang" 遊二禪師道場, *Shen Gua quanji*, vol. 1, 173. For *Brush Talks* items, see MXBT, items 284 and 351, 17,121, 20.143–44, among a few others. For a brief analysis of these items, see Halperin, *Out of the Cloister*, 87.

21. Shen, "Yunzhou Xingguo si Chanyue tang ji," CXJ 22.3a.

22. Shen, "Xuanzhou Shi'ang si Chuandeng ge ji" 宣州石盎寺傳燈閣記, CXJ 22.6b.

23. Nonduality, *advaya* in Sanskrit or *bu'er* in Chinese, was a fundamental Mahāyāna Buddhist doctrine. A nondual state was one that transcended conventional dichotomies, such as right and wrong, good and evil. See Buswell and Lopez, *The Princeton Dictionary of Buddhism*, 18–19.

24. Shen, "Xuanzhou Shi'ang si Chuandeng ge ji," CXJ 22.6b. For a full translation, see appendix 3. For an analysis of the passage and an appraisal of Shen's stance on Buddhism, see Halperin, *Out of the Cloister*, 87–92.

25. Shen, "Suizhou Fayun chanyuan foge zhong ming" 隨州法雲禪院佛閣鐘銘, 24.9b.

26. Shen, "Suizhou Fayun chanyuan foge zhong ming," 24.10a.

27. Shen, "Suizhou Fayun chanyuan foge zhong ming," 24.10a.

28. Shen, "Zi zhi," in Lu Xian, *Jiading Zhenjiang zhi*, 11.447.

29. Shen, "Zi zhi," in Lu Xian, *Jiading Zhenjiang zhi*, 11.447.

30. Shen, "Xiaoxiao Tang ji," CXJ 24.8b. Zou and Chou were regional states in the Spring and Autumn period (771–476 BCE).

31. Shen, "Xiaoxiao Tang ji," CXJ 24.9a.

32. Shen, "Xiaoxiao Tang ji," CXJ 24.9a.

33. Shen, "Xiaoxiao Tang ji," CXJ 24.9a.

34. Shen, "Xiaoxiao Tang ji," CXJ 24.9a.

35. This text was supposed to be the sequel to Shen's earlier work, titled *Recollections of My Mountain Life* (*Huai shan lu* 懷山錄). See appendix 2, entries 13 and 17. Given the putative continuity between the texts, I translate *wanghuai* as "forgetting and recollecting" (rather the conventional "forgetting the longings"), consistently rendering the polysemic *huai* as "recollection."

36. For information on this text, see appendix 2, entry 17; for a few translated examples of its content, see appendix 3.

37. See appendix 2, entry 28.

38. See appendix 2, entry 30, and appendix 3 for a selection of translated items.

39. Zhu Yu, *Ping zhou ke tan*, 3.62.

40. MXBT, 15.

41. MXBT, 15.

42. MXBT, 15.

43. MXBT, 15.

44. The other two were, respectively, *li de* 立德 (establishing virtue) and *li gong* 立功 (establishing deeds). For the provenance of this reference, see *Chunqiu Zuo Zhuan zhengyi*, 35.277.

45. Fu Daiwie, for instance, analyzes the text in connection with Shen's experience in local administration. See Fu Daiwie, "World Knowledge and Local Administrative Techniques."

46. For instance, see MXBT, item 168, 9.76, item 180, 9.80–81, item 260, 14.109, BBT, item 582, 3.228, item 584, 3.228, XBT, item 609, 239, and item 609, 239.

47. Zhu Yizun, *Jingyi kao*, 183.10a. For an annotated list of Shen's other works, see appendix 2. For each of the texts mentioned below, appendix 2 provides detailed bibliographical information.

48. All six treatises are lost. See appendix 2, entries 19–24.

49. See appendix 2, entries 14 and 12.

50. See appendix 2, entries 6–10.

51. See appendix 2, entry 18.

52. See appendix 2, entry 27.

53. See appendix 2, entries 29 and 30.

54. For instance, in item 103, Shen claimed that he had discussed the system "incorporation of pitch names" (*nayin* 納音) at length in his earlier treatise *On Music*, so he decided to gloss over it this time. MXBT, item 103, 5.48.

55. The original text is no longer extant. For parts recovered from other sources, see Wang Anshi, *Wang Anshi Zi Shuo ji*.

56. Huihong, *Leng zhai yehua*, 10.79.

57. Su Zhe, *Luancheng hou ji*, in *Su Zhe ji*, 22.1127. For Su's anti-Wang sentiments, see Chao, *Junzhai dushu zhi jiaozheng*, 1.58.

58. Inspired by Pomata and Siraisi's literal (rather than Geertzian) use of "thick description." See Pomata and Siraisi, "Introduction," 26.

59. Modern scholars have reached the consensus that this structure was assigned by Shen himself rather than later editors. See Hu Daojing's analysis in Shen, *Xin jiaozheng Mengxi bitan*, edited and annotated by Hu Daojing, 3–4. In addition to the main text, there are two later additions, *Supplement to Brush Talks* (*Bu bitan* 補筆談) and *Sequel to Brush Talks* (*Xu bitan* 續筆談). For information on them, see Hu Daojing's introduction in Shen, *Xin jiaozheng*, edited and annotated by Hu Daojing, 4–5, and appendix 1. The seventeen categories were organized into twenty-six chapters. Most scholars agree that the original *Brush Talks* had twenty-six chapters. There was a different version of thirty chapters recorded in Ming bibliographies, but no solid evidence exists to show that the longer version preceded the twenty-six-chapter edition. For critical reflections on the debate, see Fu, "A Contextual and Taxonomic Study," 34–35, and Chiang, "Shen Gua zhushu kao," 22–24. For information on the historical editions of *Brush Talks*, see appendix 1.

60. Scholars have conducted in-depth analyses of the meanings of these categories. For instance, see Fu Daiwie's analysis of "Miscellaneous," in Fu Daiwie, "On *Mengxi bitan*'s World." Fu and Lei also examine "Numinous Marvels" and "Strange Occurrences" in context. See Fu Daiwie, "A Contextual and Taxonomic Study," and Lei and Fu, "Mengxi bitan li de yuyan yu xiangsixing."

61. Fu Daiwie discusses the item-by-item format from the perspective of the history of the book. See Fu Daiwie, "The Flourishing of *Biji*," 108–13.

62. Inspired by Lorraine Daston's discussion of fact as a "nugget of experience detached from theory." Daston, "The Factual Sensibility," 465.

63. MXBT, item 303, 18.128, and item 251, 14.106–7.

64. For a detailed analysis of Shen's discussion of causation, see chapter 9.

65. MXBT, item 307, 18.130. For a full translation of this item, see appendix 3.

66. MXBT, items 220 and 221, 12.94–95.

67. For a critical assessment of the term *leishu* and its translations, see Bretelle-Establet and Chemla, "Qu'était-ce qu'écrire une encyclopédie en Chine?," especially 9–11. For basic information on *leishu* as a textual tradition, see Sun Yongzhong, *Leishu yuanyuan yu tili*, and Drège, "Des Ouvrages Classés par Catégories." For a comprehensive study of *leishu* in the Song, see Zhang Weidong, *Songdai leishu*.

68. For details on classification in *Taiping Collectanea*, see Guo Bogong, *Song si da shu kao*, 17–25.

69. For details on the editorial arrangements, see Kurz, "The Politics of Collecting Knowledge," 295–301, and "The Compilation and Publication of the *Taiping yulan*," 44–56.

70. Intellectual neutrality, however, does not exclude external purposes. The *Taiping Collectanea* served an ostensibly political goal: it was a textual simulation of the political unification of the Song empire, and its categorical structure fulfilled a metaphor of control. See Kurz, "The Politics of Collecting Knowledge." Nevertheless, the political purpose remained external to the intellectual content of the book. Unlike *Brush Talks*, the *Taiping Collectanea* did not intend for an intellectual voice intimately intertwined with the internal details of the text.

71. *Taiping yulan*, 15.76a–78b.

72. For ways the literati used encyclopedias, see Tang Guangrong, *Tangdai leishu*, 5–12, and De Weerdt, "The Encyclopedia as Textbook," especially 79–90.

73. Ronald Egan notices Shen's selectivity. See Egan, "Shen Kuo Chats with Ink Stone," 134.

## 9. Building a Nonsystem

1. For examples and analyses, see chapter 11.

2. For recent comprehensive delineations of the *biji* phenomenon, see Ellen Cong Zhang, "To Be 'Erudite in Miscellaneous Knowledge'," 46–52, and De Weerdt, *Information, Territory, and Networks*, 281–394. De Weerdt's study specifically sheds new light on ways of studying the *biji* genre in its entirety.

3. De Weerdt discusses *biji* as a genre by identifying four common characteristics: a broad and free choice of subject matters, a nonstructured mixture of topics, personal verification of the content, and itemized presentation. See De Weerdt, *Information, Territory, and Networks*, 285. My discussion is largely consistent with her summary.

4. For the diversity of topics covered in *biji*, see Ellen Cong Zhang, "To Be 'Erudite in Miscellaneous Knowledge'," 51–52.

5. For *biji* as private histories, see Franke, "Some Aspects of Chinese Private Historiography," 116–17. Owen and De Weerdt argue that "supplemental" personal records in *biji* often functioned to destabilize the authority of official historiography. See Owen, "Postface," and De Weerdt, *Information, Territory, and Networks*, 374–75. For a discussion of *biji* in conjunction with "lesser writings" (*xiaoshuo* 小說) and "accounts

of anomalies" (*zhiguai* 志怪), see Inglis, *Hong Mai's Record of the Listener*, 108–9. For the history of anomaly accounts heralding Song *biji*, see Campany, *Strange Writing*. For a discussion of *biji* as a single-/multisubject technical treatise, see Fu Daiwie, "The Flourishing of *Biji*," 104–8 and 116–22.

6. Another good example is Ouyang Xiu, who at some point felt compelled to defend his engagement in *biji*. See Egan, *The Problem of Beauty*, 64–65.

7. Ellen Cong Zhang discusses the two practices as intimately interlocked. See Ellen Cong Zhang, "To Be 'Erudite in Miscellaneous Knowledge'."

8. Zhang Bangji, *Mozhuang manlu*, 281.

9. Zhang Bangji, *Mozhuang manlu*, 281.

10. Zhang Bangji, *Mozhuang manlu*, 281.

11. *Brush Talks*, for example, included such sections as "Numinous Marvels" and "Strange Occurrences" as well.

12. In the preface to the *Records from the Dongzhai Study* (*Dong zhai jishi* 東齋記事), Fan Zhen mentioned that he included content on "ghosts and spirits, dream and divination" and followed with an apologetic explanation that he did so only to remind people of discretions they should take against such information. See Fan Zhen, *Dong zhai jishi*, 1. Ouyang Xiu emphasized in the postscript of *Records from Returning to the Fields* (*Guitian lu* 歸田錄), that in this *biji* he avoided anything that concerned "kharma," "ghosts and spirits," "dreams and divination," and matters "on one's private life." Ouyang, *Guitian lu*, 269.

13. Shen used *xin* in a number of cases, for example, see MXBT, item 74, 4.36, item 81, 4.38, and item 357, 21.357. For *yan* and *biran*, see MXBT, item 116, 7.53, and item 430, 24.166.

14. My choice of the term "reliability" draws great inspiration from the discussion of reliability and reliabilism in philosophy. Reliabilism is a mode of justification, normally contrasted with evidentialism. The reliabilist approach justifies a belief by identifying it as the product of a reliable process, rather than supplying evidence for this belief (evidentialism). I invoke this philosophical definition of reliability to highlight the fact that supplying evidential reasons is not the only justificatory mechanism. In Shen's case, he made abundant descriptions of reliable procedures, an effort to attain good knowledge that should not be neglected although it went beyond the evidentialist model. For the definition of reliabilism and its distinction with evidentialism, see Goldman, "Reliabilism," 68–94, and "Toward a Synthesis of Reliabilism and Evidentialism," 123–50. My rendition of "epistemological reliability" in this historical context, however, is broader and not equivalent to reliability as in reliablism. In defining this concept, I appeal to the generic sense of the word, which refers to good, grounded knowledge in general. In my use, reliability provides an extensive spectrum that includes Shen's reliabilist impulse (dependence on reliable processes) and his effort to seek evidential reasons (a tendency he also saliently demonstrated, later picked up by evidential scholars). Thus, the "methods for reliability" in my discussion include evidence seeking and reliable procedures that involved no evidential reasoning. Alister Inglis aptly uses "reliability" in the study of Hong Mai and defines term mainly as "historical factuality." See Inglis, *Hong Mai's Record of the Listener*, 123–51, cited 123. My invocation of "reliability" is related, yet not limited to historical credibility, a point I elaborate later in this chapter and chapter 11.

15. MXBT, item 148, 8.67.

16. Also discussed by Fu Daiwie in his case study of these two sections. Fu argues that Shen contained the fantastic within the conceptual framework of the entire text, an insight this study concurs with. My interpretation of the entire conceptual framework, however, is different from Fu's. See Fu Daiwie, "A Contextual and Taxonomic Study."

17. MXBT, item 344, 20.141.

18. MXBT, item 346, 20.142.

19. MXBT, item 349, 20.143.

20. MXBT, item 391, 22.156.

21. MXBT, item 391, 22.156.

22. MXBT, item 169, 9.76.

23. SS 269.9244.

24. Jack Chen, "Introduction," 2.

25. Some scholars notice this feature and characterize it in different language. For instance, Joël Brenier calls it a preference of "comparisons and analogies" over "formal discursive articulation." See Brenier et al., "Shen Gua et les sciences," 350.

26. BBT, item 523, 1.202.

27. For a full translation of this item, see appendix 3.

28. Zhuangzi, *Zhuangzi jishi*, 1.4.

29. MXBT, item 66, 3.32.

30. MXBT, item 66, 3.32.

31. MXBT, item 66, 3.32.

32. BBT, item 523, 1.202.

33. MXBT, item 363, 21.149.

34. MXBT, item 357, 21.147.

35. MXBT, item 357, 21.147.

36. MXBT, item 360, 21.148.

37. MXBT, item 486, 26.186.

38. MXBT, item 53, 3.30; MXBT, item 491, 26.187.

39. For a survey of Shen's achievements in mathematics, see Sivin, "Shen Kua," 13–15.

40. MXBT, item 301, 18.126–27.

41. For an introduction to this method, see Needham and Wang, *Science and Civilisation in China*, vol. 3, 39, and Martzloff, *A History of Chinese Mathematics*, 328–29.

42. For introductions to this method, see Needham and Wang, *Science and Civilisation in China*, vol. 3, 142–43; Sivin, "Shen Kua," 14; Bréard, "Shen Gua's Cuts"; and Tian Miao, "The Westernization of Chinese Mathematics," 46.

43. A system to count off days in a cycle of 60. See Sivin, *Granting the Seasons*, 63.

44. For a full translation of this item, see appendix 3.

45. For the antiquarian-cosmic scheme in Song sonic world, see Zuo, "Keeping Your Ear to the Cosmos."

46. MXBT, item 97, 5.63.

47. MXBT, item 44, 3.27–28. For a full, annotated translation of this item, see Graham and Sivin, "A Systematic Approach to the Mohist Optics," 145–47.

48. MXBT, item 44, 3.27.

49. MXBT, item 347, 20.142.

50. MXBT, item 347, 20.142.

51. MXBT, item 347, 20.142.

52. My discussion of commonsense epistemic virtues is inspired by Alvin Goldman's "epistemic folkways." See Goldman, "Epistemic Folkways and Scientific Epistemology."

53. The issue of "nature" in premodern China was complex because there was no exact counterpart of Western "nature," only some approximations. For a survey of them, see Harbsmeier, "Towards a Conceptual History of Some Concepts of Nature." It is worth noting that none of these terms (including *tiandao*, which I choose to focus on) designated a natural realm rigorously demarcated from the human world. My current argument is not a reiteration of the idea that Chinese cosmology was characterized by a "pre-given, continuous cosmos" denying a nature-culture break, an argument Michael Puett brilliantly disputes. See Puett, *The Ambivalence of Creation*. The old "continuity" thesis asserts that culture (passively) fulfilled natural patterns. Continuity took the form of homogeneous repetitions of patterns, first in Heaven and Earth and then among humans. To echo Puett's analysis, my study provides another example (in a later period) in which the human realm featured humans' own creative work and a discontinuity from Heaven. More important, the Song thinkers did not view such discontinuity as disruption of a larger unitary order (the *dao*). My objection to a nature-culture demarcation is thus the by-product of the historical observation, that in the Song world, unity (continuity) did not necessarily depend on self-same repetitions between Heaven and humans.

54. For a discussion of the topic in the classical times, see Graham, *Studies in Chinese Philosophy*, 54–57.

55. Cheng and Cheng, *Er Cheng ji*, Yishu, 18.182.

56. See, respectively, MXBT, item 127, 7.57, item 126, 7.57, item 133, 7.60, item 304, 18.128–29, and item 150, 8.68.

57. See, respectively, MXBT, item 307, 18.130 (for a full translation of this item, see appendix 3), item 299, 18.125, and item 308, 18.130–31.

58. *Bo* can be translated as "erudite," "erudition" when referring to a human quality.

59. Ellen Cong Zhang views *bo* as a new scholarly ideal that distinguished itself from the more highbrow classical studies, an argument I endorse with the caveat that *bo* did not necessarily turn away from textual learning. See Ellen Cong Zhang, "To Be 'Erudite in Miscellaneous Knowledge'."

60. See Zhang Lei, *Mingdao zazhi*, 14; Zhu Xi, *Zhu Zi yulei*, 92.2342; and Ma Duanlin, *Wenxian tongkao*, 216.10a.

61. Su Shi, *Su Shi wenji*, 10.340.

62. For negative connotations of *bo*, see also Bol, "The Rise of Local History," 58–59.

63. For a biographical account of Su, see Needham, Wang Ling, and Price, *Heavenly Clockwork*, 5–9. Sivin initiates the comparison between Su and Shen. See Sivin, "Shen Kua," 31–32.

64. For example, see Su Xiangxian, *Chengxiang Weigong tanxun*, 3.1133.

65. For details of Su's astronomical clock tower, see Needham, Wang Ling, and Price, *Heavenly Clockwork*; Wang Zhenduo, "Songdai shuiyun yixiang tai"; Li Zhichao, *Shuiyun yixiang zhi*, 76–101; and Su Song, *Xin yixiang fayao*, 18–40.

66. Su Song, *Su Weigong wenji*, 56.996–97, 998.

67. For Su's partisan association, see Wang Ruilai, "Su Song lun," 120–22, and Jin Qiupeng, "Lue lun Su Song," 5–7.

68. A number of political incidents indicated a potentially uneasy relationship between Su and Shen. For one thing, Su strongly opposed Wang's appointment of the controversial Li Ding (see chapter 3). For Su's protesting memorials, see Su Song, *Su Weigong wenji*, 16.220–24. Also, Su Song was a close friend with Su Shi, and both were victims of the Crow Terrace Poetry Case, for which Shen was possibly responsible (see chapter 6). For a detailed analysis, see Guan and Wang, "Su Song yu Su Shi jiaoyi."

69. For examples of plants and food, see Su Shi, *Su Shi wenji*, 73.2361 and 73.2371. For his medical writings and recipes, see Su and Shen, *Su Shen liangfang*.

70. See Bol, "A Literati Miscellany."

71. Zhang Lei, *Mingdao zaozhi*, 14, 16.

72. Zhang Lei, *Mingdao zaozhi*, 22.

73. Zhang Lei, *Zhang Lei ji*, 48.758. Translation after Bol with changes, see Bol, "A Literati Miscellany," 130–31.

74. Ellen Cong Zhang, "To Be 'Erudite in Miscellaneous Knowledge'."

75. Li Zhao, *Tang guoshi bu*, 3.

76. Fan Zhen, *Dong zhai jishi*, 1.

77. Ouyang, *Guitian lu*, 269.

78. Wang Pizhi, *Mianshui yantan lu*.

79. For a detailed analysis of this trend, see Jiang Mei, "Songdai biji," 147–48.

80. Qian Yi, *Nanbu xinshu*, 1.

81. For the development of *biji* in the age of factional struggles, see Jiang Mei, "Songdai biji," 148–51. For a detailed study of the influence factional struggles exerted on historical composition at the imperial court, see Liang, "Beisong houqi dangzheng yu shixue."

82. See Qi Xia's analysis, Qi, "Fan Zhongyan jituan," 2.

83. See the preface to this text, Liu Yanshi, *Sun Gong tan pu*, 139.

84. Similar with what Pomata and Siraisi call an "empirisme érudit." See Pomata and Siraisi, "Introduction," 17.

85. Fu Daiwie proposes a "two-tier theory of knowledge" to characterize Shen's epistemological stance, which is another way to frame this complexity. See Fu Daiwie, "When Shen Gua Encountered the 'Natural World'."

86. "Experience" assumed a complex role in medical practice. Farquhar and Sivin argue that in "experience" praxis and conceptual knowledge are inseparable from each other, a conceptualization similar to what I call the relationship between sensory knowing and deep orders. See Farquhar, *Knowing Practice*, 2–3, and Sivin, "Text and Experience in Classical Chinese Medicine," 195–98.

87. MXBT, item 314, 18.132. Also mentioned in the preface to *Liangfang*, Su and Shen, *Su Shen liangfang*, 3.

88. Su and Shen, *Su Shen liangfang*, 4.

89. Su and Shen, *Su Shen lifangfang*, 1–2. Translation after Sivin, *Health Care in Eleventh-Century China*, 83. For the medical use of the Course-*Qi* system, see Despeux, "The System of the Five Circulatory Phases," 139–40, and Goldschmidt, *The Evolution of Chinese Medicine*, 175–76.

90. Su and Shen, *Su Shen liangfang*, 1. Translation after Sivin, *Health Care in Eleventh-Century China*, 82.

91. Su and Shen, *Su Shen liangfang*, 1. Translation after Sivin, *Health Care in Eleventh-Century China*, 82–83.

92. Sivin, *Health Care in Eleventh-Century China*, 83, n.81.

93. Furth, *A Flourishing Yin*, 23. For an introduction to the organ system, see Unschuld, *Huang Di nei jing su wen*, 129–36.

94. *Huangdi neijing suwen*, 3.165. For the subtle meanings of color in Chinese medicine, see Kuriyama, *The Expressiveness of the Body*, 153–92.

95. Su and Shen, *Su Shen liangfang*, 1.

## 10. Farewell to System

1. Except in sporadic cases, Shen respectfully cited Cheng Yi's words. For instance, in his essay on eclipses, Shen quoted Cheng's grand characterization of the sun and moon as the consummation of yin and yang. See Shen, "Lun ri yue" 論日月, CXJ 3.14a.

2. The concept of *li* in neo-Confucianism has provoked long-term debates regarding its translation. In the past fifty years, English-speaking scholars have respectively rendered it as "reason," "law," "pattern," "principle," "form," and "coherence," among a few other choices. For critical evaluations of these translations, see Peterson, "Another Look at Li," 13–14, and Ziporyn, "Form, Principle, Pattern, or Coherence." The current study adopts "Coherence" with a capital C as the one-word English translation of *li*. My choice builds on generations of scholarly work. Since Peterson's first proposal of "coherence," multiple scholars from various angles have reinforced the validity of this translation. Peter Bol first applies it in his historical account of neo-Confucianism. See Bol, *Neo-Confucianism*, 163–68. Brook Ziporyn and Stephen Angle provide thoroughgoing philosophical reasoning for this reading. See Ziporyn, "Form, Principle, Pattern, or Coherence" and *Beyond Oneness and Difference*, 321–44, and Angle, *Sagehood*, 31–50. My definition of Coherence primarily depends on these three scholars in addition to Peterson. I follow Angle's example in adopting the capital C in Coherence. In his conversation with Justin Tiwald, Angle explains that the capitalization highlights that "speaking of Coherence is to make a significant metaphysical claim about the structuring of the universe, rather than a deflationary view according to which *li* is whatever one happens to find coherent." See Angle, "Reply to Justin Tiwald," 239. I endorse Angle's rejection of the deflationary "coherence-only" idea by adding a further historical reason: *li* was a deep fact about the structuring of the universe, and the basic content of these structures was historically grounded in the written culture hitherto accumulated (see my discussion of the cultural repertoire in chapter 7). From a historical point of view, *li* could never be whatever one found coherent in a purely subjective sense.

3. Peterson, "Another Look at Li," 14.

4. Su Shi, *Su Shi wenji*, 1.1.

5. Angle, Review of *Neo-Confucianism in History*, 348.

6. Cheng and Cheng, *Er Cheng ji*, Yishu, 18.193.

7. Cheng and Cheng, *Er Cheng ji*, Wenji, 9.609.

8. Angle, *Sagehood*, 48.

9. Cheng and Cheng, *Er Cheng ji*, Yishu, 2A.34.

10. Cheng and Cheng, *Er Cheng ji*, Yishu, 18.189.

11. The "overwriting" process was mainly a matter of metaphysical adjustment so the rise of *li* did not eliminate the discursive presence of other orders.

12. Cheng and Cheng, *Er Cheng ji*, Yishu, 2B.51.

13. Cheng and Cheng, *Er Cheng ji*, Yishu, 2A.16–17. Translation after Ziporyn with minor changes. See Ziporyn, *Beyond Oneness and Difference*, 322.

14. Cheng and Cheng, *Er Cheng ji*, Yishu, 15.144.

15. Cheng and Cheng, *Er Cheng ji*, Yishu, 18.237.

16. Cheng and Cheng, *Er Cheng ji*, Yishu, 2A.36.

17. Cheng and Cheng, *Er Cheng ji*, Jing shuo, 1.1028.

18. My summary of the distinction is based on Wyatt and Adler's discussions. See Wyatt, *The Recluse of Loyang*, 91–92, and Adler, "Chu Hsi and Divination," 174, n.20. For a book-length exposition of representative figures in the two schools, see Zhu Bokun, *Yixue zhexue shi*, vol. 2.

19. For a detailed study of Cheng Yi's interpretative approach, see Hon, "Redefining the Civil Governance."

20. Wyatt, *The Recluse of Loyang*, 88–89.

21. Cheng and Cheng, *Er Cheng ji*, Wenji, 9.615.

22. Cheng and Cheng, *Er Cheng ji*, Cuiyan, 2.1227.

23. Kasoff, *The Thought of Chang Tsai*, 137. Although Zhang and Cheng differed in terms of assigning priority, they both agreed that *li* and *qi* could not exist independently from each other, an ontological argument later inherited by Zhu Xi. See Yong Huang, *Why Be Moral*, 296, n.16.

24. For Zhu's stipulation of the *li-qi* relation, see Peterson, "Another Look at Li," 18–20, and Angle, *Sagehood*, 38–44.

25. Peterson, "Another Look at Li," 21–22.

26. Zhu Xi, *Zhu Zi Yulei*, 9.156.

27. For a similar analysis, see Angle, *Sagehood*, 49.

28. Partly because that "function" belongs to the sensory in modern conceptualization.

29. I borrow this term from Angle and Tiwald but use it without making the distinction between what they call type-two and type-three knowing. See Angle and Tiwald, *Neo-Confucianism*, 122–26.

30. To do justice to Shao Yong, his later thinking under the rubric of "observation of things" invested extensive attention to the heart-mind. See Wyatt, "Shao Yong's Numerological-Cosmological System," 25–29. However, it is not clear to me that the new line of thinking was oriented to address the normativity issue in the number system.

31. This was why later in Zhang Zai and Cheng Yi's phrasing, modeling was known as "virtuous nature's knowing." For an introduction to human nature and its connection with the heart-mind, see Graham, *Two Chinese Philosophers*, 44–60.

32. Another discourse associated with this argument was the dichotomy between "the heart-mind of humans" (*renxin* 人心) and "the heart-mind of the *dao*" (*daoxin* 道心). The latter referred to the undisturbed heart-mind embodying the perfect cosmic perspective, and the former stood for the human condition susceptible to the vagaries of

passions and intentions. A number of eleventh-century literati participated in this discourse, for example, Su Shi (see Egan, *Word, Image, and Deed*, 84–85) and Cheng Yi (see Graham, *Two Chinese Philosophers*, 64).

33. Wang Anshi, *Linchuan*, 65.691.

34. Wang Anshi, *Linchuan*, 65.691.

35. For a full translation and analysis, see Wyatt, *The Recluse of Loyang*, 99–100.

36. For a comprehensive analysis of Zhang's thinking on the heart-mind, see Kasoff, *The Thought of Chang Tsai*, 85–91.

37. Zhang Zai, *Zhang zai ji*, Zhengmen, 24, Liku, 266, Liku, 275, and Yulu, 2.325.

38. Cheng and Cheng, *Er Cheng ji*, Yishu, 21B.276.

39. Cheng and Cheng, *Er Cheng ji*, Yishu, 18.204.

40. Cheng and Cheng, *Er Cheng ji*, Yishu, 18.197.

41. Cheng and Cheng, *Er Cheng ji*, Yishu, 2A.22. Translation after Graham with minor changes. Graham, *Two Chinese Philosophers*, 81.

42. Cheng and Cheng, *Er Cheng ji*, Yishu, 21A.271–72. Translation after Smith with changes; Smith, "Sung Literati Thought and the *I Ching*," 217.

43. Cheng and Cheng, *Er Cheng ji*, Yishu 2B.57.

44. For Shao's possible solution, see Smith, "Sung Literati Thought and the *I Ching*," 216.

45. Bol, *This Culture of Ours*, 326.

46. Cheng and Cheng, *Er Cheng ji*, Yishu, 15.163.

47. A number of scholars believe that Cheng Hao, in contrast, placed less emphasis on the external world. See Graham, *Two Chinese Philosophers*, 95, and Bol, *Neo-Confucianism*, 214. The difference between the Chengs later evolved into the divergence between "Learning of Coherence" (*lixue* 理學) and "Learning of the Heart-mind" (*xinxue* 心學). The latter school focused on exploring one's heart-mind at the expense of investigating the external world. For this development, see de Bary, *Neo-Confucian Orthodoxy*.

48. James Liu first proposes the "inward turn" in the sense that at the transition from the Northern Song to the Southern Song, a pursuit of the perfection of the self through moral cultivation replaced that of the perfection of the social order through good government. See James Liu, *China Turning Inward*. A few other scholars make similar observations without necessarily invoking the term "inward." Peter Bol argues that in contrast to building a "political, social, economic, and cultural system," neo-Confucians devised "a serious program of self-transformation." See Bol, *Neo-Confucianism*, cited 271 and 273. Some scholars disagree. For instance, Yü Ying-shih questions the "inward" thesis by pointing out that the neo-Confucians in the twelfth century were still active in participating in court politics. He thus argues for the coexistence of the concern for self-cultivation (what Yü phrases as "internal sageliness") and that for effective governance ("external kingliness") in neo-Confucianism. See Yü, *Zhu Xi de lishi shijie*. I agree that the definition of "inwardness" can be further refined and self-cultivation and statecraft did not have to be mutually exclusive commitments. There was, however, a turn toward the "self," in the sense that the thinking on the internality of a person—including his heart-mind and nature—grew in sophistication and the self started to demand more intellectual and political authority. This change did not necessarily invalidate the intention to make outward efforts to order the world, yet might affect the forms of political

participation. My narrative on Cheng Yi's reorientation of the guiding function of his system provides one example of this refined "inwardness."

49. Shen, "Xie fu qiju sheren chong Longtu ge daizhi biao" 謝復起居舍人充龍圖閣待制表, CXJ 14.1b.

50. MXBT, item 137, 7.62, item 82, 5.40–42, item 432, 24.167, and item 430, 24.166.

51. Drawing on examples from *Brush Talks*, Sakade also notices that Shen was not committed to a unitary *li*, especially in comparison to Cheng Yi. See Sakade, "Shin Katsu no shizenkan ni tsuite," 82–84.

52. Shen, "Da tongren shu," CXJ 19.4b. Yang Weisheng regards this sentence to be misplaced from another passage, "Da Li Yanfu xiucai shu," with which I concur. See *Shen Gua quanji*, CXJ, 46.

53. MXBT, item 437, 24.169.

54. MXBT, item 432, 24.167.

55. MXBT, item 128, 7.58.

56. This is not to affirm that the reason Shen proposed was the only cause of the glitch. For other possible antecedents, see Guo Shengchi, "Shen Gua faxian de louhu chiji," 208–9.

57. Given that the variation of the length of the apparent solar day is small (fifty-one seconds maximum), scholars have different opinions as to whether Shen captured this difference through experiential observation or theoretical deduction. For reasons to support the former interpretation, see Qian Jingkui, "Guanyu Shen Gua" and Hua, *Zhongguo louke*, 94–95.

58. MXBT, item 44, 3.27.

59. BBT, item 537, 1.206.

60. MXBT, item 485, 26.185–86. For a full translation of this item, see appendix 3.

61. For a survey of the changing meanings of the subject and subjectivity, see Rorty, "The Vanishing Subject," 35–45. For the purpose of making a focused comparison, I limit my definition of the subject to the basic parameters of the Cartesian self. For an introduction to the Cartesian self, see Dicker, *Descartes*, 39–90.

62. Cheng and Cheng, *Er Cheng ji*, Yishu, 2A.16.

63. Cheng and Cheng, *Er Cheng ji*, Yishu, 9.247.

64. Heidegger, *The Question Concerning Technology*, 128.

65. Cheng and Cheng, *Er Cheng ji*, Yishu 1.9.

66. Wang Anshi, *Linchuan*, 56.608–9. For an analysis of both Cheng and Wang's views, see Lin Sufen, *Beisong zhongqi ruxue*, 438–39.

67. For a systematic study of the issue in the classical period, see Makeham, *Name and Actuality*.

68. Makeham, *Name and Actuality*, 46; Allen, *Vanishing into Things*, 31.

69. Shao Yong, *Huangji*, 11.493. Translation after Wyatt, "The Transcendence of the Past," 212. For Shao's relativism, see Wyatt, "The Transcendence of the Past," 212–15.

70. SS 427.12728.

71. SS 427.12728.

72. MXBT, item 350, 20.143.

73. MXBT, item 144, 8.66. For another similar example regarding the *Change*, see BBT, item 551, 2.215. See also Sivin's brief analysis, "Shen Kua," 35–36.

74. MXBT, item 128, 7.58.

75. MXBT, item 314, 18.132. Also see Su and Shen, *Su Shen liangfang*, 4.

## *11. Reverberating in the World (1100–1800)*

1. MXBT, Mao edition, 26.12b.

2. Zhu Xi, *Zhu Zi yulei*, 92.2343.

3. For Zhu's appreciation of Shen, see Hu Daojing, "Zhu Zi dui Shen Gua kexue xueshuo de zuanyan." In recent years, scholars who study Zhu Xi's engagement with natural studies often discuss Zhu's opinions on Shen. For instance, see Kim, *The Natural Philosophy of Chu Hsi* and Le, *Zhu Xi de ziran yanjiu*.

4. Zhu Xi, *Zhu Zi yulei*, 94.2367 and 64.1597.

5. Zhu Xi, *Zhu Zi yulei*, 23.536. For a detailed analysis, see Kim, *The Natural Philosophy of Chu Hsi*, 147–48.

6. Zhu Xi, *Hui'an xiansheng Zhu Wen gong wenji*, 45.710. For a detailed analysis, see Kim, *The Natural Philosophy of Chu Hsi*, 149.

7. Ma Mingheng, *Shangshu yiyi*, 4.30a.

8. Zhu Xi, *Zhu Zi yulei*, 2.19.

9. Zhu Xi, *Zhu Zi yulei*, 92.2342. Zhu made this comment in the particular context of music and harmonics; therefore, *qi* specifically referred to musical instruments, and *shu* to calculations of harmonics.

10. For another discussion of Zhu's interest in this dyad in the context of classical exegesis, see De Weerdt, *Competition over Content*, 240.

11. See Elman, *On Their Own Terms*, 5–6.

12. Zhu Xi, *Zhu Zi yulei*, 11.188.

13. Zhu Xi, *Zhu Zi yulei*, 11.188.

14. Zhu Xi, *Zhu Zi yulei*, 11.188.

15. A number of scholars have paid attention to this development. For instance, Alister Inglis employs "reliability" to characterize Hong Mai's *biji* writing. See Inglis, *Hong Mai's Record of the Listener*, 123–51. Hilde De Weerdt notes that personal verification of content became an important concern of *biji* authors in the Southern Song. See De Weerdt, *Information, Territory, and Networks*, 285. Fu Daiwie addresses some Song *biji* as *kaoju* 考據, "evidential research," emphasizing the authors' proclivity to produce grounded knowledge. See Fu Daiwie, "The Flourishing of *Biji*," 109 and 115.

16. Hong, *Rong zhai suibi*, 980.

17. Hong, *Rong zhai suibi*, 980.

18. Hong, *Rong zhai suibi*, 980.

19. Tanzi, allegedly the teacher of Kongzi, used to name official titles after birds. See *Chunqiu Zuo zhuan zhengyi*, 48.4523–26.

20. For the story of Bozong, see *Chunqiu Zuo Zhuan zhengyi*, 24.4097–98.

21. Zha, the third son of the Duke of Wu, once observed the performance of ritual music at the Zhou court and made abundant comments. See *Chunqiu Zuo Zhuan zhengyi*, 39.4355–61. For the whole quotation, see Zhou Mi, *Qi dong yeyu*, 1.

22. Gong Yizheng, *Jieyin biji*, 121. For a full translation of this preface and analysis, see De Weerdt, *Information, Territory, and Networks*, 285–86.

23. Zhang Shinan, *Youhuan jiwen*, 95. Translation after Cong Zhang with changes. See Ellen Cong Zhang, "To Be 'Erudite in Miscellaneous Knowledge'," 68.

24. Wu Zeng, *Nenggai zhai manlu*, 8–48.

25. Wu Zeng, *Nenggai zhai manlu*, 76–206.

26. The de facto leader of the state immediately under the emperor.

27. Hong, *Rong zhai suibi*, 4.53–54. For Shen's original record, see MXBT, item 175, 9.78.

28. MXBT, item 392, 22.157.

29. Hong, *Rong zhai suibi*, 4.54.

30. Hong, *Rong zhai suibi*, 4.54.

31. MXBT, item 321, 19.135.

32. Wu Zeng, *Nenggai zhai manlu*, 3.72.

33. Wu Zeng, *Nenggai zhai manlu*, 6.159.

34. The Wenmin brothers referred to Hong Mai and his two brothers, Hong Shi 洪適 (1117–1184) and Hong Zun 洪遵 (1120–1174).

35. Hong, *Rong zhai suibi*, 984.

36. For the official biographies of the three Liu, see SS 319.10383–90. These biographies were grouped together with those of Ouyang Xiu and Zeng Gong, which demonstrated that at least in the context of SS, the three Liu were valued for their classical, historical, and literary work, similarly to Ouyang and Zeng.

37. Peterson, "Confucian Learning in Late Ming Thought," 779. For a comprehensive discussion of the term, see Elman, *From Philosophy to Philology*, 72–122.

38. For a definition of Song Learning as inquiries centered on "meanings and principles," see Elman, *From Philosophy to Philology*, 46–47. For a comprehensive survey of the meanings of the term, see Zhu and Wang, "'Song xue de lishi," specifically 80–83.

39. The term *kong* (empty, hollow) was frequently invoked by evidential scholars in criticizing the Song learning. Here I cite Dai Zhen's *zaokong* as an example. See Dai Zhen, *Dai Zhen quanshu*, vol. 6, 505.

40. "Tiyao," in Zheng Fangkun, *Jingbai*, 1b–2a. This is a remark presumably made by two compilers of the *Four Treasures*. Most editors involved in this project were advocates of the evidential movement. See Elman, *From Philosophy to Philology*, 100–102. Elman also notices that Qing scholars looked to the genre *biji* for the early history of evidential scholarship. See Elman, *From Philosophy to Philology*, 212 and 242.

41. Qian Daxin, *Qianyan tang wenji*, in *Jiading Qian Daxin xiansheng quanji*, vol. 9, 25.405. Cheng was famous for his work on historical geography and he authored *A Geographical Map of Mountains and Rivers from the Tribute of Yu* (*Yugong lun shanchuan dili tu* 禹貢論山川地理圖). Sun authored a *biji* entitled *Collection of Instructions for [My] Sons at the Lüzhai Study* (*Lü zhai shi'er bian* 履齋示兒編).

42. Qian Daxin, *Qianyan tang wenji*, 25.405.

43. Qian Daxin, *Qianyan tang wenji*, 25.405.

44. Hong Jing's postscript, in Hong, *Rong zhai suibi*, 986.

45. Xue et al., *Jiu Wudai shi*, 40b.

46. Zheng Fangkun, *Jingbai*, 2b.

47. Yü Ying-shih notices that the Qing evaluation of Song learning (under the rubric of "meanings and principles") only partially, and to a considerable extent inaccurately,

reflected the content of Song scholarship due to an overriding concern to accentuate the Song-Qing distinction on their own terms. See Yü, *Zhongguo sixiang chuantong*, 172–73.

48. Zheng Fangkun, *Jingbai*, 2b.

49. Qian Daxin, *Qianyan tang wenji*, 25.405.

50. Another important reason Qing scholars fixated on book learning when faulting Song learning was that the object of criticism they had in mind was often the "learning of the heart-mind" in its post-Ming form (the part of "Song learning" that went beyond the Song period). Learning of the heart-mind placed more emphasis on one's internality vis-à-vis the external world, and such a penchant became more salient after Wang Yangming's proposal of a new concept, "good knowing" (*liangzhi* 良知). "Good knowing" was supposed to be a faculty of the heart-mind and the first guide for learning; it posed a contrast to hearing, seeing, and rote book learning, which were subsidiary sources often involved in "vulgar learning" (*suxue* 俗學). These Ming changes prompted many debates on this issue to focus on a contrast between book learning and a literal reliance on one's heart-mind (speculation), hence the evidential scholars' criticism that Song scholars did not read. For an analysis of "good knowing" in regard to epistemology, see Angle and Tiwald, *Neo-Confucianism*, 121. For a detailed analysis of Qing scholars' emphasis on book learning, see Yü, *Zhongguo sixiang chuantong*, 174–86.

51. For a detailed study of this trend, see Jami, "'European Science in China' or 'Western Learning'?"

52. Ruan, *Chouren zhuan*, 2b. For more on Ruan's rendition of Shen, see Martzloff, "French Research in to the *Mengxi bitan*," 45–46.

53. For a few conventional ways of distinguishing "theoretical knowledge" and "practical knowledge," see Fantl, "Knowing How."

54. The distinction between "knowing how" and "knowing that," a thesis popularized by Gilbert Ryle, has inspired a rich body of philosophical arguments. Here I only draw on the most basic meaning of this distinction for the sake of keeping my discussion in focus. For Ryle's initial discussions, see "Knowing How and Knowing That" and *The Concept of Mind*. For a recent survey of positions on this issue, see Bengson and Moffett, "Two Conceptions of Mind and Action."

55. For a survey of the term from the Song through the late Ming, see Elman, *On Their Own Terms*, 5–9. For a detailed discussion in the late imperial period, see Elman, "The Investigation of Things."

56. Sivin, "Why the Scientific Revolution Did Not Take Place in China," 45. Scholars of Chinese science have developed thoughtful critiques of the Needham question and its implication on the "universality" of sciences across different cultures. For some examples, see Bray, *Technology and Gender*, 8–9; Hart, "Beyond Science and Civilization"; and Chemla, "The Dangers and Promises of Comparative History of Science."

# Bibliography

Adler, Joseph A. "Chu Hsi and Divination." In Kidder Smith et al., *Sung Dynasty Uses of the I Ching*, 169–205. Princeton, NJ: Princeton University Press, 1990.

Allen, Barry. *Vanishing into Things: Knowledge in Chinese Tradition*. Cambridge, MA: Harvard University Press, 2015.

Amelung, Iwo. "Historiography of Science and Technology in China." In *Science and Technology in Modern China, 1880s–1940s*, edited by Jing Tsu and Benjamin A. Elman, 39–65. Leiden: Brill, 2014.

Ames, Roger T. "Meaning as Imaging: Prolegomena to a Confucian Epistemology." In *Culture and Modernity: East-West Philosophic Perspectives*, edited by Eliot Deutsch, 227–44. Honolulu: University of Hawai'i Press, 1991.

An Zuozhang 安作璋 and Wang Zhimin 王志民. *Qi Lu wenhua tongshi: Song Yuan juan* 齊魯文化通史: 宋元卷. Beijing: Zhonghua shuju, 2004.

Ang, Isabelle, and Pierre-Étienne Will, eds. *Nombres, astres, plantes et viscères: sept essais sur l'histoire des sciences et des techniques en Asie orientale*. Paris: Collège de France, Institut des Hautes Études Chinoises, 1994.

Angle, Stephen C. "Reply to Justin Tiwald." *Dao: A Journal of Comparative Philosophy* 10 (2011): 237–239.

———. Review of *Neo-Confucianism in History* by Peter K. Bol. *Journal of Chinese Studies* 50 (2010): 345–52.

———. *Sagehood: The Contemporary Significance of Neo-Confucian Philosophy*. Oxford: Oxford University Press, 2009.

Angle, Stephen C., and Justin Tiwald. *Neo-Confucianism: A Philosophical Introduction*. Cambridge: Polity, 2017.

Aoyama Sadao 青山定雄. "Sōdai ni okeru kanan kanryō no keifu nitsuite: Toku ni Yōsukō ryūiki o chūshin to shite" 宋代における華南官僚の系譜にいつて: 特に揚子江流域お中心として. *Chūō daigaku bungakubu kiyō* 72 (1974): 51–76.

Araki Toshikazu 荒木敏一, "Sōdai no dōkin: tokuni Ō Anseki no dōkin teppai no jijyō ni tsuite" 宋代の銅禁：特に王安石の銅禁撤廢の事情について. *Tōyō shi kenkyū* 4.1 (1938): 1–29.

Arrault, Alain. *Shao Yong (1012–1077): Poète et Cosmologue*. Paris: Collège de France, Institut des Hautes Études Chinoises, 2002.

Asmis, Elizabeth. "Epicurean Epistemology." In *The Cambridge History of Hellenistic Philosophy*, edited by Keimpt Algra, Jonathan Barnes, Jaap Mansfeld, and Malcolm Schofield, 260–94. Cambridge: Cambridge University Press, 1999.

Ayers, Michael. *Locke: Epistemology and Ontology*. Oxford: Oxford University Press, 1991.

Azuma Jūji 吾妻重二. "Sōdai no keireikyū: Dōkyō saishi to jūka saishi no kōsa" 宋代の景靈宮について— 道教祭祀と儒家祭祀の交差. In *Dōkyō no saihō girei no shisōshiteki kenkyū* 道教の斎法儀禮の思想史的研究, edited by Kobayashi Masami 小林正美, 286–93. Tokyo: Chisen Shokan, 2006.

Bacon, Francis. *Francis Bacon: The New Organon*. Compiled by Lisa Jardine and Michael Silverthorne. Cambridge: Cambridge University Press, 2000.

Bao Weimin 包偉民. "Shen Gua shiji xianyi liu ze" 沈括事跡獻疑六則. In *Song shi yanjiu jikan* 宋史研究集刊, edited by Hangzhou daxue lishi xi Song shi yanjiu shi 杭州大學歷史系宋史研究室, 306–15. Hangzhou: Zhejiang guji chubanshe, 1986.

———. "Shen Gua yanjiu lunzhu suoyin" 沈括研究論著索引. In *Shen Gua Yanjiu* 沈括研究, edited by Hangzhou daxue Song shi yanjiu shi 杭州大學宋史研究室, 322–36. Hangzhou: Zhejiang renmin chubanshe, 1985.

Barbieri-Low, Anthony J. *Artisans in Early Imperial China*. Seattle: University of Washington Press, 2007.

Bengson, John, and Marc A. Moffett. "Two Conceptions of Mind and Action: Knowing How and the Philosophical Theory of Intelligence." In *Knowing How: Essays on Knowledge, Mind, and Action*, edited by John Bengson and Marc A. Moffett, 3–57. Oxford: Oxford University Press, 2011.

Biard, Joël. "Nominalism in the Late Middle Ages." In *The Cambridge History of Medieval Philosophy*, edited by Robert Pasnau, 661–73. Cambridge: Cambridge University Press, 2009.

Billeter, Jean François. "Florilège des notes du Ruisseau des Rêves (Mengxi bitan) de Shen Gua (1031–1095)." *Asiatische Studien, Zeitschrift der Schweizerischen Gesellschaft für Asiengesellschaft. Revue de la Société Suisse-Asie* 47.3 (1993): 389–451.

Birdwhistell, Anne D. *Transition to Neo-Confucianism: Shao Yung on Knowledge and Symbols of Reality*. Stanford, CA: Stanford University Press, 1989.

Bloom, Irene. "Mencian Arguments on Human Nature (Jen-hsing)." *Philosophy East and West* 44.1 (1994): 19–53.

———. *Mencius*. Edited and with an introduction by Philip J. Ivanhoe. New York: Columbia University Press, 2009.

Bo Juyi 白居易. *Bo Juyi ji jianjiao* 白居易集箋校, annotated by Zhu Jincheng 朱金城. Shanghai: Shanghai guji chubanshe, 1988.

Bol, Peter K. "Chu Hsi's Redefinition of Literati Learning." In *Neo-Confucian Education: The Formative Stage*, edited by William Theodore de Bary, 151–85. Berkeley: University of California Press, 1989.

———. "Government, Society, and State: The Political Visions of Ssu-ma Kuang and Wang An-shih." In *Ordering the World: Approaches to State and Society in Sung Dy-*

*nasty China*, edited by Robert P. Hymes and Conrad Schirokauer, 128–92. Berkeley: University of California Press, 1993.

———. "A Literati Miscellany and Sung Intellectual History: The Case of Chang Lei's Ming-tao tsa-chih." *Journal of Sung-Yuan Studies* 25 (1995): 121–51.

———. *Neo-Confucianism in History*. Cambridge, MA: Harvard University Asia Center, 2008.

———. "On Shao Yong's Method for Observing Things." *Monumenta Serica* 61 (2013): 287–99.

———. "On the Problem of Contextualizing Ideas." *Journal of Song-Yuan Studies* 34 (2004): 59–79.

———. "Reconceptualizing the Order of Things in Northern and Southern Sung." In *The Cambridge History of China Volume 5: Sung China, 960–1279 AD, Part 2*, edited by John W. Chaffee and Denis Twitchett, 665–726. Cambridge: Cambridge University Press, 2015.

———. "The Rise of Local History: History, Geography, and Culture in Southern Song and Yuan Wuzhou." *Harvard Journal of Asiatic Studies* 61.1 (2001): 37–76.

———. *This Culture of Ours: Intellectual Transitions in T'ang and Sung China*. Stanford, CA: Stanford University Press, 1992.

———. "Wang Anshi and the *Zhouli*." In *Statecraft and Classical Learning: The Rituals of Zhou in East Asian History*, edited by Benjamin A. Elman and Martin Kern, 227–51. Leiden: Brill, 2009.

Bossler, Beverly J. *Powerful Relations: Kinship, Status, and the State in Sung China (960–1279)*. Cambridge, MA: Harvard University Council on East Asian Studies, 1998.

Bray, Francesca. *Technology and Gender: Fabrics of Power in Late Imperial China*. Berkeley: University of California Press, 1997.

Bréard, Andrea. "Shen Gua's Cuts." In *Zhongguo keji dianji yanjiu* 中國科技典籍研究, edited by Hua Jueming 華覺明, 149–63. Zhengzhou: Daxiang chubanshe, 1997.

Brenier, Joël, Colette Diény, Jean Claude Martzloff, and Wladislaw de Wieclawik. "Shen Gua (1031–1095) et les sciences." *Revue d'histoire des sciences* 42.4 (1989): 333–51.

Bretelle-Establet, Florence, and Karine Chemla. "Qu'était-ce qu'écrire une encyclopédie en Chine?" In *Qu'était-ce qu'écrire une encyclopédie en Chine*, edited by Florence Bretelle-Establet and Karine Chemla, 7–18. Paris: Presses Universitaires de Vincennes, 2007.

Brindley, Erica Fox. *Music, Cosmology, and the Politics of Harmony in Early China*. Albany: State University of New York Press, 2013.

Brokaw, Cynthia J. "On the History of the Book in China." In *Printing and Book Culture in Late Imperial China*, edited by Cynthia J. Brokaw and Kai-Wing Chow, 1–53. Berkeley: University of California Press, 2005.

Buswell, Robert E., Jr., and Donald S. Lopez Jr. *The Princeton Dictionary of Buddhism*. Princeton, NJ: Princeton University Press, 2013.

Cai Shangxiang 蔡上翔. *Wang Jinggong nianpu kaolue* 王荊公年譜考略. In *Wang Anshi nianpu sanzhong* 王安石年譜三種, edited by Pei Rucheng 裴汝誠, 165–744. Beijing: Zhonghua shuju, 2006.

Cai Tao 蔡絛. *Tienwei shan congtan* 鐵圍山叢談. Beijing: Zhonghua shuju, 1983.

Campany, Robert F. *Making Transcendents: Ascetics and Social Memory in Early Medieval China*. Honolulu: University of Hawai'i Press, 2009.

———. *Strange Writing: Anomaly Accounts in Early Medieval China*. Albany: State University of New York Press, 1996.

Cao Tingdong 曹庭棟. *Yangsheng biji* 養生筆記. Shanghai: Shanghai shudian, 1981.

Cao Wanru 曹婉如. "Lun Shen Gua zai ditu xue fangmian de gongxian" 論沈括在地圖學方面的貢獻. In *Keji shi wenji*, vol. 3 科技史文集 (三), edited by Ziran kexue shi yanjiusuo 自然科學史研究所, 81–84. Shanghai: Shanghai kexue jishu chubanshe, 1980.

Chaffee, John W. "Sung Education: Schools, Academics, and Examinations." In *The Cambridge History of China Volume 5: Sung China, 960–1279 AD, Part 2*, edited by John W. Chaffee and Denis Twitchett, 286–320. Cambridge: Cambridge University Press, 2015.

———. *The Thorny Gates of Learning in Sung China: A Social History of Examinations*. Cambridge: Cambridge University Press, 1985.

Chao Gongwu 晁公武. *Jun zhai dushu zhi jiaozheng* 郡齋讀書志校證, annotated by Sun Meng 孫猛. Shanghai: Shanghai guji chubanshe, 1990.

Chemla, Karine. "The Dangers and Promises of Comparative History of Science." *Sartoniana* 27 (2016): 174–98.

Chen Guangyi 陳光隘, Li Guangyu 李光羽, and Xie Baogeng 謝寶耿. *Zhongguo gudai kexue jia, xia ji: Shen Gua* 中國古代科學家，下集：沈括. Shanghai: Shanghai renmin meishu chubanshe, 1978.

Chen Hao 陳昊. "*Zhenglei bencao* yu Beisong shiqi dui yaowu zhi wei de renshi" 《證類本草》與北宋時期對藥物之味的認識. Paper presented at the Annual Meeting for the Society of Song History, Guangzhou, China, August 19–21, 2016.

Chen, Jack W. "Introduction." In *Idle Talk: Gossip and Anecdote in Traditional China*, edited by Jack W. Chen and David Schaberg, 1–15. Berkeley: University of California Press, 2014.

Chen Jiujin 陳久金. "Fengyuan li shuping" 奉元曆述評. In *Shen Gua yanjiu wenji* 沈括研究文集, edited by Zhenjiang shi wenwu guanli weiyuanhui bangongshi 鎮江市文物管理委員會辦公室, 408–12. Hong Kong: Xianggang wenxue baoshe chuban gongsi, 2002.

Chen Meidong 陳美東. "Wo guo gudai louhu de lilun yu jishu: Shen Gua de Fulou yi ji qita" 我國古代漏壺的理論與技術：沈括的《浮漏議》及其它. *Ziran kexue shi yanjiu* 1 (1982): 21–33.

———. *Zhongguo kexue jishu shi: tianwen xue juan* 中國科學技術史：天文學卷. Beijing: Kexue chubanshe, 2003.

Chen Shuguo 陳戍國. *Zhongguo lizhi shi: Song Liao Jin Xia juan* 中國禮制史：宋遼金夏卷. Changsha: Hunan jiaoyu chubanshe, 2001.

Chen Si 陳思 and Chen Shilong 陳世隆. *Liang Song mingxian xiao ji* 兩宋名賢小集. SKQS edition. Taipei: Taiwan shangwu yinshuguan, 1983–86.

Chen Xiaoshan 陳曉珊. "Beisong baojia fa zhiding yu shishi guocheng zhong de quyu chayi" 北宋保甲法制定與實施過程中的區域差異. *Shixue yuekan* 6 (2013): 49–56.

Chen Yuanlong 陳元龍. *Gezhi jingyuan* 格致鏡原. SKQS edition. Taipei: Shangwu yin-shuguan, 1983–86.

Chen Zhen 陳振. "Lue lun baoma fa de yanbian: jian lun Ma Duanlin dui baoma fa de wujie ji yingxiang" 略論保馬法的演變：兼論馬端臨對保馬法的誤解及影響. *Xueshu yanjiu jikan* 1 (1980): 16–28.

Chen Zhensun 陳振孫. *Zhi zhai shulu jieti* 直齋書錄解題, edited by Xiang Jiada 向家達. 1774.

Chen Zhi 陳直. *Shouqin yanglao xinshu* 壽親養老新書. *Congshu jicheng xubian* 叢書集成續編, vol. 43. Taipei: Xinwenfeng chuban gongsi, 1989.

Chen Zhi'e 陳植鍔. *Beisong wenhua shi shulun* 北宋文化史述論. Beijing: Zhongguo she-hui kexue chubanshe, 1992.

Chen Zungui 陳遵媯. *Zhongguo tianwen xue shi* 中國天文學史. Shanghai: Shanghai renmin chubanshe, 1980.

Cheng, Chung-ying. "Categories of Creativity in Whitehead and Neo-Confucianism." *Journal of Chinese Philosophy* 6 (1979): 251–74.

Cheng Hao 程顥 and Cheng Yi 程頤. *Er Cheng ji* 二程集. Beijing: Zhonghua shuju, 2004.

Cheng Minsheng 程民生. *Songdai wujia yanjiu* 宋代物價研究. Beijing: Renmin chu-banshe, 2008.

Chia, Lucille, and Hilde De Weerdt, eds. *Knowledge and Text Production in an Age of Print: China, 900–1400*. Leiden: Brill, 2010.

Chiang Hsiang-Ling 蔣湘伶. "Shen Gua zhushu kao" 沈括著述考. Master's thesis, Chi-nese Culture University, 2011.

Chien, Cecilia Lee-fang. *Salt and State: An Annotated Translation of the Songshi Salt Monopoly Treatise*. Ann Arbor: Center for Chinese Studies at the University of Mich-igan, 2004.

*Chunqiu Zuo zhuan zhengyi* 春秋左傳正義. In *Shisanjing zhushu* 十三經注疏, compiled by Ruan Yuan 阮元, 1697–2188. Reprint. Beijing: Zhonghua shuju, 1982.

Clark, Hugh R. "Overseas Trade and Social Change." In *The Emporium of the World, Maritime Quanzhou, 1000–1400*, edited by Angela Schottenhammer, 47–94. Leiden: Brill, 2001.

———. *Portrait of a Community: Society, Culture, and the Structures of Kinship in the Mulan River Valley (Fujian) from the Late Tang through the Song*. Hong Kong: Chi-nese University Press, 2007.

Conti, Alessandro D. "Realism." In *The Cambridge History of Medieval Philosophy*, ed-ited by Robert Pasnau, 647–60. Cambridge: Cambridge University Press, 2009.

Crisciani, Chiara. "Histories, Stories, Exempla, and Anecdotes: Michele Savonarola from Latin to Vernacular." In *Historia: Empiricism and Erudition in Early Modern Europe*, edited by Gianna Pomata and Nancy G. Siraisi, 295–324. Cambridge, MA: MIT Press, 2005.

Cua, Antonio S. "Ti and Yong: Substance and Function." In *Encyclopedia of Chinese Phi-losophy*, edited by Antonio S. Cua, 718–26. London: Routledge, 2013.

Cui Yuqian 崔玉謙. "Beisong Shaanxi lu zhizhi Xie yan si kaolun" 北宋陝西路制置解鹽司考論. *Xi Xia yanjiu* 1 (2015): 59–65.

Cullen, Christopher. *Astronomy and Mathematics in Ancient China: The Zhou bi suan jing*. Cambridge: Cambridge University Press, 2006.

———. *The Foundations of Celestial Reckoning: Three Ancient Chinese Astronomical Systems*. Oxford: Taylor and Francis, 2016.

Dai Yixuan 戴裔煊. *Songdai chaoyan zhidu yanjiu* 宋代鈔鹽制度研究. Shanghai: Shangwu yinshuguan, 1957.

Dai Zhen 戴震. *Dai Zhen quanshu* 戴震全書. Vols. 1–7. Hefei: Huangshan shushe, 1994.

Daston, Lorraine. "The Factual Sensibility." *Isis* 79 (1988): 452–67.

Daston, Lorraine, and Peter Galison. *Objectivity*. Brooklyn: Zone Books, 2007.

De Bary, Wm. Theodore. *Neo-Confucian Orthodoxy and the Learning of the Mind-and-Heart*. New York: Columbia University Press, 1981.

De Weerdt, Hilde. *Competition over Content: Negotiating Standards for the Civil Service Examinations in Imperial China (1127–1279)*. Cambridge, MA: Harvard University Asia Center, 2007.

———. "The Encyclopedia as Textbook: Selling Private Chinese Encyclopedias in the Twelfth and Thirteenth Centuries." In *Qu'était-ce qu'écrire une encyclopédie en Chine*, edited by Florence Bretelle-Establet and Karine Chemla, 77–101. Paris: Presses Universitaires de Vincennes, 2007.

———. *Information, Territory, and Networks: The Crisis and Maintenance of Empire in Song China*. Cambridge, MA: Harvard University Asia Center, 2016.

Deng Guangming 鄧廣銘. *Beisong zhengzhi gaigejia Wang Anshi* 北宋政治改革家王安石. Beijing: Renmin chubanshe, 1997.

———. "Buyao zai wei Shen Gua jinshangtianhua le" 不要再為沈括錦上添花了. In *Shen Gua Yanjiu*, edited by Hangzhou daxue Song shi yanjiu shi, 16–26. Hangzhou: Zhejiang renmin chubanshe, 1985.

Despeux, Cathrine. "The System of the Five Circulatory Phases and the Six Seasonal Influences (wuyun liuqi), A Source of Innovation in Medicine under the Song (960–1279)," translated by Janet Lloyd. In *Innovation in Chinese Medicine*, edited by Elisabeth Hsu. 121–66. Cambridge: Cambridge University Press, 2001.

Diao Peijun 刁培俊. "Songdai 'baojia fa si ti'" 宋代"保甲法"四題. *Zhongguoshi yanjiu* 1 (2009): 69–81.

Dicker, Georges. *Descartes: An Analytic and Historical Introduction*. Oxford: Oxford University Press, 2013.

Diény, Colette. "On Some Trends in Contemporary Critiques of Shen Gua and His Works." In *Historical Perspectives on East Asian Science, Technology and Medicine*, edited by Alan K. L. Chan, Gregory K. Clancey, and Hui-Chieh Loy, 560–69. Singapore: Singapore University Press, 2002.

Ding Chao 丁超. "Jin tu kaimi: Zhongguo ditu xue shi shang de 'zhi tu liuti' yu Pei Xiu ditu shiye" 晉圖開秘：中國地圖學史上的'製圖六體'與裴秀地圖事業. *Zhongguo lishi dili luncong* 30.1 (2015): 5–18.

Dong Yuyu 董煜宇. "Cong Fengyuan li gaige kan Beisong tianwen guanli de jixiao" 從《奉元曆》改革看北宋天文管理的績效. *Ziran kexueshi yanjiu* 27.2 (2008): 203–12.

Dong Yuyu and Guan Zengjian 關增建. "Songdai de tianwenxue wenxian guanli" 宋代的天文學文獻管理. *Ziran kexueshi yanjiu* 23.4 (2004): 345–55.

Dong Zhongshu 董仲舒. *Chunqiu fanlu yizheng* 春秋繁露義證, collated by Su Yu 蘇輿 and annotated by Zhong Zhe 鍾哲. Beijing: Zhonghua shuju, 1992.

Drège, Jean-Pierre. "Des Ouvrages Classés par Catégories: Les Encyclopédies Chinoises." In *Qu'était-ce qu'écrire une encyclopédie en Chine*, edited by Florence Bretelle-Establet and Karine Chemla, 19–38. Paris: Presses Universitaires de Vincennes, 2007.

Duan Chengshi 段成式. *Youyang zazu* 酉陽雜俎, annotated by Fang Nansheng 方南生. Beijing: Zhonghua shuju, 1981.

Dunnel, Ruth W. "The Hsi Hsia." In *The Cambridge History of China Vol. 6: Alien Regimes and Border States, 907–1368*, edited by Herbert Franke and Denis C. Twichett, 154–214. Cambridge: Cambridge University Press, 1994.

Ebrey, Patricia B. *The Aristocratic Families of Early Imperial China: A Case Study of the Po-ling Ts'ui Family*. Cambridge: Cambridge University Press, 1978.

———. "The Dynamics of Elite Domination in Song China." *Harvard Journal of Asiatic Studies* 48.2 (1988): 493–519.

———. *The Inner Quarters: Marriage and the Lives of Chinese Women in the Sung Period*. Berkeley: University of California Press, 1993.

Ebrey, Patricia, Anne Walthall, and James Palais. *East Asia: A Cultural, Social, and Political History*. Wadsworth: Cengage Learning, 2009.

Egan, Ronald C. *The Problem of Beauty: Aesthetic Thought and Pursuits in Northern Song Dynasty China*. Cambridge, MA: Harvard University Asia Center, 2006.

———. "Shen Kuo Chats with Ink Stone and Writing Brush." In *Idle Talk: Gossip and Anecdote in Traditional China*, edited by Jack W. Chen and David Schaberg, 132–53. Berkeley: University of California Press, 2014.

———. *Word, Image, and Deed in the Life of Su Shi*. Cambridge, MA: Harvard University Asia Center, 1994.

Elman, Benjamin A. *From Philosophy to Philology: Intellectual and Social Aspects of Change in Late Imperial China*. Second edition. Los Angeles: UCLA Asian Pacific Monograph Series, Asia-Pacific Institute, 2001.

———. "The Investigation of Things (*Gewu* 格物): Natural Studies (*Gezhixue* 格致學), and Evidential Studies (*Kaozhengxue* 考證學) in Late Imperial China, 1600–1800." In *Concepts of Nature: A Chinese-European Cross-Cultural Perspective*, edited by Hans Ulrich Vogel, Guenter Dux, and Mark Elvin, 368–99. Leiden: Brill, 2010.

———. *On Their Own Terms: Science in China, 1550–1900*. Cambridge, MA: Harvard University Press, 2005.

*Erya zhushu* 爾雅注疏. In *Shisanjing zhushu* 十三经注疏, compiled by Ruan Yuan 阮元, 5575–783. Beijing: Zhonghua shuju, 1982.

Fan Chengda 范成大. *Wu Jun zhi* 吳郡志. In *Shoushan Ge congshu* 守山閣叢書. Taipei: Yiwen chubanshe, 1968.

Fan Jiawei 范家偉. *Beisong jiaozheng yishu ju xin tan* 北宋校正醫書局新探. Hong Kong: Zhonghua shuju, 2014.

Fan Zhen 范鎮. *Dong zhai jishi* 東齋記事. Beijing: Zhonghua shuju, 1980.

Fang Yaozhong 方藥中 and Xu Jiasong 許家松. *Huangdi neijing suwen yunqi qipian jiangjie* 黃帝內經素問運氣七篇講解. Beijing: Remin weisheng chubanshe, 1984.

Fantl, Jeremy. "Knowing How." *The Stanford Encyclopedia of Philosophy*, Spring 2016 edition, edited by Edward N. Zalta, http://plato.stanford.edu/archives/spr2016/entries/knowledge-how/ (accessed January 8, 2018).

Farquhar, Judith. *Knowing Practice: The Clinical Encounter of Chinese Medicine.* Boulder, CO: Westview Press, 1996.

Feng Hanyong 馮漢鏞. "Bi Sheng huozi jiaoni wei liuyi ni kao" 畢昇活字膠泥為六一泥考. *Wen shi zhe* 3 (1983): 84–85.

Fish, William. *Philosophy of Perception.* London: Routledge, 2010.

Forage, Paul Christopher. "Science, Technology, and War in Song China: Reflections in the *Brush Talks from the Dream Creek* by Shen Kuo (1031–1095)." PhD diss., University of Toronto, 1991.

———. "The Sino-Tangut War of 1081–1085." *Journal of Asian History* 25 (1991): 1–28.

Franke, Herbert. "Some Aspects of Chinese Private Historiography in the Thirteenth and Fourteenth Centuries." In *Historian of China and Japan*, edited by W. G. Beasley and E. G. Pulleyblank, 115–34. Oxford: Oxford University Press, 1961.

Fraser, Chris. "Knowledge and Error in Early Chinese Thought." *Dao: A Journal of Comparative Philosophy* 10 (2011): 127–48.

Fu Daiwie. "A Contextual and Taxonomic Study of the 'Divine Marvels' and 'Strange Occurrences' in the *Mengxi Bitan*." *Chinese Science* 11 (1993–1994): 3–35.

———. "The Flourishing of *Biji* or Pen-Notes Texts and Its Relations to History of Knowledge in Song China (960–1279)." *hors série, Extrême-Orient Extrême-Occident* 27 (2007): 103–30.

———. "On *Mengxi bitan*'s World of Marginalities and 'South-Pointing Needles': Fragment Translation vs. Contextual Translation." In *Current Perspectives in the History of Science in East Asia*, edited by Kim Yung Sik and Francesca Bray, 52–66. Seoul: Seoul National University Press, 1999.

———. "When Shen Gua Encountered the 'Natural World': A Preliminary Discussion on the *Mengxi bitan* and the Concept of Nature." In *Concepts of Nature: A Chinese-European Cross-Cultural Perspective*, edited by Hans Ulrich Vogel, Guenter Dux, and Mark Elvin, 285–309. Leiden: Brill, 2010.

———. "World Knowledge and Local Administrative Techniques: Literati's *Biji* Experience in Some Song *Biji*." In *Songdai guojia wenhua zhong de kexue* 宋代國家文化中的科學, edited by Sun Xiaochun 孫小淳 and Zeng Xiongsheng 曾雄生, 253–67. Beijing: Zhongguo kexue jishu chubanshe, 2007.

Fu Xinian 傅熹年. "Shanxi sheng Fanzhi xian Yanshan si nandian Jindai bihua suo hui jianzhu de chubu fenxi" 山西省繁峙縣岩山寺南殿金代壁畫所繪建築的初步分析. In Fu Xinian, *Zhongguo gudai jianzhu shi lun* 中國古代建築十論, 246–95. Shanghai: Fudan daxue chubanshe, 2004.

Fujiki Toshirō 藤木俊郎. "Gomi no ōyō no hensen" 五味の応用の変遷. In Fujiki Toshirō, *Shinkyugaku genryukō: Sommon igaku no sekai II* 鍼灸医学源流考：素問医学の世界 II, 62–74. Tokyo: Sekibundo, 1979.

Furth, Charlotte. *A Flourishing Yin: Gender in China's Medical History, 960–1665.* Berkeley: University of California Press.

Gao Congming 高聰明. *Songdai huobi yu huobi liutong yanjiu* 宋代貨幣與貨幣流通研究. Baoding: Hebei daxue chubanshe, 1998.

Gardner, Daniel K. *Chu Hsi and the Ta-hsueh: Neo-Confucian Reflection on the Confucian Canon.* Cambridge, MA: Harvard University Council on East Asian Studies, 1986.

Geaney, Jane. *On the Epistemology of the Senses in Early Chinese Thought.* Honolulu: University of Hawai'i Press, 2002.

Geng Hu 耿虎. "Hedi, pingdi guanxi zai tan: jian yu Yuan Yitang xiansheng shangque" 和糴，平糴關系再談：兼與袁一堂先生商榷. *Zhongguo jingji shi yanjiu* 2 (2002): 48–56.

Gill, Mary Louise. "Problems for Forms." In *A Companion to Plato,* edited by Hugh H. Benson, 184–98. Oxford: Blackwell, 2006.

Golas, Peter. "Rural China in the Song." *Journal of Asian Studies* 39.2 (1980): 291–325.

Goldman, Alvin I. "Epistemic Folkways and Scientific Epistemology." *Philosophical Issues* 3 (1993): 271–85.

———. "Reliabilism." In Alvin I. Goldman, *Reliabilism and Contemporary Epistemology: Essays,* 68–94. Oxford: Oxford University Press, 2012.

———. "Toward a Synthesis of Reliabilism and Evidentialism." In Alvin I. Goldman, *Reliabilism and Contemporary Epistemology: Essays,* 123–50. Oxford: Oxford University Press, 2012.

Goldschmidt, Asaf. *The Evolution of Chinese Medicine: Song Dynasty, 960–1200.* London: Routledge, 2009.

Gong Yanming 龔延明. *Songdai guanzhi cidian* 宋代官制辭典. Beijing: Zhonghua shuju, 1997.

Gong Yizheng 龔頤正. *Jieyin biji* 芥隱筆記, annotated by Li Guoqiang 李國強. In *Quan Song biji* 全宋筆記, series 5, vol. 2. Zhengzhou: Daxiang chubanshe, 2012.

Graham, Angus C. *Disputers of the Tao: Philosophical Argument in Ancient China.* La Salle, IL: Open Court, 1989.

———. *Studies in Chinese Philosophy and Philosophical Literature.* Albany: State University of New York Press, 1986.

———. *Two Chinese Philosophers.* Second edition. La Salle, IL: Open Court, 1992.

———. *Yin-Yang and the Nature of Correlative Thinking.* Singapore: Institute of East Asian Philosophies, 1986.

Graham, Angus C., and Nathan Sivin. "A Systematic Approach to the Mohist Optics (ca. 300 B.C.)." In *Chinese Science: Explorations of an Ancient Tradition,* edited by Shigeru Nakayama and Nathan Sivin, 105–52. Cambridge, MA: MIT Press, 1973.

Gu Liwei 古麗巍. *"Da youwei" zhi zheng* "大有為" 之政. Beijing: Shehui kexue wenxian chubanshe, forthcoming.

Guan Chengxue 管成學 and Wang Xingwen 王興文. "Su Song yu Su Shi jiaoyi kaoshu" 蘇頌與蘇軾交誼考述. *Tsinghua daxue xuebao* (*zhexue shehui kexue ban*) 2.17 (2002): 85–89.

Guan, Yuzhen. "A New Interpretation of Shen Kuo's Ying Biao Yi." *Archive for History of Exact Sciences* 64.6 (2010): 707–19.

Guo, Bogong 郭伯恭. *Song si da shu kao* 宋四大書考. Taipei: Taiwan shangwu yinshuguan, 1967.

Guo Shaoyu 郭紹虞. *Beisong shihua kao* 北宋詩話考. Hong Kong: Chongwen shudian, 1971.

Guo Shengchi 郭盛熾. "Shen Gua faxian de louhu chiji he taiyang zhounian shiyun-dong de bu junyun xing" 沈括發現的漏壺遲疾和太陽週年視運動的不均勻性. *Zhongguo kexueyuan shanghai tianwentai niankan* 2 (1980): 202–10.

Guo Zhengzhong 郭正忠. "Beisong qianqi Xie yan de 'quejin' yu tongshang" 北宋前期解鹽的"権禁"與通商. *Beijing shiyuan xuebao (shehui kexue ban)* 2 (1981): 62–70.

Hall, David L., and Roger T. Ames. *Thinking from the Han: Self, Truth, and Transcendence in Chinese and Western Culture.* Albany: State University of New York Press, 1998.

Halperin, Mark. *Out of the Cloister: Literati Perspectives on Buddhism in Sung China, 960–1279.* Cambridge, MA: Harvard University Asia Center, 2006.

Han Wo 韓偓. *Yushan qiaoren ji* 玉山樵人集. SBCK edition. Shanghai: Shangwu yin-shuguan, 1929.

Han Yü 韓愈. *Han Changli wenji jiaozhu* 韓昌黎文集校注, annoted by Ma Qichang 馬其昶and Ma Maoyuan 馬茂元. Shanghai: Shanghai guji chubanshe, 1986.

Harbsmeier, Christoph. "Conceptions of Knowledge in Ancient China." In *Epistemological Issues in Classical Chinese Philosophy*, edited by Hans Lenk and Paul Gregor, 11–30. Albany: State University of New York Press, 1993.

———. "Towards a Conceptual History of Some Concepts of Nature in Classical Chinese: *Ziran* 自然 and *Ziran zhi li* 自然之理." In *Concepts of Nature: A Chinese-European Cross-Cultural Perspective*, edited by Hans Ulrich Vogel, Guenter Dux, and Mark Elvin, 220–54. Leiden: Brill, 2010.

Hart, Roger. "Beyond Science and Civilization: A Post-Needham Critique." *East Asian Science, Technology, and Medicine*, 16 (1999): 88–114.

Hartman, Charles. *Han Yü and the T'ang Search for Unity.* Princeton, NJ: Princeton University Press, 1986.

———. "Poetry and Politics in 1079: The Crow Terrace Poetry Case of Su Shih." *Chinese Literature: Essays, Articles, Reviews (CLEAR)* 12 (1990): 15–44.

Hartwell, Robert. "Demographic, Political, and Social Transformations of China, 750–1550." *Harvard Journal of Asiatic Studies* 42.2 (1982): 365–442.

He Yongqiang 何勇強. "Shen Gua yu Wang Anshi de guanxi xintan" 沈括與王安石的關系新探. In *Songxue yanjiu jikan* 宋學研究集刊, edited by Zhejiang daxue Songxue yanjiu zhongxin 浙江大學宋學研究中心, 137–61. Hangzhou: Zhejiang daxue chuban-she, 2008.

Heidegger, Martin. *The Question Concerning Technology and Other Essays*, translated and introduction by William Lovitt. New York: Harper Perennial, 2013.

Henderson, John B. "Cosmology and Concepts of Nature in Traditional China." In *Concepts of Nature: A Chinese-European Cross-Cultural Perspective*, edited by Hans Ulrich Vogel, Guenter Dux, and Mark Elvin, 181–97. Leiden: Brill, 2010.

———. *The Development and Decline of Chinese Cosmology.* Taipei: Windstone Press, 2011.

Heng, Chye Kiang. *Cities of Aristocrats and Bureaucrats: The Development of Medieval Chinese Citiescapes.* Honolulu: University of Hawai'i Press, 1999.

Herrmann, Konrad. *Pinsel-Unterhaltungen am Traumbach: Das gesamte Wissen des alten China.* Munich: Eugen Diederichs, 1997.

Higashi Ichio 東一夫. *Ō Anseki to Shiba Kō* 王安石と司馬光. Tokyo: Chusekisha, 1980.

Hinrichs, T. J. "The Song and Jin Periods." In *Chinese Medicine and Healing*, edited by T. J. Hinrichs and Linda L. Barnes, 97–127. Cambridge, MA: Belknap Press of Harvard University Press, 2013.

Holzman, Donald. "Shen Kua and His Meng-ch'i pi-t'an." *T'oung Pao*, second series 46.3 (1958): 260–92.

Hon, Tze-ki. "Redefining the Civil Governance: The *Yichuan Yizhuan* of Cheng Yi." *Monumenta Serica* 52 (2004): 199–219.

Hong Mai 洪邁. *Rong zhai suibi* 容齋隨筆. Beijing: Zhonghua shuju, 2005.

Hu Daojing 胡道靜. "Guanyu Shen Gua pingzhuan" 關於沈括評傳. In Hu Daojing, *Shen Gua yanjiu: keji shi lun*, 144–46. Shanghai: Shanghai renmin chubanshe, 2011.

———. "Gudai ditu cehui jishu shang de 'qifa' wenti" 古代地圖測繪技術上的 "七法" 問題. *Zhonghua wenshi luncong* 6 (1964): 236. Reprinted in Hu Daojing, *Shen Gua yanjiu, keji shi lun*, 246–47. Shanghai: Shanghai renmin chubanshe, 2011.

———. "*Mengxi bitan* buzheng" 《夢溪筆談》補證. *Zhonghua wenshi luncong* 3 (1979): 111–35. Reprinted in Hu Daojing, *Shen Gua yanjiu, keji shi lun*, 35–56. Shanghai: Shanghai renmin chubanshe, 2011.

———. "Shen Gua de nongxue zhuzuo Mengxi wanghuai lu" 沈括的農學著作《夢溪忘懷錄》. *Wenshi* 3 (1963): 221–25. Reprinted in Hu Daojing, *Shen Gua yanjiu, keji shi lun* 沈括研究，科技史論, 17–34. Shanghai: Shanghai renmin chubanshe, 2011.

———. "Shen Gua junshi sixiang tanyuan: Lun Shen Gua yu qi jiufu Xu Dong de shicheng guanxi" 沈括軍事思想探源：論沈括與其舅父許洞的師承關係. In *Shen Gua yanjiu*, edited by Hangzhou daxue Song shi yanjiu shi, 27–35. Hangzhou: Zhejiang renmin chubanshe, 1985.

———. *Shen Gua yanjiu: keji shi lun* 沈括研究：科技史論. Shanghai: Shanghai renmin chubanshe, 2011.

———. "*Su Shen neihan liangfang* chu shu pan: fenxi benshu meige fang, lun suoshu de zuozhe: 'Shen fang' yi wei 'Su fang'" 蘇沈內翰良方楚蜀判：分析本書每個方、論所屬的作者：'沈方' 抑為'蘇方'. *Shehui kexue zhanxian* 3 (1980): 195–209.

———. "Zhu Zi dui Shen Gua kexue xueshuo de zuanyan yu fazhan" 朱子對沈括科學學說的鑽研與發展. In Hu Daojing, *Shen Gua yanjiu: keji shi lun*, 147–52. Shanghai: Shanghai renmin chubanshe, 2011.

Hu Daojing and Wu Zuoxin 吳作忻. "Mengxi wanghuai lu go chen" 梦溪忘懷錄鉤沉. *Hangzhou daxue xuebao* 11.1 (1981): 40–55.

Hu Jinyin 胡勁茵. "Cong Da'an dao Dasheng" 從大安到大晟. PhD diss., Zhongshan University, 2010.

Hua Tongxu 華同旭. *Zhongguo louke* 中國漏刻. Hefei: Anhui kexue jishu chubanshe, 1991.

Huang I-long 黃一農. "Jixing yu gudu kao" 極星與古度考. *Tsing Hua Journal of Chinese Studies* 22.2 (1992): 93–117.

Huang Shiheng 黃士恆. "Wang Anshi Laozi zhu de dao lun yu tianren guanxi" 王安石老子注的道論與天人關係. *Tsinghua zhongwen xuebao* 2 (2009): 17–44.

Huang, Yong. *Why Be Moral: Learning from the Neo-Confucian Cheng Brothers*. Albany: State University of New York Press, 2015.

*Huangdi neijing suwen jiaozhu* 黃帝內經素問校注, annotated by Guo Aichun 郭靄春. Beijing: Renmin weisheng chubanshe, 1992.

Hucker, Charles O. *A Dictionary of Official Titles in Imperial China*. Stanford, CA: Stanford University Press, 1985.

Huihong 惠洪. *Leng zhai yehua* 冷齋夜話. Beijing: Zhonghua shuju, 1988.

Hymes, Robert P. *Statesmen and Gentlemen: The Elite of Fu-chou, Chiang-hsi, in Northern Song and Southern Song*. Cambridge: Cambridge University Press, 1986.

Inglis, Alister David. *Hong Mai's Record of the Listener and Its Song Dynasty Context*. Albany: State University of New York Press, 2006.

Ivanhoe, Philip J. *Confucian Moral Self Cultivation*. Indianapolis: Hackett, 2000.

Jami, Catherine. "'European Science in China' or 'Western Learning'? Representations of Cross-Cultural Transmission, 1600–1800." *Science in Context* 12.3 (1999): 413–34.

Ji, Xiao-bin. *Politics and Conservatism in Northern Song China: The Career and Thought of Sima Guang (A.D. 1019–1086)*. Hong Kong: Chinese University Press, 2005.

Jia Daquan 賈大泉. *Songdai Sichuan jingji shulun* 宋代四川經濟述論. Chengdu: Sichuan shehui kexueyuan chubanshe, 1985.

Jiang Mei 江湄. "Songdai biji, lishi jiyi yu shiren shehui de lishi yishi" 宋代筆記、歷史記憶與士人社會的歷史意識. *Tianjin shehui kexue* 4 (2016): 146–55.

Jiang Shaoyu 江少虞. *Song chao shishi leiyuan* 宋朝事實類苑. Shanghai: Shanghai guji chubanshe, 1981.

Jin Liangnian 金良年 and Hu Xiaojing 胡小靜. *Mengxi bitan quanyi* 夢溪筆談全譯. Guiyan: Guizhou renmin chubanshe, 1998.

Jin Qiupeng 金秋鵬. "Lue lun Su Song de zhengzhi shengya" 略論蘇頌的政治生涯. *Ziran kexue shi yanjiu* 1 (1991): 1–7.

Kaptchuk, Ted. *The Web That Has No Weaver: Understanding Chinese Medicine*. New York: Rosetta Books, 2010.

Kasoff, Ira. *The Thought of Chang Tsai (1020–1077)*. Cambridge: Cambridge University Press, 2002.

Kelly, David. "Between System and History." In *Historia: Empiricism and Erudition in Early Modern Europe*, edited by Gianna Pomata and Nancy G. Siraisi, 211–40. Cambridge, MA: MIT Press, 2005.

Kim, Yung Sik. *The Natural Philosophy of Chu Hsi (1130–1200)*. Philadelphia: American Philosophical Society, 2000.

Kinugawa Tsuyoshi 衣川強. *Songdai wenguan fengji zhidu* 宋代文官俸給制度, translated by Zheng Liangsheng 鄭樑生. Taipei: Taiwan shangwu yinshuguan, 1977.

Kumamoto Takashi 熊本崇. "Chūsho kenseikan: Ō An-seki seiken kaku no ninaitetachi" 中書檢正官：王安石政権閣のにないてたち. *Tōyō shi kenkyū* 47.1 (1988): 54–80.

Kuriyama, Shigehisa. *The Expressiveness of the Body and the Divergence of Greek and Chinese Medicine*. Brooklyn, NY: Zone Books, 2002.

Kurz, Johannes L. "The Compilation and Publication of the *Taiping yulan* and the *Cefu yuangui*." In *Qu'était-ce qu'écrire une encyclopédie en Chine*, edited by Florence Bretelle-Establet and Karine Chemla, 39–76. Paris: Presses Universitaires de Vincennes, 2007.

———. "The Politics of Collecting Knowledge: Song Taizong's Compilation Projects." *T'oung Pao* 87 (2001): 289–316.

Lam, Joseph S. C. "Huizong's Dashengyue, a Musical Performance of Emperorship and Officialdom." In *Emperor Huizong and Late Northern Song China: The Politics of Culture and the Culture of Politics*, edited by Patricia B. Ebrey and Maggie Bickford, 418–27. Cambridge, MA: Harvard University Asia Center, 2006.

Lamouroux, Christian. "Geography and Politics: The Song-Liao Border Dispute of 1074/75." In *China and Her Neighbours: Borders, Visions of the Other, Foreign Policy 10th to 19th Century*, edited by Sabine Dabringhaus and Roderich Ptak, 1–28. Wiesbaden: Harrassowitz, 1997.

Le Aiguo 樂愛國. "Beisong ruxue beijing xia Shen Gua de kexue yanjiu" 北宋儒學背景下沈括的科學研究. *Zhejiang shifan daxue xuebao* 32.6 (2007): 9–15.

———. *Zhu Xi de ziran yanjiu* 朱熹的自然研究. Shenzhen: Haitian chubanshe, 2014.

Lei Hsianglin 雷祥麟 and Fu Daiwie. "Mengxi bitan li de yuyan yu xiangsixing" 夢溪筆談裡的語言與相似性. *Ch'ing-hua hsueh-pao* 清華學報 23 (1993): 31–60.

Lei, Sean Hsiang-lin. "How Did Chinese Medicine Become Experiential? The Political Epistemology of *Jingyan*." *Positions* 10.2 (2002): 333–64.

———. *Neither Donkey Nor Horse: Medicine in the Struggle over China's Modernity*. Chicago: University of Chicago Press, 2014.

Lenk, Hans. "Introduction: If Aristotle Had Spoken and Wittgenstein Known Chinese: Remarks Regarding Logic and Epistemology, A Comparison between Classical Chinese and Some Western Approaches." In *Epistemological Issues in Classical Chinese Philosophy*, edited by Hans Lenk and Paul Gregor, 1–11. Albany: State University of New York Press, 1993.

Levine, Ari Daniel. *Divided by a Common Language: Factional Conflict in Late Northern Song China*. Honolulu: University of Hawai'i Press, 2008.

Li Cuicui 李翠翠. *Kexue juxing: Shen Gua* 科學巨星：沈括. Changchun: Jilin wenshi chubanshe, 2011.

Li Fang 李昉 et al. *Taiping guangji* 太平廣記. 1522.

———. *Taiping yulan* 太平御覽. Taipei: Taiwan shangwu yinshuguan, 1968.

Li Geng 李更. *Songdai guange jiaokan yanjiu* 宋代館閣校勘研究. Nanjing: Fenghuang chubanshe, 2006.

Li Jinshui 李金水. *Wang Anshi jingji bianfa yanjiu* 王安石經濟變法研究. Fuzhou: Fujian renmin chubanshe, 2007.

Li Kan 李衎. *Zhu pu xianglu* 竹譜詳錄. SKQS edition. Taipei: Taiwan shangwu yinshuguan, 1983–86.

Li Shizhen 李時珍. *Bencao gangmu* 本草綱目, annotated by Liu Hengru 劉衡如 and Liu Shanyong 劉山永. Beijing: Huaxia chubanshe, 1998.

Li Tao 李燾. *Xu Zizhi tongjian changbian* 續資治通鑒長編. Beijing: Zhonghua shuju, 2004. (XCB)

Li Wenze 李文澤 and Wu Hongze 吳洪澤. *Mengxi bitan quanyi: wenbai duizhao* 夢溪筆談全譯：文白對照. Chengdu: Bashu shushe, 1996.

Li, Xiaobing. *China at War: An Encyclopedia*. Santa Barbara, CA: ABC-CLIO, 2012.

Li Xiaocong 李孝聰. *Lishi chengshi dili* 歷史城市地理. Jinan: Shandong jiaoyu chubanshe, 2007.

Li Yong 李勇. "Zhongguo guli jingshuo shuju de huifu ji yingyong" 中國古曆經朔數據的恢復及應用. *Tianwen xuebao* 46.4 (2005): 474–84.

Li Yumin 李裕民. *"Mengxi bitan yu Shen Gua Liangfang yanjiu"* 《夢溪筆談》與沈括 《良方》研究. In Li Yumin, *Songshi xintan* 宋史新探, 290–310. Xi'an: Shaanxi shi-fan daxue chubanshe, 1999.

———. "Shen Gua de qinshu, jiaoyou ji yizhu" 沈括的親屬、交游及佚著. In *Songshi xintan*, 274–89. Xi'an: Shaanxi shifan daxue chubanshe, 1999.

Li Zhao 李肇. *Tang guoshi bu* 唐國史補. Shanghai: Shanghai guji chubanshe, 1979.

Li Zhichao 李志超. "Fulou yi kaoshi" 浮漏議考釋. *Zhongguo kexue jishu daxue xuebao,* zengkan 6 (1982): 33–39.

———. "Huangdao hunyi ji Xining hunyi: Kaozheng yu fuyuan" 黃道渾儀及熙寧渾 儀：考證與復原. In *Tianren guyi: Zhongguo kexueshi lun gang* 天人古義：中國科學 史論綱, 187–94. Zhengzhou: Daxiang chubanshe, 1998.

———. "'Hunyi yi' pingzhu" 渾儀議評註. In *Tianren guyi: Zhongguo kexueshi lun gang*, 195–210. Zhengzhou: Daxiang chubanshe, 1998.

———. "Shen Gua de tianwen yanjiu er: rishi he xingdu" 沈括的天文研究 (二)：日 食和星度. *Zhongguo kexue jishu daxue xuebao* 10.1 (1980): 51–56.

———. "Shen Gua de tianwen yanjiu yi: kelou yu tuofa" 沈括的天文研究 (一)：刻漏 與妥法. *Zhongguo kexue jishu daxue xuebao* 1 (1978): 9–14.

———. *Shuiyun yixiang zhi: Zhongguo gudai tianwen zhong de lishi* (*fu* Xin yixiang fayao *yijie*) 水運儀象志：中國古代天文鐘的歷史 (附《新儀象法要》譯解). Hefei: Zhongguo keji daxue chubanshe, 1997.

Liang Sile 梁思樂. "Beisong houqi dangzheng yu shixue: yi Shenzong pingjia ji Zhe-zong jiwei wenti wei zhongxin" 北宋後期黨爭與史學：以神宗評價及哲宗繼位問題 為中心. Paper presented at the Annual Meeting for the Society of Song History, Guangzhou, China, August 19–21, 2016.

Lin Cen 林岑. "Lue lun Shen Gua yu Wang Anshi de guanxi" 略論沈括與王安石的關 系. *Beijing shifan xueyuan xuebao* 4 (1980): 45–51.

Lin Ruihan 林瑞翰. "Songdai baojia" 宋代保甲. In *Song Liao Jin shi yanjiu lunji* 宋遼 金史研究論集, *Dalu zazhi shixue congshu* 大陸雜誌史學叢書, series 1, volume 5, 14–20. Taipei: Dalu zazhi she, 1960.

Lin Sufen 林素芬. *Beisong zhongqi ruxue dao lun leixing yanjiu* 北宋中期儒學道論類型 研究. Taipei: Liren shuju, 2008.

Liu An 劉安. *Huainanzi jiaoshi* 淮南子校釋, annotated by Zhang Shuangli 張雙棣. Bei-jing: Beijing daxue chubanshe, 1997.

Liu Mu 劉牧. *Yi shu gouyin tu* 易數鉤隱圖. SKQS edition. Taipei: Shangwu yinshu-guan, 1983–86.

Liu Shangheng 劉尚恆. "Ye tan Wanchun wei de xingjian: shi yu Deng Guangming xiansheng shangque" 也談萬春圩的興建：試與鄧廣銘先生商榷. *Xueshu yuekan* 8 (1979): 77–81.

Liu Tzu-Chien 劉子健 (James T. C. Liu). *China Turning Inward: Intellectual-Political Changes in the Early Twelfth Century*. Cambridge, MA: Harvard University Council on East Asian Studies, 1989.

———. *Ouyang Xiu de zhixue yu congzheng* 歐陽修的治學與從政. Hong Kong: New Asia Institute of Advanced Chinese Studies, 1963.

———. *Reform in Sung China: Wang An-shih (1021–1086) and His New Policies*. Cam-bridge, MA: Harvard University Press, 1959.

———. "The Sung Views on the Control of Government Clerks." *Journal of Economic and Social History of the Orient* 10 (1967): 317–44.

Liu, William Guanglin. *The Chinese Market Economy 1000–1500*. Albany: State University of New York Press, 2015.

Liu Yanshi 劉延世. *Sun Gong tan pu* 孫公談圃. In *Quan Song biji*, series 2, vol. 1. Zhengzhou: Daxiang chubanshe, 2006.

Lo, Winston W. *An Introduction to the Civil Service of Sung China*. Honolulu: University of Hawai'i Press, 1987.

Locke, John. *An Essay Concerning Human Understanding*, edited by Peter H. Nidditch. Oxford: Clarendon Press, 1975.

Lorge, Peter. "The Great Ditch of China and the Song-Liao Border." In *Battlefronts Real and Imagined: War, Border, and Identity in the Chinese Middle Period*, edited by Don J. Wyatt, 59–74. New York: Palgrave Macmillan, 2008.

Lü Buwei 呂不韋. *Lü shi chunqiu jishi* 呂氏春秋集釋, annotated by Xu Weijue 許維遹. Beijing: Zhongguo shudian, 1985.

Lu Xian 盧憲. *Jiading Zhenjiang zhi* 嘉定鎮江志. In *Song Yuan difang zhi songchu* 宋元地方志叢書, vol. 5. Taipei: Zhongguo dizhi yanjiu hui, 1978.

*Lunyu* 論語 (*Analects*). Standard text and chapter divisions.

Ma Duanlin 馬端臨. *Wenxian tongkao* 文獻通考. Beijing: Zhonghua shuju, 1986.

Ma Mingheng 馬明衡. *Shangshu yiyi* 尚書疑義. Edition of SKQS. Taipei: Taiwan shangwu yinshuguan, 1983–86.

Makeham, John. *Name and Actuality in Early Chinese Thought*. Albany: State University of New York Press, 1994.

Martzloff, Jean-Claude. "French Research into the *Mengxi bitan*: Its Past, Present, and Predictable Future." In *Current Perspectives in the History of Science in East Asia*, edited by Kim Yung Sik and Francesca Bray, 42–51. Seoul: Seoul National University Press, 1999.

———. *A History of Chinese Mathematics*. New York: Springer, 1987.

McRae, John R. *Seeing through Zen: Encounter, Transformatoin, and Genealogy in Chinese Chan Buddhism*. Berkeley: University of California Press, 2004.

Meng Qi 孟祺 et al. *Nongsang jiyao jiaozhu* 農桑輯要校注, annotated by Shi Shenghan 石聲漢. Beijing: Nongye chubanshe, 1982.

Meng Yuanlao 孟元老. *Dongjing meng hua lu jianzhu* 東京夢華錄箋注. Beijing: Zhonghua shuju, 2006.

*Mengzi* 孟子 (*Mencius*). Standard text and chapter divisions.

Metzger, Thomas. *A Cloud across the Pacific: Essays on the Clash between China and Western Political Theory Today*. Hong Kong: Chinese University of Hong Kong Press, 2005.

Miao Shumei 苗書梅. *Songdai guanyuan xuanren he guanli zhidu* 宋代官員選任和管理制度. Kaifeng: Henan daxue chubanshe, 1996.

Mihelich, Mira Ann. "Polders and the Politics of Land Reclamation in Southeast China during the Northern Sung Dynasty (960–1126)." PhD diss., Cornell University, 1979.

Miyazawa Tomoyuki 宮澤知之. *Sōdai Chūgoku no kokka to keizai* 宋代中國の国家と経済. Tokyo: Sōbunsha, 1998.

Morgan, Daniel Patrick. *Astral Sciences in Early Imperial China: Observation, Sagehood and Society.* Cambridge: Cambridge University Press, 2017.

Munro, Donald J. "Unequal Human Worth." In *The Philosophical Challenge from China*, edited by Brian Bruya, 121–58. Cambridge, MA: MIT Press, 2015.

Nagase Mamoru 長瀬守. "Sōdai ni okeru suiri kaihatsu: toku ni Inken to sono shūiki o chushin to shite" 宋代における水利開発：特に鄞県とその周域を中心として. In *Aoyama Hakushi Koki Kinen Sodaishi Ronsō* 青山博士古稀記念宋代史論叢, edited by Aoyama Hakushi Koki Kinen sodaishi Ronsō Kankōkai 青山博士古稀記念宋代史論叢刊行会, 315–37. Tokyo: Seishin Shobo, 1974.

Nakamura Hajime 中村元. *Bukkyōgo daijiten* 仏教大辞典. Tokyo: Tōkyō Shoseki, 1975.

Narens, Louis. *Theories of Meaningfulness.* Mahwah, NJ: Lawrence Erlbaum Associates, 2014.

Needham, Joseph. *The Grand Titration: Science and Society in East and West.* Toronto: University of Toronto Press, 1969.

Needham, Joseph, Ho Ping-yü, Lu Gwei-djen, and Wang Ling. *Science and Civlisation in China, Volume 5: Chemistry and Chemical Technology, Part 7: Military Technology; The Gunpowder Epic.* Cambridge: Cambridge University Press, 1986.

Needham, Joseph, and Wang Ling. *Science and Civilisation in China, Volume 2: History of Scientific Thought.* Cambridge: Cambridge University Press, 1956.

———. *Science and Civilisation in China, Volume 3: Mathematics and the Sciences of the Heavens and the Earth.* Cambridge: Cambridge University Press, 1995.

Needham, Joseph, Wang Ling, and Derek J. de Solla Price. *Heavenly Clockwork: The Great Astronomical Clocks of Medieval China.* Second edition, with supplement by John H. Combridge. Cambridge: Cambridge University Press, 1986.

Needham, Joseph, Wang Ling, and Kenneth Girdwood Robinson. *Science and Civilisation in China, Volume 4: Physics and Physical Technology, Part 1: Physics.* Cambridge: Cambridge University Press, 1962.

Nylan, Michael. "Yin-yang, Five Phases, and *qi*." In *China's Early Empires: A Re-appraisal*, edited by Michael Nylan and Michael Loewe, 398–414. Cambridge: Cambridge University Press, 2010.

Ong, Chang Woei. *Men of Letters within the Passes: Guanzhong Literati in Chinese History, 907–1911.* Cambridge, MA: Harvard University Asia Center, 2008.

Ouyang Xiu 歐陽修. *Guitian lu* 歸田錄. In *Quan Song biji*, series 1, vol. 5. Zhengzhou: Daxiang chubanshe, 2003.

———. *Ouyang Xiu quanji* 歐陽修全集. Beijing: Zhonghua shuju, 2001.

———. *Wenzhong ji* 文忠集, compiled by Zhou Bida 周必大. SKQS edition. Taipei: Taiwan shangwu yinshuguan, 1983–86.

Ouyang Xiu et al. *Xin Tang shu* 新唐書. Beijing: Zhonghua shuju, 1975.

Owen, Stephen. "Postface: Believe It or Not." In *Idle Talk: Gossip and Anecdote in Traditional China*, edited by Jack W. Chen and David Schaberg, 217–23. Berkeley: University of California Press, 2014.

Pelliot, Paul. *Les Débuts de l'Imprimerie en Chine*. Paris: Imprimerie nationale Librairie d'Amérique et d'Orient, 1953.

Perkins, Franklin. "What Is a Thing (wu)?" In *Chinese Metaphysics and Its Problems*, edited by Chenyang Li and Franklin Perkins, 54–68. Cambridge: Cambridge University Press, 2015.

Peterson, Willard J. "Another Look at Li." *Bulletin of Sung-Yuan Studies* 18 (1986): 13–31.

———. "Confucian Learning in Late Ming Thought." In *The Cambridge History of China Volume 8: The Ming Dynasty, Part 2: 1388–1644*, edited by Denis C. Twichett and Frederick W. Mote, 708–88. Cambridge: Cambridge University Press, 1998.

———. "Fang I-chih: Western Learning and the 'Investigation of Things.'" In *The Unfolding of Neo-Confucianism*, edited by Wm. Theodore de Bary, 369–411. New York: Columbia University Press, 1975.

———. "What Causes This?" In *Interpreting Culture Through Translation: A Festschrift for D. C. Lau*, edited by Roger T. Ames, Chan Sin-wai, and Mau-sang Ng, 185–205. Hong Kong: Chinese University Press, 1991.

Pomata, Gianna, and Nancy G. Siraisi. "Introduction," in *Historia: Empiricism and Erudition in Early Modern Europe*, edited by Gianna Pomata and Nancy G. Siraisi, 1–38. Cambridge, MA: MIT Press, 2005.

Poovey, Mary. *A History of the Modern Fact: Problems of Knowledge in the Sciences of Wealth and Society*. Chicago: University of Chicago Press, 1998.

Puett, Michael J. *The Ambivalence of Creation: Debates Concerning Innovation and Artifice in Early China*. Stanford, CA: Stanford University Press, 2001.

———. "The Haunted World of Humanity: Ritual Theory from Early China." In *Rethinking the Human*, edited by J. Michelle Molina and Donald K. Swearer, 95–111. Cambridge, MA: Center for the Study of World Religions and Harvard University Press, 2010.

Qi Xia 漆俠. "Fan Zhongyan jituan yu Qingli xinzheng: Du Ouyang Xiu 'Pengdang lun' shu hou" 范仲淹集團與慶曆新政 : 讀歐陽修《朋黨論》書後. In *Song shi yanjiu luncong* 宋史研究論叢, vol. 2, edited by Qi Xia, 1–10. Baoding: Hebei daxue chubanshe, 1993.

———. *Wang Anshi bianfa* 王安石變法. Shijiazhuang: Hebei renmin chubanshe, 2001.

Qian Baocong 錢寶琮. "Hanren yuexing yanjiu" 漢人月行研究. In Qian Baocong, *Qian Baocong kexue shi lunwen xuanji* 錢寶琮科學史論文選集, 175–92. Beijing: Kexue chubanshe, 1983.

Qian Daxin 錢大昕. *Qianyan tang wenji* 潛研堂文集. In *Jiading Qian Daxin xiansheng quanji* 嘉定錢大昕先生全集, vol. 9. Nanjing: Jiangsu guji chubanshe, 1997.

Qian Jingkui 錢景奎. "Guanyu Shen Gua yong gui, lou guance faxian zhen taiyang ri you changduan de tantao" 關於沈括用晷、漏觀測發現真太陽日有長短的探討. *Ziran kexueshi yanjiu* 1.2 (1982): 140–43.

Qian Yi 錢易. *Nanbu xinshu* 南部新書, annotated by Huang Shoucheng 黃壽成. Beijing: Zhonghua shuju, 2002.

*Qiantang xianzhi* 錢塘縣志. 1609. Reprint. *Zhongguo fangzhi congshu* 中國方志叢書. Taipei: Chengwen chubanshe, 1976.

Quan Hansheng 全漢昇. *Tang Song diguo yu yunhe* 唐宋帝國與運河. Chongqing: Shangwu yinshuguan, 1944.

Reardon-Anderson, James. *The Study of Change: Chemistry in China, 1840–1949*. Cambridge: Cambridge University Press, 1991.

Ren Yingqiu 任應秋. *Wu yun liu qi* 五運六氣. Hong Kong: Xianggang weisheng chubanshe, 1971.

Rheinberger, Hans-Jörg. *On Historicizing Epistemology: An Essay*. Translated by David Fernbach. Stanford, CA: Stanford University Press, 2010.

Rorty, Amelie Oksenberg. "The Vanishing Subject: The Many Faces of Subjectivity." In *Subjectivity: Ethnographic Investigations*, edited by João Biehl, Byron Good, and Arthur Kleinman, 34–51. Berkeley: University of California Press, 2007.

Rošker, Jana S. *Searching for the Way*. Hong Kong: Chinese University Press, 2008.

Rossi, Paolo. *Francis Bacon: From Magic to Science*. Translated by Sacha Rabinovitch. Chicago: University of Chicago, 1968.

Ruan Yuan 阮元. *Chouren zhuan* 疇人傳. Shanghai: Shangwu yinshuguan, 1955.

Ryle, Gilbert. *The Concept of Mind*. Chicago: University of Chicago Press, 1949.

———. "Knowing How and Knowing That." *Proceedings of the Aristotelian Society New Series* 46 (1945–1946): 1–16.

Sakade Yoshinobu 坂出祥伸. "Shin Katsu no shizenkan ni tsuite" 沈括の自然観について. *Tōhōgaku* 39 (1970): 74–87.

Schaberg, David. "The *Zhouli* as Constitutional Text." In *Statecraft and Classical Learning: The Rituals of Zhou in East Asian History*, edited by Benjamin A. Elman and Martin Kern, 33–63. Leiden: Brill, 2009.

Schäfer, Dagmar. *The Crafting of the 10,000 Things: Knowledge and Technology in Seventeenth-Century China*. Chicago: University of Chicago Press, 2011.

———. "Silken Strands: Making Technology Work in China." In *Cultures of Knowledge: Technology in Chinese History*, edited by Dagmar Schäfer, 45–73. Leiden: Brill, 2011.

Shao Yong 邵雍. *Huangji jingshi shu* 皇極經世書. Zhengzhou: Zhongzhou guji chubanshe, 2006.

Shen Gou 沈遘. *Xixi ji* 西溪集. In *Shenshi san xiansheng wenji* 沈氏三先生文集. SBCK edition. Shanghai: Shangwu yinshuguan, 1936.

———. *Xixi ji*. In *Shenshi san xiansheng wenji* 沈氏三先生文集, edited by Wu Yunjia 吳允嘉. Hangzhou: Zhejiang shuju, 1896.

Shen Gua 沈括. *Bu Bitan* 補筆談, edited by Chen Jiru 陳繼儒. In *Baoyan Tang miji* 寶顏堂秘集. Shanghai: Wenming shuju, 1922.

———. *Bu Bitan*, edited by Shang Jun 商濬. In *Bai hai* 稗海. Taipei: Xinxing shuju, 1968.

———. *Bu Bitan*, edited by Ma Yuandiao 馬元調. 1631.

———. *Bu Bitan*. SKQS edition. Taipei: Shangwu yinshuguan, 1983.

———. *Bu Bitan*, edited by Zhang Haipeng 張海鵬. In *Xuejin taoyuan* 學津討原. Taipei: Yiwen yinshuguan, 1965.

————. *Bu Bitan*, edited by Tao Fuxiang 陶福祥. Shanghai: Shanghai gushu liutong chu, 1922.

————. *Bu Bitan*, edited by Hu Daojing. Beijing: Zhonghua shuju, 1957.

————. *Bu Bitan*, edited by Hu Jingyi 胡靜宜. In *Quan Song biji*, series 2, vol. 3. Zhengzhou: Daxiang chubanshe, 2006.

————. *Bu Bitan*, edited by Hu Daojing. Shanghai: Shanghai renmin chubanshe, 2011. (BBT)

————. *Bu Bitan*, edited by Yang Weisheng 楊渭生. In *Shen Gua quanji* 沈括全集, vol. 2. edited by Yang Weisheng. Hangzhou: Zhejiang daxue chubanshe, 2011.

————. *Changxing ji* 長興集. In *Shenshi san xiansheng wenji*. SBCK edition. Shanghai: Shangwu yinshuguan, 1936.

————. *Changxing ji*. In *Shenshi san xiansheng wenji*, edited by Wu Yunjia. Hangzhou: Zhejiang shuju, 1896. (CXJ)

————. *Changxing ji*. In *Quan Song wen* 全宋文, vols. 78–79, edited by Zeng Zaozhuang 曾棗莊 and Liu Lin 劉琳. Shanghai: Shanghai cishu chubanshe, 2006.

————. *Changxing ji*. *Shen Gua quanji* 沈括全集, vol. 1, edited by Yang Weisheng. 2011.

————. *Mengxi bitan (Yuan kan)* 夢溪筆談 (元刊), edited by Chen Renzi 陳仁子. Beijing: Wenwu chubanshe, 1975.

————. *Mengxi bitan* 夢溪筆談, edited by Xu Bao 徐瑤. 1495.

————. *Mengxi bitan*, edited by Shen Jingkai 沈敬炌. 1602.

————. *Mengxi bitan*, edited by Shang Jun. In *Bai hai*. 1968.

————. *Mengxi bitan*, edited by Mao Jin 毛晋. In *Jindai mishu* 津逮秘書. Taipei: Yiwen yinshuguan, 1966.

————. *Mengxi bitan*, edited by Ma Yuandiao. 1631.

————. *Mengxi bitan*. SKQS edition. 1983.

————. *Mengxi bitan*, edited by Zhang Haipeng. In *Xuejin taoyuan*. 1965.

————. *Mengxi bitan*, edited by Tao Fuxiang. 1922.

————. *Mengxi bitan*, edited by Liu Shihang. 1916.

————. *Mengxi bitan*. SBCK edition. 1934.

————. *Mengxi bitan*, edited by Hu Jingyi. In *Quan Song biji*, series 2, vol. 3. 2006.

————. *Mengxi bitan jiaozheng* 夢溪筆談校證, edited and annotated by Hu Daojing. Beijing: Zhonghua shuju, 1956.

————. *Mengxi bitan jiaozheng*, edited and annotated by Hu Daojing. Shanghai: Shanghai renmin chubanshe, 2011.

————. *Mengxi bitan jiaozheng*, edited and annotated by Hu Daojing, reprint. Shanghai: Shanghai renmin chubanshe, 2016.

————. *Mengxi bitan*, edited by Yang Weisheng. In *Shen Gua quanji*, vol. 2. Hangzhou: Zhejiang daxue chubanshe, 2011.

————. *Shen Gua shici ji cun* 沈括詩詞輯存, collated by Hu Daojing. Shanghai: Shanghai shudian, 1985.

————. *Shen Gua quanji*, edited by Yang Weisheng. vols. 1–3. Hangzhou: Zhejiang daxue chubanshe, 2011.

————. *Xin jiaozheng Mengxi bitan* 新校正夢溪筆談, edited by Hu Daojing. Beijing: Zhonghua shuju, 1957.

———. *Xin jiaozheng Mengxi bitan*, edited by Hu Daojing. Shanghai: Shanghai renmin chubanshe, 2011. (MXBT)

———. *Xu Bitan* 續筆談, edited by Shang Jun. In *Bai hai*. 1602.

———. *Xu Bitan*, edited by Ma Yuandiao. 1631.

———. *Xu Bitan*. SKQS edition. 1983.

———. *Xu Bitan*, edited by Zhang Haipeng. In *Xuejin taoyuan*. 1965.

———. *Xu Bitan*, edited by Tao Fuxiang. 1922.

———. *Xu Bitan*, edited by Hu Daojing. 1957.

———. *Xu Bitan*, edited by Hu Jingyi. 2006.

———. *Xu Bitan*, edited by Hu Daojing. 2011. (XBT)

———. *Xu Bitan*, edited by Yang Weisheng, 2011.

*Shen Gua*. Eighteen episodes. Directed and written by Li Huaxin 李華新. Beijing: Shiyou gongye chubanshe, 1999.

Shen Liao 沈遼. *Yunchao ji* 雲巢集. In *Shenshi san xiansheng wenji*. SBCK edition. Shanghai: Shangwu yinshuguan, 1936.

———. *Yunchao ji*. In *Shenshi san xiansheng wenji*, edited by Wu Yunjia. Hangzhou: Zhejiang shuju, 1896.

Shen Songqin 沈松勤. *Beisong wenren yu dangzheng* 北宋文人與黨爭. Beijing: Renmin chubanshe, 1998.

Shen Zinan 沈自南. *Yilin huikao* 藝林彙考. SKQS edition. Taipei: Taiwan shangwu yinshuguan, 1983–86.

Shi Yunli 石雲里. *Zhongguo kexue shi gang: tianwen juan* 中國天文學史綱：天文卷. Shenyang: Liaoning jiaoyu chubanshe, 1997.

Sima Guang 司馬光. *Zi zhi tongjian* 資治通鑑, annotated by Hu Sanxing 胡三省. Taipei: Cui wen tang, 1975.

Sivin, Nathan. *Granting the Seasons: The Chinese Astronomical Reform of 1280, with a Study of Its Many Dimensions and an Annotated Translation of Its Records*. New York: Springer, 2008.

———. *Health Care in Eleventh-Century China*. New York: Springer, 2015.

———. "A Multi-Dimensional Approach to Research on Ancient Science." *East Asian Science, Technology, and Medicine* 23 (2005): 10–25.

———. "On the Limits of Empirical Knowledge in the Traditional Chinese Sciences." In *Time, Science, and Society in China and the West*, edited by J. T. Fraser, N. Lawrence, and F. C. Haber, 151–69. Amherst: University of Massachusetts Press, 1995.

———. "Recent Publications on Shen Kuo's *Mengxi bitan* (Brush Talks from Dream Brook)." *East Asian Science, Technology, and Medicine* 42 (2015): 93–102.

———. "Science and Medicine in Imperial China: The State of the Field." *Journal of Asian Studies* 47.1 (1988): 41–90.

———. "Shen Kua." In *Dictionary of Scientific Biography*, edited by Charles Coulston Gillispie, vol. 12, 369–93. New York: Scribner's, 1975. Reprinted as "Shen Kua: A Preliminary Assessment of his Scientific Thought and Achievements," *Sung Studies Newsletter* 13 (1977): 31–56. Reprinted and revised in *Science in Ancient China: Researches and Reflections*, 1–55. Aldershot, UK: Variorum, 1995.

———. "Text and Experience in Classical Chinese Medicine." In *Knowledge and the Scholarly Medical Traditions*, edited by Don Bates, 177–204. Cambridge: Cambridge University Press, 1995.

———. *Traditional Medicine in Contemporary China: A Partial Translation of Revised Outline of Chinese Medicine (1972): With an Introductory Study on Change in Present Day and Early Medicine*. Ann Arbor: Center for Chinese Studies, University of Michigan, 1987.

———. "Why the Scientific Revolution Did Not Take Place in China—Or Didn't It?" *Chinese Science* 5 (1982): 45–66.

Skonicki, Douglas E. "Cosmos, State, and Society: Song Dynasty Arguments Concerning the Creation of Political Order." PhD diss., Harvard University, 2007.

Smith, Kidder. "Sung Literati Thought and the *I Ching*." In Kidder Smith et al., *Song Dynasty Uses of the I Ching*, 206–35. Princeton, NJ: Princeton University Press, 1990.

Smith, Kidder, and Don Wyatt. "Shao Yung and Number." In Kidder Smith et al., *Song Dynasty Uses of the I Ching*, 100–135. Princeton, NJ: Princeton University Press, 1990.

Smith, Paul J. "Shen-tsung's Reign." In *The Cambridge History of China. Volume 5, Part 1, The Sung Dynasty and its Precursors, 907–1279*, edited by Denis Twichett and Paul J. Smith, 348–483. Cambridge: Cambridge University Press, 2008.

———. "State Power and Economic Activism during the New Policies, 1068–1085: The Tea and Horse Trade and the 'Green Sprouts' Loan Policy." In *Ordering the World: Approaches to State and Society in Sung Dynasty China*, edited by Robert P. Hymes and Conrad Schirokauer, 76–127. Berkeley: University of California Press, 1993.

———. *Taxing Heaven's Storehouse: Horses, Bureaucrats, and the Destruction of the Sichuan Tea Industry 1071–1244*. Cambridge, MA: Harvard University Council on East Asian Studies, 1991.

Sogabe Shizuo 曾我部静雄. *Sōdai zaisei shi* 宋代財政史. Tokyo: Seikatsusha, 1941.

———. "Ō Anseki no hokōhō" 王安石の保甲法. In Sogabe Shizuo, *Sōdai seikeishi no kenkyū* 宋代政経史の研究, 1–63. Tokyo: Yoshikawa Kōbunkan, 1974.

Song Lian 宋濂 et al. *Yuan shi* 元史. Beijing: Zhonghua shuju, 1976.

Song, Jaeyoon. *Traces of Grand Peace: Classics and State Activism in Imperial China*. Cambridge, MA: Harvard University Asia Center, 2015.

Song Shou 宋綬 and Song Minqiu 宋敏求. *Song da zhaoling ji* 宋大詔令集. Beijing: Zhonghua shuju, 1962.

Su Shi 蘇軾. *Dongpo zhilin* 東坡志林, annotated by Wang Songling 王松齡. Beijing: Zhonghua shuju, 1981.

———. *Shu zhuan* 書傳, edited by Ling Mengchu 凌濛初. Circa seventeenth century.

———. *Su Shi wenji* 蘇軾文集, annotated by Kong Fanli 孔凡禮. Beijing: Zhonghua shuju, 2004.

Su Shi 蘇軾 and Shen Gua. *Su Shen neihan liangfang* 蘇沈內翰良方, annotated by Song Zhenmin 宋珍民 and Li Enjun 李恩軍. Beijing: Zhongyi guji chubanshe, 2009.

Su Song 蘇頌. *Su Weigong wenji: fu Weigong tanxun* 蘇魏公文集：附魏公譚訓, annotated by Wang Tongce 王同策, Guan Chengxue 管成學, and Yan Zhongqi 顏中其. Beijing: Zhonghua shuju, 1988.

———. *Xin yixiang fayao* 新儀象法要, annotated by Hu Weijia 胡維佳. Shenyang: Liaoning jiaoyu chubanshe, 1997.

Su Xiangxian 蘇象先. *Chengxiang Weigong tanxun* 丞相魏公譚訓. In Su Song, *Su Weigong wenji: fu Weigong tanxun*, 1113–89. Beijing: Zhonghua shuju, 1988.

Su Zhe 蘇轍. *Luancheng hou ji*. In *Su Zhe ji* 蘇轍集, annotated by Chen Hongtian 陳宏天 and Gao Xiufang 高秀芳. Beijing: Zhonghua shuju, 1990.

Sudō Yoshiyuki 周藤吉之. "Ō Anseki no seibyōhō no kigen ni tsuite" 王安石の青苗法の起原について. *Tōyō gakuhō* 53.2 (1970): 133–77.

———. "Ō Anseki no seibyōhō no shikō katei ichi" 王安石の青苗法の施行過程 (一). *Tōyō daigaku daigakuin kiyō* 8 (1971): 171–99.

Sun Simiao 孫思邈. *Beiji qianjin yaofang* 備急千金要方, annotated by Liu Qingguo 劉清國 et al. Beijing: Zhongguo zhongyiyuan chubanshe, 1998.

Sun, Xiaochun. "State and Science: Scientific Innovations in Northern Song China, 960–1127." PhD diss., University of Pennsylvania, 2007.

Sun, Xiaochun, and Han Yi. "The Northern Song State's Financial Support for Astronomy." *East Asian Science, Technology and Medicine* 38 (2013): 17–53.

Sun Xiaochun 孫小淳 and Zeng Xiongsheng 曾雄生, eds. *Songdai guojia wenhua zhong de kexue* 宋代國家文化中的科學. Beijing: Zhongguo kexue jishu chubanshe, 2007.

Sun Yongzhong 孫永忠. *Leishu yuanyuan yu tili xingcheng zhi yanjiu* 類書淵源與體例形成之研究. *Gudian wenxian yanjiu jikan* 古典文獻研究集刊, series 4, vol. 3. Taipei: Huamulan wenhua chubanshe, 2007.

Sun Yuanlu 孫遠路. "Beisong de yiyong yu qiangzhuang" 北宋的義勇與強壯. Master's thesis, Hebei University, 2014.

Swidler, Ann. *Talk of Love: How Culture Matters*. Chicago: University of Chicago Press, 2001.

Tackett, Nicolas. *The Destruction of the Medieval Chinese Aristocracy*. Cambridge, MA: Harvard University Asia Center, 2014.

———. "The Great Wall and Conceptualizations of the Border under the Northern Song." *Journal of Song-Yuan Studies* 38 (2008): 99–138.

Tang Daijian 唐代劍. *Songdai daojiao guanli zhidu yanjiu* 宋代道教管理制度研究. Beijing: Xianzhuang shuju, 2003.

Tang Guangrong 唐光榮. *Tangdai leishu yu wenxue* 唐代類書與文學. Chengdu: Bashu shushe, 2008.

Tang Shenwei 唐慎微. *Chongxiu Zhenghe jingshi zhenglei beiyong bencao* 重修政和經史證類備用本草. Beijing: Renmin weisheng chubanshe, 1957.

Tao, Jing-shen. *Two Sons of Heaven: Studies in Sung-Liao Relations*. Tucson: University of Arizona Press, 1988.

Tao Zongyi 陶宗儀. *Shuofu* 說郛. Beijing: Zhongguo shudian, 1986.

Taylor, Charles. *A Secular Age*. Cambridge, MA: Harvard University Press, 2009.

———. *Sources of the Self: The Making of the Modern Identity*. Cambridge, MA: Harvard University Press, 1989.

Teraji Jun 寺地遵. "Shin Katsu no shizen kenkyū to sono haikei" 沈括の自然研究とその背景. *Hiroshima Daigaku Bungakubu kiyō* 27.1 (1967): 99–121.

Tian Kuang 田況. *Rulin gongyi* 儒林公議. In *Quan Song biji*, series 1, vol. 5. Zheng-zhou: Daxiang chubanshe, 2003.

Tian, Miao. "The Westernization of Chinese Mathematics: A Case Study of the *duoji* Method and Its Development." *East Asian Science, Technology, and Medicine* 20 (2003): 45–72.

Tillman, Hoyt Cleveland. "The Idea and The Reality of the 'Thing' during the Sung: Philosophical Attitudes toward *Wu*." *Bulletin of Sung and Yuan Studies* 14 (1978): 68–82.

Tsu, Jing, and Benjamin A. Elman. "Introduction." In *Science and Technology in Modern China, 1880s–1940s*, edited by Jing Tsu and Benjamin A. Elman, 1–14. Leiden: Brill, 2014.

Tuotuo (Toktoghan) 脫脫 et al. *Song shi* 宋史. Beijing: Zhonghua shuju, 1977. (SS)

Twitchett, Denis, and Paul J. Smith, eds. *The Cambridge History of China. Volume 5, Part 1, The Sung Dynasty and Its Precursors, 907–1279*. Cambridge: Cambridge University Press, 2008.

Umehara Kaoru 梅原郁. *Mukei hitsudan* 夢溪筆談. Tokyo: Heibonsha, 1978–1981.

———. *Sōdai kanryō seido kenkyū* 宋代官僚制度研究. Kyoto: Dōhōsha Shuppan, 1985.

Unschuld, Paul. *Essential Subtleties on the Silver Sea: The Yin-Hai Jing-Wei: A Chinese Classic on Ophthalmology*. Translated by Jürgen Kovacs. Berkeley: University of California Press, 1999.

———. *Huang Di nei jing su wen: Nature, Knowledge, Imagery in an Ancient Chinese Medical Text*. Berkeley: University of California Press, 2003.

Valenstein, Suzanne G. *A Handbook of Chinese Ceramics*. New York: Metropolitan Museum of Art, 1988.

Van Fraassen, Bas. *The Empirical Stance*. New Haven, CT: Yale University Press, 2008.

von Glahn, Richard. "Community and Welfare: Chu Hsi's Community Granary in Theory and Practice." In *Ordering the World: Approaches to State and Society in Sung Dynasty China*, edited by Robert P. Hymes and Conrad Schirokauer, 221–54. Berkeley: University of California Press, 1993.

———. "Revisiting the Song Monetary Revolution: A Review Essay." *International Journal of Asian Studies* 1.1 (2004), 159–78.

Wang Anshi 王安石. *Linchuan xiansheng wenji* 臨川先生文集. Beijing: Zhonghua shuju, 1959.

———. *Sanjing xinyi jikao huiping (3): Zhouli* 三經新義輯考彙評（3）：周禮, edited by Cheng Yuanmin 程元敏. Taipei: Guoli bianyi guan, 1987.

———. *Wang Anshi Zi Shuo ji* 王安石《字說》輯, compiled by Zhang Zongxiang 張宗祥 and annotated by Cao Jingyan 曹錦炎. Fuzhou: Fujian renmin chubanshe, 2005.

Wang Hong 王宏 and Zhao Zheng 趙崢. *Shen Kuo: Brush Talks from Dream Brook*. Reading: Path International, and Chengdu: Sichuan rennin chubanshe, 2011.

Wang Huaiyin 王懷隱. *Taiping shenghui fang* 太平聖惠方. Beijing: Renmin weisheng chubanshe, 1959.

Wang Ling. "On the Invention and Use of Gunpowder and Firearms in China." In *Science and Technology in East Asia*, edited by Nathan Sivin, 140–58. New York: Science History Publications, 1977.

Wang Meihua 王美華. "Li fa heliu yu Tang Song lizhi de tuixing" 禮法合流與唐宋禮制的推行. *Shehui kexue jikan* 4 (2008): 119–26.

Wang Meng 王蒙. "Beisong Jingling gong guoji xingxiang lue lun" 北宋景靈宮國忌行香略論. *Zongjiao xue yanjiu* 2 (2016): 268–73.

Wang Pizhi 王闢之. *Mianshui yantan lu* 澠水燕談錄. Beijing: Zhonghua shuju, 1981.

Wang, Robin R. *Yinyang: The Way of Heaven and Earth in Chinese Thought and Culture.* Cambridge: Cambridge University Press, 2012.

Wang Ruilai 王瑞來. "Su Song lun" 蘇頌論. *Zhejiang xuekan* 4 (1988): 118–23.

Wang Wendong 王文東. "Songchao qingmiao fa yu Tang Song Changping cang zhidu bijiao yanjiu" 宋朝青苗法與唐宋常平倉制度比較研究. *Zhongguo shehui jingji shi yanjiu* 3 (2006): 29–36.

Wang Yinglin 王應麟. *Kun xue ji wen* 困學紀聞. Shanghai: Shangwu yinshuguan, 1935.

———. *Yu hai* 玉海. Nanjing: Jiangsu guji chubanshe and Shanghai: Shanghai shudian, 1987.

Wang Zengyu 王曾瑜. "Songdai de fangguohu" 宋代的坊郭戶. In *Song Liao Jin shi luncong* 宋遼金史論叢, edited by Zhongguo shehuikexueyuan lishi yanjiu suo 中國社會科學院歷史研究所 and Song Liao Jin Yuan shi yanjiushi 宋遼金元史研究所, vol. 1, 64–82. Beijing: Zhonghua shuju, 1985.

Wang Zhenduo 王振鐸. "Songdai shuiyun yixiang tai de fuyuan" 宋代水運儀象臺的復原. In Wang Zhenduo, *Keji kaogu luncong* 科技考古論叢, 238–74. Beijing: Wenwu chubanshe, 1989.

Will, Pierre-Etienne, et al. "Translation of the *Mengxi bitan* (in French)." Manuscript.

Williams, Paul. *Mahāyāna Buddhism: The Doctrinal Foundations.* London: Routledge, 2008.

Wright, David C. *From War to Diplomatic Parity in Eleventh-Century China.* Leiden: Brill, 2005.

Wu Bing 吳兵. "Beisong Dingzhou junshi ruogan wenti yanjiu" 北宋定州軍事若干問題研究. Master's thesis, Hebei University, 2014.

Wu Hui 吳慧. "Song Yuan de duliangheng" 宋元的度量衡. *Zhongguo shehui jingji shi yanjiu* 1 (1994): 16–23, 7.

Wu Tai 吳泰. "Songdai 'baojia fa' tanwei" 宋代《保甲法》探微. In *Song Liao Jin shi luncong* 宋遼金史論叢, edited by Zhongguo shehuikexueyuan lishi yanjiu suo 中國社會科學院歷史研究所 and Song Liao Jin Yuan shi yanjiushi 宋遼金元史研究所, vol. 2, 178–200. Beijing: Zhonghua shuju, 1991.

Wu Zeng 吳曾. *Nenggai zhai manlu* 能改齋漫錄. In *Quan Song biji*, series 5, vol. 3. Zhengzhou: Daxiang chubanshe, 2012.

Wu Zuoxin 吳佐忻. "Shen Gua de *Lingyuan Fang*" 沈括的《靈苑方》. *Zhonghua wenshi luncong* 3 (1978): 78.

Wuhan shuili dianli xueyuan 武汉水利电力学院 and Shuili shuidian kexue yanjiuyuan 水利水电科学研究院. *Zhongguo shuili shigao* 中國水利史稿, vol. 2. Beijing: Zhongguo shuili dianli chubanshe, 1989.

Wyatt, Don J. *The Recluse of Loyang: Shao Yung and the Moral Evolution of Early Sung Thought*. Honolulu: University of Hawai'i Press, 1996.

———. "Shao Yong's Numerological-Cosmological System." In *Dao Companions to Chinese Philosophy*, edited by Yong Huang, 17–37. New York: Springer, 2010.

———. "The Transcendence of the Past: Objectivity, Relativism, and Moralism in the Historical Thought of Shao Yong." *Monumenta Serica* 61 (2013): 203–26.

Xie Jin 解縉 et al. *Yongle dadian* 永樂大典. Beijing: Zhonghua shuju, 1986.

Xu Gui 徐規. "Shen Gua shiji biannian" 沈括事跡編年. In *Liu Tze-chien boshi songshou jinian Song shi yanjiu lunji* 劉子健博士頌壽紀念宋史研究論集, edited by Tsuyoshi Kinugawa 衣川強, 85–94. Kyoto: Dōhōsha, 1989.

Xu Song 徐松. *Song Huiyao ji gao* 宋會要輯稿. Taipei: Xinwenfeng chuban gongsi, 1976. (SHY)

Xue Juzheng 薛居正 et al. *Jiu Wudai shi* 舊五代史. SKQS edition. Taipei: Taiwan shangwu yinshuguan, 1983–86.

Yabuuchi Kiyoshi 薮内清, ed. *Sō Gen jidai no kagaku gijutsushi* 宋元時代の科学技術史. Kyoto: Kyōto Daigaku Jinbun Kagaku Kenkyūjo kenkyū hōkoku, 1967.

Yan Dunjie 嚴敦傑. "Fengyuan li (fusuan)" 奉元曆 (複算). In Hu Daojing, *Shen Gua yanjiu, keji shi lun*, 165–70. Shanghai: Shanghai renmin chubanshe, 2011.

Yan Jia 閻嘉 and Zhou Xiaofeng 周曉風. *Mengxi bitan baihua quan yi* 夢溪筆談白話全譯. Chengdu: Bashu shushe, 1995.

Yang Weisheng 楊渭生. "Shen Gua Xining shi Liao tu chao ji jian" 沈括《熙寧使遼圖抄》輯箋. In *Shen Gua Yanjiu*, edited by Hangzhou daxue Song shi yanjiu shi, 297–321. Hangzhou: Zhejiang renmin chubanshe, 1985.

Yang Weisheng, ed. *Shen Gua quanji* 沈括全集. Vols. 1–3. Hangzhou: Zhejiang daxue chubanshe, 2011.

Yang Yinliu 楊蔭瀏. *Zhongguo gudai yinyue shigao* 中國古代音樂史稿. Beijing: Renmin yinyue chubanshe, 2004.

Yang Zhishui 揚之水. *Tang Song jiaju xunwei* 唐宋家具尋微. Hong Kong: Zhonghe chuban youxian gongsi, 2015.

Ye Tan 葉坦. "Lun Beisong 'qian huang'" 論北宋《錢荒》. *Zhongguo shi yanjiu* 2 (1991): 20–30.

Yoshinobu Yoshioka 吉岡義信. "Hoku sō shoki ni okeru nanjin no shinshutsu" 北宋初期における南人官僚の進出. *Suzugamine jyoshi tandai kenkyo shyōhō* 2 (1955): 24–37.

You Biao 游彪. *Songdai yinbu zhidu yanjiu* 宋代蔭補制度研究. Beijing: Zhongguo shehui kexue chubanshe, 2001.

Yü, Ying-shih 余英時. "Intellectual Breakthroughs in the T'ang-Sung Transition." In *The Power of Culture: Studies in Chinese Cultural History*, edited by Willard J. Peterson, Andrew H. Plaks, and Ying-shih Yü, 158–70. Hong Kong: Chinese University Press, 1994.

———. *Zhongguo sixiang chuantong de xiandai quanshi* 中國思想傳統的現代詮釋. Nanjing: Jiangsu renmin chubanshe, 2006.

———. *Zhu Xi de lishi shijie: Songdai shidafu zhengzhi wenhua de yanjiu* 朱熹的歷史世界：宋代士大夫政治文化的研究. Taipei: Yunchen wenhua shiye gufen youxian gongsi, 2004.

Yuan Yitang 袁一堂. "Beisong qian huang: cong bizhi dao liutong tizhi de kaocha" 北
宋錢荒：從幣制到流通體制的考察. *Lishi yanjiu* 4 (1991): 129–40.
————. "Songdai shidi zhidu yanjiu" 宋代市糴制度研究. *Zhongguo jingji shi yanjiu*
3 (1994): 139–46.

Zagzebski, Linda T. *Epistemic Authority: A Theory of Trust, Authority, and Autonomy in
Belief.* Oxford: Oxford University Press, 2012.
Zeng Gong 曾鞏. *Zeng Gong ji* 曾鞏集, edited by Chen Xingzhen 陳杏珍 and Chao
Jizhou 晁繼周. Beijing: Zhonghua shuju, 1984.
Zeng Gongliang 曾公亮 and Ding Du 丁度. *Wujing zongyao* 武經總要. In *Zhongguo
bingshu jicheng* 中國兵書集成, vols. 3–5. Beijing: Jiefangjun chubanshe, Liaoshen
shushe, 1987.
Zeng Zaozhuang 曾棗莊 and Liu Lin 劉琳, eds. *Quan Song wen* 全宋文. Shanghai:
Shanghai cishu chubanshe, 2006.
Zhang Bangji 張邦基. *Mozhuang manlu* 墨莊漫錄, annotated by Kong Fanli 孔凡禮.
Beijing: Zhonghua shuju, 2002.
Zhang, Ellen Cong. "To Be 'Erudite in Miscellaneous Knowledge': A Study of Song
(960–1279) *Biji* Writing." *Asia Major Third Series* 25.2 (2012): 43–77.
Zhang Fangping 張方平. *Zhang Fangping ji* 張方平集, annotated by Zheng Han 鄭涵.
Zhengzhou: Zhongzhou guji chubanshe, 2000.
Zhang Jiaju 張家駒. *Shen Gua* 沈括. Second edition. Shanghai: Shanghai renmin chu-
banshe, 1978.
Zhang Lei 張耒. *Mingdao zaozhi* 明道雜誌. In *Quan Song biji*, series 2, vol. 7, 5–30.
Zhengzhou: Daxiang chubanshe, 2005.
————. *Zhang Lei ji* 張耒集. Beijing: Zhonghua shuju, 2005.
Zhang Shinan 張世南. *Youhuan jiwen* 遊宦紀遊. Beijing: Zhonghua shuju, 1981.
Zhang Weidong 張圍東. *Songdai leishu zhi yanjiu* 宋代類書之研究. *Gudian wenxian
yanjiu jikan*, series 1, vol. 5. Taipei: Huamulan wenhua chubanshe, 2005.
Zhang Xiumin 張秀民. *Zhongguo yinshua shi* 中國印刷史. Shanghai: Shanghai renmin
chubanshe, 1989.
Zhang Zai 張載. *Zhang Zai ji* 張載集. Beijing: Zhonghua shuju, 2006.
Zheng Fangkun 鄭方坤. *Jingbai* 經稗. SKQS edition. Taipei: Taiwan shangwu yinshu-
guan, 1983–86.
Zhongguo kexue jishu daxue 中國科學技術大學 and Hefei gangtie gongsi 合肥鋼鐵
公司. *Mengxi bitan yizhu: ziran kexue bufen* 夢溪筆談譯注：自然科學部分. Hefei: An-
hui kexue jishu chubanshe, 1979.
Zhou Chunsheng 周春生. "Shen Gua qinshu kao" 沈括親屬研究. In *Shen Gua Yanjiu*,
edited by Hangzhou daxue Song shi yanjiu shi, 50–63. Hangzhou: Zhejiang renmin
chubanshe, 1985.
*Zhou li zhushu* 周禮注疏. In *Shisanjing zhushu* 十三经注疏, compiled by Ruan Yuan
阮元, 631–937. Beijing: Zhonghua shuju, 1982.
Zhou Mi 周密. *Qi dong yeyu* 齊東野語, annotated by Zhang Maopeng 張茂鵬. Beijing:
Zhonghua shuju, 1997.
*Zhouyi zhengyi* 周易正義. In *Shisanjing zhushu*, compiled by Ruan Yuan, 10–228. Beijing:
Zhonghua shuju, 1982.

Zhu Bokun 朱伯崑. *Yixue zhexue shi* 易學哲學史. Vols. 1–4. Beijing: Huaxia chuban-she, 1995.

Zhu Gong 朱肱. *Huo ren shu* 活人書, annotated by Wan Lanqing 萬蘭清, Wan Yousheng 萬友生, Wan Shaoju 萬少菊, and Wang Yumen 王魚門. Beijing: Renmin weisheng chubanshe, 1993.

Zhu Hanmin 朱漢民 and Wang Qi 王琦. "'Song xue' de lishi kaocha yu xueshu fenshu" 《宋學》的歷史考察與學術分疏. *Zhongguo zhexue shi* 4 (2015): 80–86.

Zhu Kezhen 竺可楨. "Shen Gua duiyu dixue zhi gongxian yu jishu" 沈括對於地學之貢獻與記述. *Kexue* 11.6 (1926): 792–807. Reprinted, revised, and annotated in *Shen Gua Yanjiu*, edited by Hangzhou daxue Song shi yanjiu shi, 1–15. Hangzhou: Zhe-jiang renmin chubanshe, 1985.

Zhu Mu 祝穆. *Gujin shi wen leiju* 古今事文類聚. SKQS edition. Taipei: Taiwan shangwu yinshuguan, 1983–86.

Zhu Xi 朱熹. *Hui'an xiansheng Zhu Wen gong wenji* 晦庵先生朱文公文集. SBCK edi-tion. Taipei: Taiwan shangwu yinshuguan, 1979.

———. *Zhu Zi yulei* 朱子語類, compiled by Li Jingde 黎靖德, annotated by Wang Xingxian 王星賢. Beijing: Zhonghua shuju, 1986.

Zhu Yi 朱溢. "Tang zhi Beisong shiqi de huangdi qinjiao" 唐至北宋時期的皇帝親郊. *Guoli zhengzhi daxue lishi xuebao* 34 (2010): 1–52.

Zhu Yizun 朱彝尊. *Dianjiao buzheng Jingyi kao* 點校補正經義考, annotated by Xu Weiping 許維萍, Feng Xiaoting 馮曉庭, and Jiang Yongchuan 江永川. Taipei: Zhongyang yanjiu yuan Zhongguo wenzhe yanjiu suo choubei chu, 1997.

Zhu Yu 朱彧. *Ping zhou ke tan* 萍洲可談. Shanghai: Shanghai guji chubanshe, 1989.

Zhuangzi 莊子. *Zhuangzi jishi* 莊子集釋, compiled by Guo Qingfan 郭慶藩. Beijing: Zhonghua shuju, 1961.

Ziporyn, Brook. *Beyond Oneness and Difference: Li and Coherence in Chinese Buddhist Thought and Its Antecedents*. Albany: State University of New York Press, 2014.

———. "Form, Principle, Pattern, or Coherence? Li 理 in Chinese Philosophy." *Phi-losophy Compass* 3 (2008): 1–50.

———. *Ironies of Oneness and Difference: Coherence in Early Chinese Thought; Prole-gomena to the Study of Li*. Albany: State University of New York Press, 2013.

Zu Hui 祖慧. *Shen Gua pingzhuan* 沈括評傳. Nanjing: Nanjing daxue chuban she, 2004.

———. "Shen Gua yu Wang Anshi guanxi yanjiu" 沈括與王安石關係研究. *Xueshu yuekan* 10 (2003): 52–59.

Zuo, Ya. "Keeping Your Ear to the Cosmos: Coherence as the Standard of Good Music in the Northern Song (960–1127) Music Reforms." Forthcoming in *Standards of Va-lidity in Late Imperial China*, edited by Ari D. Levine, Joachim Kurtz, and Martin Hofmann.

———. "*Ru* versus *Li*: The Divergence between the Generalist and the Specialist in the Northern Song (960–1127)." *Journal of Song-Yuan Studies* 44 (2014): 83–137.

# Index

Page numbers for figures and tables are in italics.

# Harvard-Yenching Institute Monograph Series

## (titles now in print)

# Harvard-Yenching Institute Monograph Series

# Harvard-Yenching Institute Monograph Series

# Harvard-Yenching Institute Monograph Series